Manual of
Professional Remodeling

by
Jack P. Jones

Craftsman Book Company

6058 Corte del Cedro, P.O. Box 6500, Carlsbad, CA 92008

Acknowledgements

The author wishes to express his appreciation to the following companies and organizations for furnishing materials used in the preparation of various portions of this book.

Kohler Co., Kohler, Wisconsin
Masonite Corp., Laurel, Mississippi
Celotex Corp., Tampa, Florida
Georgia-Pacific Corp., Atlanta, Georgia
Armstrong Cork Co., Lancaster, Pennsylvania
Magic Chef, Inc., Cleveland, Tennessee
Asphalt Roofing Manufacturers Association, Washington, D.C.
Tub-Master Corp., Orlando, Florida
Elk Corporation, Tuscaloosa, Alabama
Benjamin Moore & Co., Montvale, New Jersey
Majestic Co., Huntington, Indiana
American Plywood Association, Tacoma, Washington

This book is dedicated to Claude H. Jones, B.A., M.S., who went the extra mile

Library of Congress Cataloging in Publication Data

Jones, Jack Payne 1928-
    Manual of professional remodeling.

    Includes index.
    1. Dwellings--Remodeling--Handbooks, manuals,
etc.   I. Title.
TH4816.J647   1982   690'.837'0286   82-14222
ISBN 0-910460-98-1
©1982 Craftsman Book Company

Fourth printing 1989

Edited by Sam Adrezin

Contents

1 Who Makes It, and How............5

2 The Art of Evaluating...............7
   Evaluating a Building...................8
   Fireplaces and Chimneys.............10
   Structural Supports..................11
   Wall Framing.........................12
   Roof Framing.........................13
   Windows and Doors...................14
   Walls and Ceilings....................16
   Flooring..............................17
   Insulation and Ventilation............17
   Plumbing and Electrical..............18
   Heating and Cooling..................19
   Destructive Insects...................20
   Evaluation Check List.................21

3 The Art of Planning................24
   Space Requirements...................24
   Kitchen Storage.......................27
   Ceiling Height and Basement Space......27
   Traffic Flow..........................28
   Relocating and Removing Partitions.....29
   Window Arrangement..................32
   Garages and Carports—Minimum.......33
   General Exterior Design...............34

4 Repairing Structural Defects........35
   Repositioning Girders and Sills.........35
   Installing New Pillars..................36
   Replacing Sills........................36
   Correcting Sagging Girders............38
   Eliminating Floor Squeaks.............39
   Correcting Sagging Roofs..............40
   Correcting Sagging Roof Sheathing......41
   Correcting Masonry Foundation
      and Pillars..........................42

5 The Right Material for the Job.......44
   Concrete Mix..........................45

Framing Lumber......................45
Plywood, Glass and Nails..............50
Termite Protection....................51

6 Bathrooms........................55
   The Modern Bathroom................56
   Tubs.................................56
   Shower Coves and Cabinets............57
   Lavatories............................58
   Commodes............................67
   Bidets, Whirlpool Baths and Spas.......68
   Vanities and Cabinets.................76
   Walls and Floor Covering..............77
   Bathroom Plans and Specifications......78
   Bathroom Checklist...................82
   Job Specifications....................84

7 Kitchens..........................91
   The Present Concept..................91
   U- and L-Shaped Kitchens.............92
   The Corridor Kitchen.................93
   The Sidewall Kitchen.................94
   Kitchen Cabinets......................95
   Drop-In Range Installation............98
   Kitchen Planning.....................109
   Kitchen Checklist....................110

8 Flooring..........................117
   Wood................................117
   Sheet Vinyl Flooring.................119
   Vinyl Sheet and Tile Removal..........126
   Sub-Floors...........................126
   Vinyl-Asbestos Tile..................127
   Hard Flooring........................129
   Carpet...............................130

9 Walls and Ceilings................133
   Removing a Partition.................133
   Adding a Partition...................134

Interior Wall Finish.....................136
Paneling a Basement.....................144
Ceiling Tile............................150
Suspended Ceilings.....................155
Interior Trim...........................165

10 Doors and Windows...............167
Doors...................................167
Windows.................................168
Guide to Window Selection...............175

11 Exterior Siding.................179
Wall Construction......................179
Masonry Construction...................180
Hardboard Shakes.......................181
Estimating Siding Requirements........182
Step-By-Step Procedures...............195
Painting and Staining..................196
Vinyl Siding...........................196
Wood Siding............................204
Aluminum Siding........................208

12 Insulation......................211
Insulation Materials...................211
Which is the Right Material?..........214
The R-20 Wall..........................215
Masonry Systems........................216
Insulating Ceilings and Walls.........218
Insulating Floors......................221
Weatherstripping and Caulking.........225
Ventilation 227

13 Asphalt Roofing................235
Shingles...............................235
Coverage and Exposure..................236
Estimating Materials...................239
Strip Shingling on New Construction....245
Valley Flashing........................247
Flashing...............................255
Applying Strip Shingles................260
Removing Old Roofing...................263
Roll Roofing...........................267
Wood Shingles and Shakes...............274

14 Fireplaces and Chimneys.........277
Fireplace Repairs......................277
Proper Construction....................278
Manufactured Fireplaces................280
Installing the Built-In Fireplace......282
Basement Installation..................289
Determining Chimney Height............296
Framing and Finishing..................300
Conventional Chimneys..................302
Flue Liners............................304
Factory-Built Inserts or Forms.........306

15 Stairways......................309
Interior Stairs........................309
Exterior Stairs........................310
Stair Construction.....................312

16 Skylights.......................315
Curb Mounted Skylights.................316
Self Flashing Skylights................319
Framing Dimension Chart................321
Clustering Skylights...................322

17 Paint and Painting.............327
The Effects of Color...................328
Paint Materials........................329
Exterior Surface Preparation..........331
Painting Specifications................334
Plaster and Drywall....................336
Surface Problems.......................337
Primer and Finish Coat Selection.......344
Calculating Paint Quantities..........345
Painting the Outside of a House.......345
Painting Walls, Ceilings and Trim......348

18 Basement Conversions...........353
Basement Construction..................353
Foundation Drains and Dampproofing...354

19 Attic Conversions..............359
Gable and Shed Dormers.................360
Adding a Second Story..................361
Skylights for Attic Conversions........361

20 Adding Outward.................365
Footings...............................365
Slabs..................................367
Framing................................367
Framing to Wood Girders................373
Exterior Wall Framing..................376
Plywood Underlayment Installation.....377
Wall Sheathing.........................380
Ceiling Joists and Rafters.............382
Using the Carpenter Square.............383

21 Basic Plumbing and Wiring.......387
Basic Plumbing.........................387
Basic Wiring...........................387

22 Job Estimating.................389
Labor..................................389
Material Estimate List.................390
Total Cost.............................392
Site Preparation.......................392
Overhead...............................393

Index............................395

1

Who Makes It, And How

This book is about professional home remodeling. It has the basic information, practical tips, and helpful reference data every contractor needs to survive and thrive in the construction business.

What you read here is based on 25 years of experience in construction. My background includes building custom and spec houses, extensive remodeling work, and building additions of all types. I'm a general contractor. I've spent the last 25 years doing the estimates, planning the job, drafting the contract, digging the footings, laying the block, sawing the lumber, driving the nails, putting on the roof, designing and building fireplaces, plumbing, and wiring.

I've handled everything from small add-ons to erecting mansions. Everything in this book is based on my own experience.

There's always demand for the builder who can do quality remodeling and additions. The American homeowner always wants to modernize his house or add on more space. He doesn't mind paying for a professional job. He will pay even more for a job done by a builder with a reputation for quality performance. Slipshod contractors don't last long in this business. And that's the way it should be.

Quality builders are always hard to find. And they're always busy. They're busy in the cities, in the small towns, and in the rural areas. It's a good living and a challenging occupation. It's hard work, but so is working for wages on someone else's payroll.

Many general contractors, builders or remodelers, merely *manage* a building or remodeling business. These people seldom lift anything heavier than a pencil. They do the estimating, get the jobs, draft the contracts, order the materials, and perhaps supervise some of the work. If you fit into this category, you realize that a reliable crew is essential. You've got to line up plenty of work in advance to keep them busy and generate the profit margin you need to carry your overhead. That's not always so easy.

If you're one of these "paper contractors," you've probably thought you could earn more money by hiring a couple of good men and doing the work yourself—actually working on the job along with the men you supervise. You could be right. This book will show you how. In addition to earning a salary, you get the contractor's cut. You'd be a workhorse, but you'll be paid for it.

The workhorse type of builder has a lot in his favor. Since he does the work himself, he knows exactly what materials are required, when to schedule deliveries, and how long a particular job will take. Most important, he knows that every task is done right.

It all hinges on doing the job right and having not only a satisfied customer, but one who will sing your praises until the roof sags! A builder like that never catches up with the demand for his work.

5

People will stand in line waiting for him.

There is only one rule if you want to stay in business: Do the job right and do what you promise. *Let your last job and every job you take on be your best recommendation.*

In this book you are going to learn something. If you're the type of contractor who uses a pickup truck as his office (as I do), you're exactly the person I'm writing this for. Stick with me for awhile. I'm going to lay it all out in the open.

By the time you finish this manual, you'll know everything that I've learned about additions and remodeling in the last 25 years. And none of this is technical stuff. I don't consider myself a genius. So you won't have any trouble following what I explain.

A brick mason can teach you how to lay bricks. But the fellow at the furnace usually doesn't know how to mix mortar. A carpenter can show you the best way to frame an addition. But only a builder knows how to put the whole job together.

Building something right is pure art. It takes an artist to transform a beat-up, out-of-date bathroom into a comfortable modern convenience the owner will brag about. This book will help you

begin to see yourself as an artist—or maybe the word is *craftsman*. You're the craftsman who takes pride in his accomplishments and wants the world to identify him with what he creates the same as a famous artist wants to be known by his masterpiece.

A word about money. Money is the reason people work. If you work, you should get paid for it. The high bidder with a good reputation is usually chosen over the low bidder with a questionable reputation. Know what the job is worth and charge accordingly.

There are contractors with small crews doing four or five jobs a year and putting money in the bank. There are contractors fighting for 10 to 20 jobs a year and having to borrow money to stay afloat. Know your capabilities, and use them to your benefit.

Over the last quarter century I've seen many big-time contractors with carpeted and paneled offices go bankrupt. I've witnessed the bankruptcy of many subdivision developers. But I've seldom seen a hardworking contractor go under. Can you succeed in this business? In the next chapter I'll begin to show you how.

2

The Art of Evaluating

Remodeling isn't that tricky. It's all a question of your attitude and approach. If you're about to embark on your first remodeling job, just remember to use care and common sense. Everything will fall into place with a little planning.

Mr. Brown is the owner of this home. It was built in the thirties and it's run-down. But it's in a good neighborhood, and a number of houses on the street have been repaired and repainted recently. It looks is if the area is taking on a new pride. The high price of gas and transportation has encouraged some to move back into the hub of the city. And the high cost of building or buying a new home has encouraged others to repair or remodel their existing dwellings. Consequently, areas that were once becoming abandoned or run-down are now taking on new life. And that means work for the contractor who is able and willing to take on repair and remodeling jobs.

You may be on Mr. Brown's street for quite a while. Perhaps when others see what you've done for Brown's house, they'll want you to do their work. That's one of the best—and cheapest—ways of advertising your services.

Brown wants a lot of work done: a new and bigger bathroom, and the kitchen done over completely with new cabinets and built-in appliances. He wants a partition wall knocked out and a family room built adjacent to the kitchen. Also, he wants to know if it would be better to remove all of the old, cracked plaster on the walls and ceilings. Should he put up sheetrock instead of trying to patch the crumbling mess?

He wants your opinion of the best method for replacing the wood floor on the front porch. Should it be concrete or wood? He prefers wood.

"Look at those windows," he says, following you around the house. "What should I do about those things? They look rotten."

And the heating system? He wants to know if the fireplace is still safe to use. Space gas heaters are scattered throughout the house. "What about a central system?" he asks. "How expensive would it be to have one installed?"

"And the floors need leveling. Can you fix that?"

You've only got a three-man crew. But that's all you'll need. Any more would get in each other's way. Also, the Browns don't want to move out while the house is being remodeled. They want the work done with them living in the home and with the least inconvenience possible. This is just the job for you.

Customers

As a builder gains experience, he automatically hones his ability to size-up his clients or customers. There are some people who decide overnight that it would be nice to transform their carport into a den or family room. They get on the phone the next morning, call a builder, and ask for a price ten minutes after the builder arrives. Next week they'll stop to look at a new car and dicker about a trade. They never buy anything; they just like to wheel and deal.

These people just waste your time. When you suspect that one is on the other end of your telephone line, make an appointment to meet with them in two or three days. Call them before you drive out there to keep the appointment. You'll save a lot of time and gasoline.

And then there's the cost-conscious penny pincher. No one can look with disfavor on thrift, but studs are no longer 50 cents. Skilled tradesmen don't work for minimum wage. This man wants an item-by-item cost breakdown before he can even decide whether he wants the work done. He constantly wants to know if there isn't an alternative method that is cheaper. He thinks that a 16 x 20 addition should be finished and ready for occupancy in a week. Anyway, he's just shopping for prices. When he really gets serious about adding that room, he'll search a year for the lowest bid. Let him search.

There are, of course, the dreamers. They want four rooms added to their house when, in fact, they can't afford a lean-to and have no way of borrowing the money. Such people are harmless enough and should be left to their dreams.

Then there are the building inspector types. They know more about how something should be done than the FHA. They go around talking about soffits, crawl space, headers, girders, and 4/12 pitch. The problem is that they don't know which is which. If they give you enough room to work, they're all right. If they throw too many questions at you, just start grunting your answers.

The silent inspector is another matter. He walks and looks. So long as he isn't saying anything, you're all right. When he does say something, pay attention. He's probably right. The way to keep him happy and silent is to do the job right.

And then there's the construction site lawyer. He's always suspicious. You're his main suspect. He's always looking for a breach of contract. Don't blame him. Most of your customers will be like him until your reputation is strengthened by quality work and satisfied customers. This type of customer makes the best reference, though. Once he decides that you can be trusted, he'll tell others about you.

There are, of course, other types and you'll meet them all eventually. Essentially, most of your customers will be hardworking and pleasant people.

Let's get back to Mr. Brown. You've decided that he's sincere about wanting his house remodeled. The bank is willing to loan him the extra cash he needs for the project. Now he wants some idea as to what it all will cost.

Evaluating a Building

Before determining how much work has to be done, we must first know what is required. The house was built over fifty years ago and has been neglected. We already know by a casual examination and by talking with Brown that the floor is uneven in places and that many windows will need replacing. Also, there's a sag in the center of the roof at the ridge, a few 6-inch clapboard siding boards need replacing, and the paint is peeling. But, on the whole, the house looks to be pretty solid. But is it?

The best way to find out is by a thorough, systematic, and patient examination. Don't ever agree to nail even as much as a single board onto a structure until you know that there's something solid to nail it to, and that the nailing doesn't violate the building code. Not knowing what is under the floor or behind a wall or its condition can cost the remodeler time and money. Don't ever agree that a floor can be leveled until you know *why* it isn't level. Don't nod your head in an "I-know-what-I'm-doing" agreement that the sag in the roof can be straightened out until you know why the thing sags.

By now you realize why I decided to remodel Brown's house. Everything is wrong with it. If we can remodel this home, we can remodel any house. Of course, Brown's house in New York is different from houses down South or out West. Every area of the country is different. But they don't differ that much, and building materials are about the same everywhere in the U.S. and Canada.

The procedure for checking out a foundation footing in Florida's ocean front sand bed is different from checking the footing of a house sitting on a big rock in Colorado. And checking out colonial mansions built before the Civil War will be different from evaluating a brick veneer bungalow built in 1950.

Attention to code requirements is essential—the powers that be won't tolerate the slightest code infringement. So it would help to have a building official come out and perform whatever inspection is required to determine code compliance in restoring the house. That probably won't be possible. So we have to do the next best thing—go over the house ourselves.

Let's make an inspection of Brown's house. We'll start at the bottom and work up.

Foundation

Brown's house sits on brick pillars. There's no

Number of Stories	Masonry or Masonry Veneer		Frame	
	Min. Thickness (inches)	Projection each side of wall (inches)	Min. Thickness (inches)	Projection each side of wall (inches)
One Story:				
Basement	6	4	6	3
No basement	6	3	6	2
Two Story:				
Basement	8	5	6	4
No basement	6	4	6	3

Footing Sizes
Figure 2-1

basement. It has what is called a crawl space; there's supposed to be room enough between the ground and the house to crawl around. Some houses are built very low to the ground. No matter. It's still crawl space construction. If you have to crawl around on your belly and work while lying on your back, adjust the cost for labor accordingly.

In most areas codes govern the height of foundations. There is usually at least 18 inches from the bottom of floor joists to the ground and at least 12 inches to the bottom of the girders.

While we're getting under Brown's house (which we are delighted to see is about 3 feet off the ground), let's talk about footings. A poor footing is trouble! It won't hold the house up. If a corner sinks, you've got a droop-eared house. If any part of the footing sinks under the weight, brick and finish walls will crack. Windows and doors will fall out of square and neither will work properly. Of course, floors will dip and be unlevel.

Figure 2-1 gives you the sizes of footings that will do the job for conventionally loaded wall footings on soil whose average bearing value is 2000 psf or better.

Under the house, notice that a foundation curtain wall of brick was added between the perimeter pillars sometime after the house was built. It looks good since there are no cracks in the masonry, and the mortar is hard.

Examine the pillars. Is the mortar falling out? Is it crumbly? Use a screwdriver and probe the joints. If you can scrape out the mortar without much effort then the pillars will probably have to be repaired or replaced, depending on the extent of deterioration. A number of old houses like Brown's have concrete pillars. These have to be checked for cracks and crumbling. Hairline cracks are to be expected in such pillars and in concrete foundation walls and don't affect the structural soundness. Open cracks are a different matter. They may get worse or they may not. If there is no sink or settlement in the structure, there probably is no problem.

In certain areas of the country you will find pier-type wooden posts. Check them carefully. Although these pillars were treated with a preservative, insect damage and decay are still possible. Under Brown's house you find the cause of the unlevel floor and the sag in the roof. Near the center of the house under a main girder are two pillars that have deteriorated, allowing the girder to drop down and sag over this span. The girder will have to be jacked back into position and the two pillars replaced.

Don't worry about it. There's a way to do it. We'll get back to that later. Right now there's another problem. Every builder comes face to face with it sooner or later so we might as well face it now and get it over with.

Brown, at one time, had a termite infestation. He had the house treated when he discovered it, but a lot of damage had already been done. You find it as you're probing the perimeter beam (or sill) with a sharp instrument (an ice pick is good for this) and discover that the sill was almost eaten up by the termites. The damage is over a pillar and runs about four feet in both directions along the sill.

It doesn't look like the bugs reached the joists or sub floor so perhaps the sole plate and studs are all right, too. You can check that out by removing a board on the outside or, as in this case, from the inside when you remove the plaster from the walls.

We'll tackle the problem of replacing that sill when we come back to jack up the girder to replace the pillars. We'll want to do both of these jobs while we're under the house. No one wants to spend any more time under a house than he has to.

Suppose Brown had a basement? What would you as a remodeler look for?

Basements

The time to waterproof a basement is when it is built. There are specific requirements to be met in the construction of a basement, depending on the area of the country. Some builders just don't know how to build one. Properly built, waterproofed basements are expensive to construct, and building them is a field all its own. We'll just touch on it here with an eye to what to look for when remodeling.

If anything can go wrong to cause a basement to leak or be damp, it will. Anything below ground level is in danger of contact with water—unless, of course, it's in an area where it never rains. Clogged drains, cracked walls, faulty or clogged downspouts and the lack of slope of the finish grade away from the house are a few things that allow water or dampness into the basement. The best time to check a basement is a few hours after a heavy rain.

Check the grading around the house. Is it such that water, including that coming off the roof, is directed away from the foundation? It should be. If it isn't, it could cause dampness.

In some areas the water table is higher than the basement floor. If this is the reason for the dampness, then there isn't very much that can be done about the problem.

As a remodeler, the less time you spend on faulty basements, the better. A properly constructed, waterproof basement is a delight. You can do wonders with it. A leaky, damp one is nothing but trouble.

Fireplaces and Chimneys

While we're still under Brown's house, let's take a look at the wires, pipes, and fireplace foundation. Let's find out what you're supposed to be looking for.

Somewhere along the line there was a popular notion that any brick or stone mason could build a fireplace and chimney and that it would work. Well, there are a lot of fireplaces around that won't draw, and more smoke enters the room than goes up the chimney. The notion was wrong. Later on we'll cover fireplaces. But right now we're at the base of the chimney.

How does it look? Check the joints with your screwdriver. Are they soft? Can you rake the mortar out of the joints? Tap the bricks lightly in several places with that 13-ounce hammer you brought along. Does the brick tap solidly with a nice little bounce, or does it crunch, crumble, or flake off? If it's in bad shape, it should be replaced.

Next, look for cracks in the base. Large cracks indicate that the footing has shifted or settled. A masonry chimney must always be supported by its own footing. When you get outside, you'll want to see if there are any cracks in the chimney or if it has shifted away from the house or toward it. When you crawl up into the attic, check the chimney there, too. Older houses may not have flue liners. A cracked chimney in this case is a fire hazard and should be repaired or the fireplace opening sealed.

The roof and ceiling framing should be no closer than 2 inches from the chimney. You'll find that this is not always the case. The wood often rests against the masonry. If the chimney is sound, the owner may decide to leave things as they are. But point these things out to him anyway.

Another thing you'll want to check when we get out from under the house is the fireplace damper. You probably won't find one. Brown may or may not want one installed. A damper serves several purposes and these are covered in the section on fireplaces. It's enough to say here that a damper can be closed when the fireplace is not in use. That prevents heating and cooling dollars from going up the chimney.

Unless you can find a damper made expressly for remodeling work, installing one in an existing fireplace is an expensive and major project. An alternative is to install a glass door unit on the fireplace opening and forget the damper. The owner gets more decor for his money and you don't run the risk of fouling up a nicely drawing fireplace. An add-on damper might alter the important dimensions of the existing fireplace.

While under the house you should check the adequacy of foundation vents for air circulation, the condition of the floor joists and subfloor, the condition and layout of the plumbing, the condition of the wiring, and the condition of the sills and girders. In other words, take a look at everything you can see from under the house.

Brown wants you to tell him what's wrong with the house and what has to be done to remodel it. But don't rush your inspection and evaluation. This is the ground work on which a number of decisions will be made.

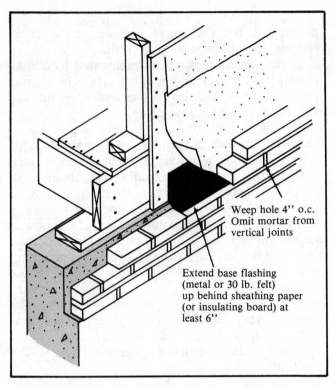

Weep hole 4" o.c.
Omit mortar from
vertical joints

Extend base flashing
(metal or 30 lb. felt)
up behind sheathing paper
(or insulating board) at
least 6"

**Weep Joints
Figure 2-2**

Brick, Stone, and Block Walls

The reason camels have big feet is because they're always standing or walking in sand beds. If their feet were like those of a deer, they'd probably sink up to their bellies. A house with "little feet" will do the same thing.

That's the problem when a masonry wall or pillar sinks. The feet (footings) are too small for the job.

If every part of a house would settle the same amount, at the same time, there would be no great problem. Few houses are so accommodating however, and masonry walls crack. Of course, small cracks can be grouted, the joints repointed and the appearance of the wall can be reasonably restored. Since mortar is hard to match, the more cracks there are, the worse the wall is going to look. A torch applied to the repaired joint after it is cured will give it an aged look.

Masonry walls soak up water like a sponge if they haven't been treated with a suitable paint or sealer. Since few exterior brick walls are painted or sealed, adequate roof overhang with the proper diversion of water from the roof is important. If there is no overhang, a clear sealer applied to the brick or stone will repel the water. Just make certain, however, that water can't be trapped behind the brick.

In new construction of brick veneer walls much of the problem of water seeping through the brick can be eliminated with weep joints. (See Figure 2-2.)

Structural Supports

Equally important with adequate footings are adequate structural supports. A stud is a post, but a post is not always a stud. When extreme weights are involved over a sizable span, as in basement and crawl space construction, the size and strength of girders, sills, posts, and pillars must be sufficient to handle the load. Figure 2-3 gives size requirements for girders over different spans.

Width of Structure	Girder Size solid or built-up	Maximum Span			
		Supporting Bearing Partition 1 story	1½ - 2 story	Supporting Non-bearing partition	Intermediate girders (other than main girder)
Up to 26' wide	4 x 6	–	–	5'-6"	7'-6"
	4 x 8	–	–	5'-6"	9'-6"
	6 x 8	7'-0"	6'-0"	9'-0"	12'-0"
	6 x 10	9'-0"	7'-6"	11'-6"	–
	6 x 12	10'-6"	9'-0"	12'-0"	–
26' to 32' wide	4 x 6	–	–	–	6'-6"
	4 x 8	–	–	7'-0"	8'-6"
	6 x 8	6'-6"	5'-6"	8'-6"	10'-6"
	6 x 10	8'-0"	7'-0"	10'-6"	13'-6"
	6 x 12	10'-0"	8'-0"	11'-6"	–

**Girder Spans
Figure 2-3**

In a house with a basement, steel posts are better. They can support wood or steel girders. A steel post set or bolted into the concrete floor, which has sufficient footing value, is about as solid and permanent as you can get unless the basement is constantly flooded with sea water. Where wood posts are used, they should be supported by a pedestal. If a post is set on or into the concrete floor, it will eventually decay due to moisture infiltration. Anytime wood is in direct contact with masonry, decay is probable. Check for decay where girders rest on exterior walls.

Sag can occur in steel as well as wood girders if the girder isn't strong enough. Sag is less likely in steel, but it has occurred. A little sag on permanently loaded girders (and most are loaded to the hilt) is seldom a serious problem unless it results in uneven floors, binding doors, or cracked plaster. If there is no more than a 1/2-inch sag over a 10- to 12-foot span, don't worry about it.

There is no such thing as a perfectly built house. Neither is there a perfectly built vehicle. Considering the purposes, most houses are built within closer tolerances than the transmission parts on your new truck. In both houses and vehicles the goal is to keep the tolerance within the acceptable boundaries. A pronounced sag (to the eye) should be corrected unless doing so will create other problems. It isn't worth fixing if it sets up a chain reaction that requires redoing the entire house.

Floor Framing
The floor of a house is comparable to the top of the dining table. It shouldn't dip, roll, or spring. The quality of lumber you sometimes get from the mills these days isn't too good. Use care in selecting the pieces used in both structural and appearance applications.

In older houses like Brown's, the sills, girders and joists are generally of better quality than what is available now. However, most of these houses were not given the benefit of termite shields and wood preservatives that are used today.

You have already located the problem under Brown's house. The girder has sunk because of pillar deterioration. This is not a sag in the general sense. When evaluating a house or building, look for sags in floor joists as well as in girders and sills. Partition walls running parallel to the floor joists should have double joists big enough to eliminate sag. If they don't an additional joist can be added after the sag is jacked out.

Again, don't bother if the sag is not a serious

problem. If the wall above is plaster or sheetrock, jacking may damage the wall finish.

In floor areas sags can be eliminated by adding extra joists. A jack here is fairly safe. The sagging joist can be jacked up to the desired point and another joist nailed into place.

In new construction, it's common practice to have the bows or crowns of joists angle upward. This often causes a serious problem. While the idea is sound (on completion building weight will push the bow down to its proper position), it doesn't always work. A bowed 2 x 10 floor joist requires a lot of weight for proper positioning. If the bow is set, it will not come out. The safest rule is this: don't use the joist if the bow is excessive. Anything over a 1/4-inch bow for a 10-foot span is excessive.

If the bow is excessive, saw the joist in the center until it drops into place and then nail a new joist beside it when the floor is in the proper position. Jumping up and down on the floor over the cut joist will usually return a reluctant floor to the position desired.

In built-up girders (such as two 2 x 10's nailed together), excessive bow can be used to advantage. Reverse the crowns, putting one up and one down. Pull them flush with a clamp and nail them together. The excessive crowns or bows of the individual pieces will be eliminated. If a slight crown exists in the built-up girder, place the crown up when putting it in place.

In cases where structural strength would not be compromised, the crown can be eliminated. Chalk a line down the length of the piece and saw if off. For neatness, nail the piece sawed off on the opposite side of the board. It'll fit like a piece in a puzzle. (See Figure 2-4.)

Once again, if the floor is springy when you walk across it, add extra joists or an extra girder.

Always pay close attention to the framing around stairway openings. The right way to frame this opening will be covered in the chapter on steps and stairs. It's amazing to me how many builders don't frame these openings properly and fail to support the stairs. When you are carrying a heavy piece of furniture up or down stairs, you shouldn't have to tip-toe lightly over a weak obstacle course. An unlevel or sagging floor at a stairway opening is cause for a close inspection and corrective action.

Wall Framing
In a two-story house, half of the studs in the first floor could be removed and the second story

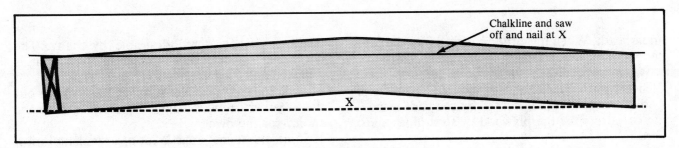

Eliminating the Crown
Figure 2-4

should still have sufficient post support to remain securely aloft. If post strength was all that was required, that would be fine. But this is not the case, as you already know. So there is no need to worry whether a house has enough studs in it to hold it up if you get a job where a second story is to be added. Of course, you may have to get an engineer's calculations to prove that before a building permit can be issued.

If the foundation sinks, so will the wall. Use a four-foot level on it. Check the squareness of the windows and doors. Do doors and windows operate properly? Are headers over wide openings sagging? If so, replacement may be the only answer, depending on the amount of sag.

It's not unusual to find an old house leaning a bit. If the floor is unlevel to the degree and in the direction of the lean, then settling is the culprit. If the floor is level and the walls lean, then the forces of nature are probably to blame. The house is not necessarily structurally unsound or unsafe.

In one seacoast town of a southern state a large number of two-story wood frame houses are noticeably leaning. They have been for a number of years. No one seems to give the matter much thought and the occupants are certainly not standing around waiting for the things to topple over. If houses lean more, the occupants will, no doubt, seek safer living quarters elsewhere.

So just because a house is slightly tipsy, don't condemn it. If the owner wants it remodeled, do it. (Think, though, before you start on the doors. Just remember that door frames have to be square *and* plumb.) Don't agree to snatch the house back upright. Some things are best left to their own angle.

Obvious defects in wall framing such as bows in or out require correcting.

When adding a room to a house, check the squareness of the house, particularly where the outside wall of the addition will be an extension of the outside wall of the house. Remember, the house is a box. If that box isn't square, then a square addition with a common outside wall won't look right. (See Figure 2-5.) The recommended procedure here is to forget about squaring the addition. Extend the common wall straight in line with the wall of the house and then work from that point.

Roof Framing
The roof of Brown's house sags as we have already said. You found the cause for that while inspecting

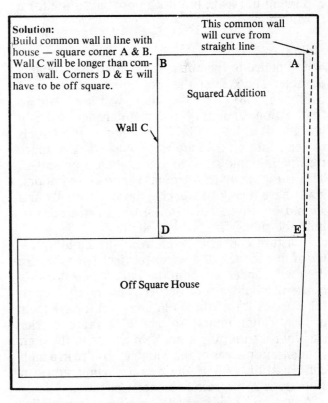

Solution:
Build common wall in line with house — square corner A & B. Wall C will be longer than common wall. Corners D & E will have to be off square.

This common wall will curve from straight line

Squared Addition

Wall C

Off Square House

Squaring the Addition
Figure 2-5

underneath. When the girder is jacked into the proper position, that problem should be solved.

When inspecting a roof, check for unevenness such as waves in the roof caused by sagging rafters. Look for sheathing sag caused by too wide a spacing of rafters for the plywood or sheathing board used.

Sometimes a ridge sag is caused by insufficient bracing or by slippage of rafters at the wall plate. This has nothing to do with settling of the house. A sag in long rafter spans can also be the result of insufficient bracing. We will cover roof framing in detail later.

From the ground you can tell if the roof is straight and uniform. In some cases irregularities and lack of a straight roof line are caused by poor construction. You can't do much to remedy the problem without a major redoing of the framing. That is hardly justified when the only objection is appearance.

A properly constructed roof ties the house together. It gives the whole structure strength. It could be said that the roof is the most important phase in house construction. Some builders think that any roof that doesn't leak is O.K. A shower cap might not leak, but it's a poor substitute for a hardhat.

Siding
Wood nailed to the outside of a house is extremely durable. The greater the protection given by roof overhang, the longer the siding will last. Add to that periodic painting and maintenance and it'll last almost as long as brick. The important thing is to keep out excessive moisture. Any crack or space that permits moisture to get in should be sealed. This is especially true around windows and doors. Nails have a way of working loose eventually and should be driven back in. Drive in extra nails to hold the siding firm at that point.

Moisture can also enter from the building interior due to lack of a vapor barrier. The moisture then will condense within the wall. When Brown's house was built, vapor barriers were rarely used in the area. A lot of these old houses still have their original siding intact. So lack of a vapor barrier usually isn't that big a problem except in the high condensation areas of the country. You're in a high condensation area if the average temperature in January is 35 degrees Fahrenheit or lower.

Any type of siding will deteriorate eventually. Builders used to say that shingle siding needed replacing if it looked ragged. Now, that ragged appearance is considered desirable by many. The weathered look is in. Even the siding of old barns is removed and used as paneling in new houses to add a rustic look. Unless the shingles fail to keep out the weather, replacement is merely a cosmetic choice for the owner.

Some old houses boast a lot of ornamental trim. Others sport a budget type trim. Whichever, trim deteriorates because it traps a lot of water. Replacement is simple, except for ginger bread trim. But it's easier to find a part for a Model A Ford than it is to match elaborate ornamental trim. Given time, most skilled carpenters can whip up a suitable replacement part for repair work.

Windows and Doors
The windows are in bad shape in Brown's house. And the front door binds when it is closed. Something will have to be done about these problems.

Windows are always a problem because they are the weakest link in the chain. They are the eye of the house. Like any eye, they can only be protected to a certain extent.

In old houses where the window unit is all wood, it's impossible to seal out the draft and still be able to open the window. Suppose you fit everything nicely so that most of the draft is sealed out but the window will still open. When the first cloud appears the window will stick! Effective repair of these old windows is practically impossible. The best thing to do is replace them.

The problem will be finding the right size replacement. Remodeling type replacement windows are now made by several manufacturers and are available through many distributors. These windows will fit nearly any size but cost considerably more than standard windows made for new construction. The only other choice is to have a custom window made to fit that job. That's a little more expensive but not nearly as expensive as redoing a whole wall.

Window sills are usually the first part of a window to go. Sills can be replaced and should be when the rest of the window is solid.

Insulated windows are available and should be used in cold or hot climates. They also help avoid condensation.

Brown's front door binds. But the door frame is not out of square by that much. The problem is that the door was not installed properly to begin with. Prehung doors are popular because few carpenters have ever learned how to hang a door properly. It seems like a simple matter to install a door, but it's not. It's easier to make a new muffler fit an old car than it is to size a door and install it right. There aren't many carpenters around who will brag about their door-fitting expertise.

You have to weatherstrip exterior doors to keep out unwanted air. This isn't a problem for most doors. But some doors are warped and won't latch properly. Sometimes the latch keeper or striker plate will have been moved four or five times. In some cases someone will have nailed the keeper or plate down for keeps with No. 16 box nails.

Suppose you have a warped door and frame with relocated and nailed down keepers. If the opening is standard size, remove the whole unit and install a prehung door.

If the door and frame aren't standard size, tread softly. Your work is cut out for you. Sometimes loose hinges are the cause for ill-fits and binds. Screws may have worked loose or wallowed out of the wood. If that's the case, longer screws will correct the problem. A little epoxy glue on the screw will help.

The front door on Brown's house hasn't been damaged by weather. But a storm door will help make the house more energy efficient. Metal storm doors can be a problem to install if the door frame is out of square. This is one of those 30-minute jobs that can take a half day. If it is out of square, reframe the opening just enough so that a good fit is possible.

A storm door has a handle or knob. So does the exterior door. Sometimes they meet in the same place and make contact, preventing the complete closing of the storm door. In such a case the handle of the storm door will have to be lowered or raised, or the storm door furred out at the frame.

Thresholds can be a problem too. Metal thresholds with a vinyl seal are not much of a problem to install, provided the wood underneath is solid. If it isn't, replacement is necessary. The metal threshold with two parts—one that fits on the bottom of the door and one that fits on the floor—is a good type to use. It eliminates the need for precision cutting the door. The vinyl seal fits on the door and is not subject to traffic wear.

Interior doors are less of a problem. They should close and open without sticking or dragging on the finished floor. They, too, are subject to warping. If the warp prevents proper fitting, replace the door. In houses with central heating and cooling, allow a little extra clearance at the bottom of interior doors. About 1/2 to 3/4 inch is enough. This promotes circulation between rooms even if the door is closed.

Porches

Brown's porch is in bad shape. The floor is wood, and while it has enough slope for water runoff and plenty of roof protection, it still shows weather damage. In checking underneath you find that the joists and sills will also have to be replaced unless Brown wants to go to concrete. Lumber treated with a preservative should be used in replacing the porch, including the wood posts which rest on the floor. Wood posts should be raised off the floor slightly to allow air flow underneath for drying. Metal pedestals are available for this purpose.

Outside Finishes

All homes with wood siding need regular repainting. It is an expensive and a recurring event. Excessive moisture in the wood is the biggest cause of finish damage and reduced paint life. Moisture can be caused by direct contact with the weather or from moisture in the walls.

Paint failure can also be caused by inferior paints, unprofessional application of quality paints on improperly prepared surfaces, and applying the wrong kind of paint over a previous paint.

Don't grab any kind of paint and slap it on an unprepared surface. Certain things have to be done the right way. Painting is one of them. If there's excessive paint build-up, you're wasting time and paint if you apply another coat. Remove paint build-up as the first step.

Roofing

Water coming into the house from any source other than the plumbing demands immediate correction. A little trickle of water through the roof or from faulty flashing can result in expensive repairs unless the problem is solved quickly. In some areas, condensation overhead is a problem. But that's not the roof's fault.

Asphalt shingles will last 15 to 20 years and are probably the most popular roof cover on residential buildings. Many different weights and thicknesses are available. The surface granules form the tough hide of the shingle that faces the weather. The granules begin coming off when the shingle is handled and continue to do so through the shingle's life. The faster the granules come off, the shorter the life. Heavy rains, wind, ice, snow, and sun contribute to the loss of these granules.

Eventually the shingles will become dry or brittle. Water and extreme temperatures also wear down shingles. The grooves between the tabs and connecting shingles become worn through, and they may not even be noticed unless you get on the roof and examine them closely.

Wood shingles made of cypress or cedar will last 25 to 30 years and are also affected by the weather. These shingles are used only on houses with a suitable pitched roof. In many areas, the wood shingle roofs have higher fire insurance rates. Wood shingles also need plenty of air, inside and out, for drying. They should not be installed on solid sheathing as asphalt shingles are. Wood shingles are used mostly for their rustic appearance. The older they get the more rustic they look. Don't tell a homeowner that his wood shingles need replacing just because they have lost that uniformity of appearance that a fairly new wood shingle roof has. It might have just the rustic tone he's been waiting for.

Broken, split, or upturned shingles might need replacing. No one can look at a wood shingle roof and tell very much about its ability to keep water out of the house. All wood shingle roofs look like they ought to leak, especially when viewed from the attic side.

Built-up roofing will be found on flat or very slightly pitched roofs. Such a roof, if properly built, will last 25 to 30 years. Bubbles, blisters, soft spots, and cracks will appear in such a roof over time. However, a new coat of tar will give the roof a little extra life. If the deterioration is excessive, a new covering is recommended.

Metal or tin roofing can be found on many old houses in some areas of the country. The popular styles are V-crimp and corrugated. Galvanized metal comes in various lengths and is about 26¾ inches wide with a 24-inch coverage. A metal roof will last 25 to 30 years. In some cases they have lasted much longer. The metal need not be replaced merely because it is rusting. A special fiber paint is available for covering metal roofs.

An easy way to cut metal roofing when repairing or replacing it is with an old power saw blade. Put it in the saw backwards and saw away. The reverse blade will make a straight and fast cut. Safety glasses or goggles are recommended during the cutting.

Other types of roofing are slate shingles and asbestos-cement shingles. Slate is laminated rock split into thin slabs and comes in different sizes and thicknesses. An average size is about 10 inches wide and 20 inches long and ranges in thickness from 3/16 to 1 inch. Asbestos-cement shingles are a composition of asbestos fiber and portland cement. They are rigid and fire resistant.

Concrete or clay tile roofs are popular on Spanish style structures. These roofing materials have a long life, as might be expected. Some have lasted beyond 40 years and look as if they're good for 40 more. Special equipment and know-how is required for extensive repairs of these roofs, so call in an expert.

Leaks usually occur around chimneys, valleys, skylights, vents, and where a roof joins a wall. Metal flashing should be checked closely. Metal corrosion is a common cause of leaks.

A leak in a roof can be a hard thing to find. Water can enter at one point and travel down the underside of the sheathing or rafters a good distance before it drops to the ceiling. The steeper the pitch, the farther water will travel. About the only safe assumption that can be made about a leak is that the water probably didn't travel up the roof. Look for stains on the sheathing or rafters to locate the entry point.

Wall and Ceilings

Brown's house, like many old houses, has plastered ceilings and walls. Gypsum board (sheetrock) is now the most popular wall and ceiling finish material. But it wasn't before the 1950's. Wood paneling was also used extensively. Many houses had at least one paneled room; some were paneled throughout.

The bad thing about plaster is that it cracks. Small hairline cracks will appear in plaster not long after it's finished. Then larger cracks will appear. Then chunks will fall out. The closer the house is to an airport, railroad tracks, or a major highway, the faster the deterioration will be because of vibration. In Brown's house the ceilings and walls have cracks and chunks out. Here we are going to have to take the plaster off the walls and ceilings. In another situation, the owner might want the plaster repaired. The best thing to do with deteriorated plaster, in old houses, is to get rid of it. Plastering is an art, and in some areas good plasterers can't be found. Sheetrock has put them out of business.

The use of wallpaper runs in cycles. Many old houses have been papered several times. If you're remodeling a room where there are several layers of wallpaper, apply paneling over it if there is something solid under it to hold a nail. Don't try to nail paneling to plaster walls; you'll end up with a mess. If the plaster is solid, paneling can be glued to the wall. The best bet is to remove the plaster and lathing first.

You don't want to keep applying wallpaper over wallpaper. If there are several layers on the wall already, strip them off. If it's an economy job, then one more layer probably won't hurt provided the walls are smooth.

Wallpaper can also be painted over, if it's reasonably smooth. The best rule is to remove the paper if the surface underneath is better for painting. If it isn't, leave the paper.

Repainting is about the easiest and most economical method of finishing a wall or ceiling. Chips, dents, and scratches will show through the paint and should be repaired first. Old, loose paint should be removed. Some painted surfaces will be incompatible with certain new paints. Determine what is on the wall (oil base, latex, etc.) and find a paint that will be compatible.

Flooring

Taste in flooring changes like everything else. The once popular hardwood floor is now seldom used in new construction. A wood floor is expensive both to install and maintain. Carpet, sheet floor covering, tile, and plywood blocks have more or less replaced hardwood. Pine (softwood) still exists in some older dwellings.

Brown's house has hardwood floors of the narrow oak board variety. Look for buckling which could have been caused by dampness or wetting. Are the boards separated? Cold will cause the boards to shrink, and heat will expand them. If the flooring is in an unheated room, there might be a thin crack between the boards due to the cold. And if the boards have contracted, the floor might squeak when walked on. Keep these things in mind while evaluating conditions during the inspection. The wider the board, the greater the shrinkage will be.

Resanding wood floors can make them new again. But there is only so much thickness available for sanding on any wood floor. Check this carefully. The first sanding might have been excessive to smooth out an uneven installation job. Under average conditions, two or three sandings are all that can be done to a wood floor; probably less to a plywood block floor. If there are wide cracks in the floor, or if it is too thin to sand, a new covering should be installed.

Check resilient flooring for broken, loose, chipped, or dented tiles. Are there ridges? Is uneven-ness in the underlayment showing through? Tile patterns and colors change with the seasons. Seldom will matching tile be found in the stores. Even if they are, they won't match the color of the aged and worn tile on the floor. A decision has to be made as to whether a new covering is needed.

Trim

The fitting of crown and base moldings and window and door trim are finish jobs which should be left to a skilled carpenter with a lot of patience. If no one is available, then do it yourself. It's the only way to learn. The joints must be smooth and tight. The trim should fit close to the wall. Replacement of trim and molding is tricky in some houses because the pattern has been discontinued and is no longer available. If such is the case, it might be more practical to take matching trim from one room to make repairs and use new trim in the cannibalized room. Sanding trim and preparing it for a new finish can cost more in manhours than replacement would cost. Keep this in mind when you're working on trim with ornamental design.

Cabinets

Cabinet doors and drawers should function properly without ill-fits. Minor adjustments can often be made in cabinet hinges and latches to correct minor problems. Binding or sticking drawers can usually be repaired by sanding. Refinishing cabinets requires considerable manhours, particularly where ornamental design or work is involved. Repainting involves less work but some preparation for the new coat is usually necessary.

Damaged countertops usually need replacing because of the nature of the covering and the difficulty in repairing them satisfactorily.

Interior doors are finished in the same way as cabinets and should fit and function properly. Most interior doors are the flush, hollow core, or paneled type.

Insulation and Ventilation

In the attic of Brown's house we find that there is no insulation. There is a vent at both ends of the attic in the gables. They appear adequate in size for the area vented. Usually 1 square inch of venting for 1 square foot of ceiling space is adequate.

A lot of winter heat and summer cooling is lost through uninsulated ceilings. There are four types of insulation: loose fill, reflective, batt, and blanket. Loose fill comes in pellet or granulated form which is poured or blown in. Mineral wool, vermiculite and the wood fiber products are good examples in this category.

Reflective insulation reflects radiant heat. A bright, metallic material such as aluminum foil is used for this purpose. Batt insulation is usually paper-backed and is made out of glass, rock wool, or minerals. Blanket insulation is wood fiber, cotton, or wool.

Six inches of insulation in the ceiling is recommended in mild areas. Colder or hotter areas require 9 to 12 inches. Mr. Brown wants the walls and floor insulated as well as the ceilings. With the high cost of energy, insulation is essential and should be included in any remodeling job. The cost will be offset by a savings in heat and cooling expenses in the first 4 or 5 years.

Plumbing

It's better to get chased by a bow-legged bull dog than tangle with a bunch of old pipes. Unless you own a 1½-ton truck and can carry four each of every fitting and pipe, then call a plumber when you need a commode moved or a sink installed. Even plumbers have to chase down fittings. And they usually carry two of everything!

Brown is on city water. The pressure is good. When the commode is flushed the water continues to flow steadily out of the faucets. The main cold and hot water distribution lines are 3/4-inch galvanized pipe. The take-offs are 1/2-inch. They seem to be in good shape, but you can never tell about galvanized pipe until you put a pipe wrench to it. Then it's too late.

When you turn the faucet off quickly the pipes hammer. Either there are no air chambers or they are waterlogged. An air chamber is merely a short extension pipe, capped, in the supply line near the fixture lines. Air chambers absorb the shock of sudden stops much the same as shock absorbing bumpers on cars absorb the shock of minor impact.

Hammering is not that much of a problem in plastic pipes. As a matter of fact, many of the problems inherent in metal pipes are eliminated by the use of plastic pipes. Plastic lines can be cut with a hack saw or hand saw and stuck together with glue. Repairing a broken line doesn't require four trips to the store and three hours of cutting threads and screwing on unions. And plastic won't rust.

Metal drain lines offer the same disadvantages as metal water lines. Happily, plastic pipe has replaced them, too.

Adequate venting of drain or sewage lines is essential. The plumbing code is the best guide here. If there is too much suction in the drain line, water in the commode and traps is pulled out. This allows sewage gases to escape into the house.

Slow drainage indicates blockage in the lines or a faulty fall in the flow line. In a private sewage system such as a septic tank and drainage field, the problem could be in the line to the tank, in the tank itself, or in the distribution field. If the line from the house to the tank is clear, the tank should be pumped out. Unless the distribution lines are clogged by tree roots, this will usually solve the problem. A dose of baking soda about once a month is a good prescription for keeping a septic tank regular.

Hot water or steam heat systems are still being used in older houses. These need to be checked for efficiency and overall condition. Hot water heaters must also be checked for performance, adequacy, and condition. When planning a bathroom addition, a larger heater may be necessary. The longer the line from the hot water source, the more hot water is needed. Hot water energy is transferred to the water line itself.

In rural areas, the water supply is often a private well. Improper pressure setting on the well controls can cause too little or too much pressure. With some above ground pumps the water tank can become water logged. Health codes specify the minimum distance between wells and sewer systems and livestock operations.

Electrical

In the 1930s a 60 amp service entrance panel was sufficient for lights and the few electrical appliances available. With modern appliances and equipment, a 150 to 200 amp service is necessary. Brown's 100 amp service panel needs replacing. He wants a central heating and cooling system with a gas-fired furnace, built-in dishwasher, and built-in oven and stove top. In addition, he wants to get a freezer later. We'll install a 200 amp panel with a 3/0 copper or 4/0 aluminum entrance cable. That will take care of a house using modern equipment.

The wiring underneath the house and in the attic looks O.K. Usually if wiring insulation is brittle, damaged, or otherwise deteriorated, replacement is a good idea. Armored cable will usually prevent the wire from being damaged. However, seriously rusted cable should be replaced. Fortunately, you won't find many cases where a rewiring job is necessary. Electricity hasn't been around that long in some areas.

Remodeling time is a good time to add those extra wall outlets that are needed. Old houses usually have too few outlets. There should be a minimum of one outlet to a wall. Long walls require more.

Every room needs a wall switch to the ceiling light. If there is no ceiling light, the wall switch should control a wall outlet for plugging in a lamp.

Faulty or damaged wiring has probably resulted in more house fires than any other cause. Compliance with codes and professional installation of wiring are essential. Many electricians like to down-size the fuse or circuit breaker to provide an additional safety margin. A 12-2 copper wire has a 20 amp capacity. They would use a 15 amp fuse or circuit breaker. The wire can't get hot because the fuse or breaker cuts out before damage can occur.

Heating and Cooling

Heating and cooling systems designed for houses are always being modified, changed, or updated. Solar energy systems are in various stages of development. Some solar systems work quite well. Most remodelers aren't too concerned with solar systems yet. But you need enough knowledge of heating and cooling systems to make a preliminary evaluation. You also need the sense to know that a professional is required to make a thorough examination.

There are three types of warm air central systems: forced air, gravity and radiant. The forced type uses a fan to force the heat through ductwork to various rooms. A return air duct sucks air out of the house, returns it back to the heat source and into the duct system again.

The gravity system may be the pipeless or piped type. The piped variety works much the same as the forced air unit, but without the fan. The furnace is located in the center of the house in the basement. Heated air rises. Cooler air enters the system at the intake at the lower part of the furnace, passes through the furnace heat chamber, and rises to flow to the rooms through the duct.

The radiant system forces warm air or water through ducts under the floor slab.

All of these systems have a furnace which uses gas, oil, or coal as its energy source.

Steam and hot water systems are still operating efficiently in older homes. The steam system converts water to steam in a gas, oil, or coal furnace. The steam travels through pipes to radiators, loses its heat, turns back into water and flows to the boiler. In the two-piped system the steam passes to the radiators and the water (after condensation has occurred) returns by a separate pipe connected to the opposite end of the radiator. In a one-pipe system the water returns by gravity through the same pipe used by the steam.

In a hot water system there is no steam. The hot water circulates from the boiler through pipes to radiators and returns to the boiler in a return pipe. The water may be circulated by a pump or by gravity.

There are many types of electric heating systems. They generally are the cheapest type to install. These include wall heaters with or without fans, baseboard heaters, heat pumps, wall panels, or ceiling cables. Electric systems are usually found in newer homes and apartments with adequate insulation in the floors, walls, and ceilings.

You'll also find houses with a combination of heating systems. One common combination is central steam with gas space heaters in rooms where quick heat-up is required.

Cooling systems can be an integral part of the forced warm air system or can be separate window units. In dry areas, an evaporator type cooling system is used to balance the humidity for comfort. It's common to integrate the cooling system with the heating unit: cooling coils are installed in the air plenum, usually adjacent to the heating unit. Cold air is circulated through the ductwork in the same manner as hot air. The thermostat has a cool and warm setting and operates both the furnace and the air conditioning unit. The air conditioning condenser is located outside the house and is connected to the cooling coil by pipes.

Window units, which can also be installed through an exterior wall, are available in one room or larger sizes. The unit is self-contained. Your only job will be installing or removing them.

House Design

The design or layout of a house is important. Some houses are poorly designed for room layout and traffic patterns. The idea is to reduce travel through rooms to get to other rooms and to avoid traffic passing through the center of a room.

Hallways should be wide enough for two-way traffic. Kitchens should be near back or side doors for easy transfer of groceries from the car. Utility rooms should be near the kitchen. Bedrooms should be isolated as much as possible from general living areas and should have convenient access to a bathroom.

The basic design is to have a place to cook, to eat, to sit, to sleep, and to bathe. From this basic plan all others derive. Forty-room mansions are mere extensions of the basic design with things like game rooms, library, powder rooms, study, office, and secret rooms thrown in.

One-room cabins are cozy, easy to heat, and easy to maintain. Few people would want to live in one, however. So, when planning step-saving, easy to heat, and cozy remodeling projects, keep in mind that anything can be overdone. Comfort, spaciousness, convenience, and economy should be meshed into a uniform package acceptable to the particular customer and his own peculiar tastes.

Destructive Insects

If it's wood, there is a bug around looking to make a meal out of it. Termites will work up through masonry walls and eat the rafters. In some areas of the southern part of the U.S., they'll start working on the lumber at the building site as soon as it's unloaded. If you don't hurry, they'll eat it up before you can nail it up. The three worst kinds of insects are termites, powderpost beetles and carpenter ants.

Termites come in two varieties: subterranean, which must have access to the ground or another water source; and non-subterranean, which do not require access to water. This latter type is usually found only in Hawaii and the narrow coastal areas of central California and Virginia.

The presence of powderpost beetles is shown by their borings which have a flour-like consistency. Look for them in damp areas near the ground. The wood will have a number of small holes around 1/8 inch in diameter.

Carpenter ants chew out a nest, shoving the sawdust out.

The best solution for these pests is a permanent one. Pest control companies will treat the house to protect against these insects.

Summary

Don't hurry your evaluation. Use a checklist to make certain that you don't miss anything. (See Figure 2-6.) Hidden problems can cost money and manhours. Look for them, and don't take anything for granted. As a remodeler you're expected to offer alternatives in design and decoration. You're expected to have the background and touch of an interior decorator and the eye of an architect. It is a challenging occupation. You can make it a successful one.

In the following chapters we'll plan for and remodel Brown's house. The house contains practically every basic problem that you will meet as a remodeler. You will learn each phase of a complete remodeling job, from computing materials to figuring manhours; from digging footings to nailing on the ridge cap.

It is not your job to decide whether any particular house in any particular location should or should not be remodeled. If the owner wants to remodel his home, that's his business, regardless of where the house is located.

You are in business to remodel. If the owner wants to spend thirty thousand dollars redoing a ten-thousand dollar house, then that's his business, too. You provide professional services at a fair profit in accord with your contract terms.

A fancy remodeling job in a ritzy neighborhood is high-class advertising. So is a good remodeling job in a worn-out neighborhood. Everyone has to live someplace. The fact that a homeowner wants a fancy bathroom in his drab little house shows his excellent taste, not an absence of good sense.

Home Evaluation Checklist

Address _____

Phone_____ Date_____

Evaluated by_____ Owner_____

Year Built_____ Year Last Remodeled_____

1. Foundation. Type _____
☐ Evidence of deterioration_____
☐ Water in basement_____
☐ Uneven settlement_____
    ☐ Windows or doors cracked _____
    ☐ Frame distorted_____
    ☐ Mortar or grout loose _____
☐ Support or pier settlement _____
☐ Decay in wood supports _____
☐ _____

2. Basement. Average depth below floor joists_____
☐ Water present_____
☐ Walls cracked_____
☐ Walls damp _____
☐ Surface water entering basement_____
☐ Waterproofing needed _____
☐ Suitability for remodeling into rooms_____
☐ _____

3. Masonry Veneer. Where located _____
☐ Cracked _____
☐ Repointing needed_____
☐ Mortar joints deteriorated _____
☐ Sandblasting needed _____
☐ Flashing rusted or missing _____
☐ Caulking needed _____
☐ Separating from wall_____
☐ _____

4. Masonry fireplace and chimney. Type _____
☐ Cracks present _____
☐ Repointing needed_____
☐ Mortar joints deteriorated _____
☐ Sandblasting needed _____
☐ Flashing rusted or missing _____
☐ Caulking needed _____
☐ Cap deteriorated _____
☐ Foundation missing or cracked _____
☐ Separating from wall_____
☐ Evidence of rain entry or smoke exist _____
☐ Fireproof lining needed _____
☐ Wood frame closer than 2'' to chimney _____
☐ Damper needed or needs repairs _____
☐ Glass door unit recommended_____

☐ Draws poorly _____
☐ _____

5. Floor Supports
☐ Girders deteriorated at bearings _____
☐ Posts not on pedestals _____
☐ Posts decayed_____
☐ Girders deflected over ½'' _____
☐ _____

6. Floor framing
☐ Sill plates deteriorated _____
☐ Decay or insect damage _____
☐ Joists deflected excessively _____
☐ Floor springy under load _____
☐ Floor deflection around stair openings _____
☐ _____

7. Wall framing
☐ Distorted _____
☐ Headers over windows or doors distorted _____
☐ _____

8. Roof framing
☐ Ridge deflected _____
☐ Rafters deflected _____
☐ Sheathing deteriorated _____
☐ Surface springy under load _____
☐ Inadequate internal support_____
☐ _____

9. Siding and Trim. Type of siding_____
☐ Inadequate roof overhang _____
☐ Decayed or deteriorated _____
☐ Vapor barrier needed _____
☐ Cracked _____
☐ Loose or warped _____
☐ Needs caulking _____
☐ Needs renailing _____
☐ Needs painting or resurfacing _____
☐ _____

10. Windows. Type _____
☐ Loose fitting _____
☐ Need weatherstripping _____
☐ Inoperative_____
☐ Decayed sash or sill _____
☐ Warped _____
☐ Hardware missing or broken _____
☐ Windows not standard size _____

Home Evaluation Checklist
Figure 2-6

☐ Need storm windows _____
☐ Need double glazing _____
☐ Inadequate window area _____
☐ Inadequate ventilation area _____
☐ Need resurfacing _____
☐ _____

11. Doors
☐ Fit properly _____
☐ Warped or damaged _____
☐ Hardware missing or broken _____
☐ Not weatherstripped _____
☐ Obsolete design _____
☐ Need resurfacing _____
☐ Fail to latch _____
☐ Need storm doors _____
☐ Frame out of square _____
☐ Threshold decayed or worm _____
☐ Trim missing or damaged _____
☐ _____

12. Porch
☐ Flooring decayed _____
☐ Supports decayed _____
☐ Steps in contact with soil _____
☐ Insect damage _____
☐ Trim damaged or missing _____
☐ Needs refinishing _____
☐ _____

13. Roofing. Type _____
☐ Evidence of leaks present _____
☐ Asphalt shingles. Year of application _____
  ☐ Losing surface granules _____
  ☐ Shingles missing or broken _____
  ☐ Worn spots _____
  ☐ Nails loose or missing _____
  ☐ Valley worn or inadequate _____
  ☐ Ridge missing or deteriorated _____
  ☐ Poorly applied _____
☐ Wood shingles. Year of application _____
  ☐ Worn or deteriorated _____
  ☐ Shingles missing or broken _____
  ☐ Shingles loose _____
  ☐ Nails loose or missing _____
  ☐ Poorly applied _____
  ☐ Valley deteriorated _____
  ☐ Ridge missing or deteriorated _____
  ☐ Warped or upturned _____
  ☐ Fungus growth _____
☐ Built-up roofing. Year of application _____
  ☐ Blistered _____
  ☐ Cracked _____
  ☐ Water accumulation _____

☐ Gravel surface deteriorated _____
☐ Flashing deteriorated _____
☐ Caulking missing or deteriorated _____
☐ Gutters and downspouts deteriorated _____
☐ Attachments or decorations on roof _____
☐ _____

14. Flooring
☐ Wood flooring. Type _____
  ☐ Buckling or cupping _____
  ☐ Separated or loose _____
  ☐ Rough surface _____
☐ Resilient. Type _____
  ☐ Loose _____
  ☐ Cracked _____
  ☐ Broken _____
  ☐ Missing _____
  ☐ Uneven surface _____
  ☐ Discolored _____
☐ Carpeting. Type _____
  ☐ Worn _____
  ☐ Spotted _____
  ☐ Needs cleaning _____
  ☐ _____

15. Walls and ceilings
☐ Plastered or wallboard. Type _____
  ☐ Cracks _____
  ☐ Holes _____
  ☐ Uneven surface _____
  ☐ Needs new finish _____
☐ Paneling. Type _____
  ☐ Chipped or cracked _____
  ☐ Discolored _____
  ☐ Needs new finish _____
☐ Wallpaper. Type _____
  ☐ Peeling or worn _____
  ☐ Discolored _____
  ☐ Needs to be removed _____
  ☐ _____

16. Cabinets and trim
☐ Need refinishing _____
☐ Obsolete style or design _____
☐ Inadequate space _____
☐ Hardware needed _____
☐ Trim damaged or missing _____
☐ Racked or distorted _____
☐ _____

17. Insulation. Type _____
☐ Needed in walls _____
☐ Needed in crawl space _____
☐ Needed in attic _____
☐ _____

Home Evaluation Checklist
Figure 2-6 (Continued)

18. Vapor Barrier. Type _____
☐ Needed on inside of walls _____
☐ Needed above ceiling_____
☐ Needed in crawl space _____
☐ _____

19. Ventilation
☐ Attic needs vents _____
☐ Crawl space needs vents _____
☐ _____

20. Plumbing. Type _____
☐ Faucet flow inadequate_____
☐ Supply inadequate_____
☐ Valves frozen _____
☐ Leaks apparent _____
☐ Water hammers _____
☐ Waste drains poorly _____
☐ Vent stacks inadequate _____
☐ Inadequate traps on fixtures _____
☐ Inadequate water heater _____
☐ Fixtures need replacement _____
☐ _____

21. Heating. Type_____
☐ Capacity inadequate _____
☐ Poor heat distribution _____
☐ System repairs needed _____
☐ Controls obsolete _____
☐ System obsolete_____

22. Cooling. Type_____
☐ Capacity inadequate _____
☐ Poor cold air distribution_____
☐ System repairs needed _____
☐ Controls obsolete _____
☐ System obsolete_____
☐ _____

23. Electrical
☐ 100 amps, ☐ 200 amps, ☐ 110 volts, ☐ 220 volts
☐ Wiring insulation deteriorated_____
☐ Inadequate electrical outlets_____
☐ Distribution equipment inadequate_____
☐ Fixtures need replacing_____
☐ Switches need replacing _____
☐ _____

24. Community considerations
☐ Adjoining properties deteriorated _____
☐ Future neighborhood improvement unlikely _____
☐ Local transportation system inadequate_____
☐ Remote from schools, shopping, etc. _____
☐ Flood or storm damage likely_____
☐ Inadequate privacy_____
☐ Offensive activities in community_____
☐ _____

25. Design considerations
Number of bedrooms _____
Number of bathrooms _____
Approximate square footage _____
Size of garage_____
Unusual features _____
☐ Room size adequate _____
☐ Poor traffic patterns _____
☐ Bedrooms not remote from kitchen _____
☐ Kitchen poorly located or designed_____
☐ Kitchen too small or too large _____
☐ Bathrooms poorly located _____
☐ Bathrooms too small_____
☐ Inadequate closet space _____
☐ Inadequate ceiling height _____
☐ Excessive ceiling height_____
☐ Architectural style unattractive _____
☐ Cluttered, unattractive exterior _____
☐ Landscaping inadequate or unattractive_____
☐ _____

Home Evaluation Checklist
Figure 2-6 (Continued)

3

The Art of Planning

Who designs room additions and remodeling projects? If you've been in the remodeling business very long, you know that most of the time the builder does the design work. That's seldom true in new construction. But it makes sense to have the man who will do the work plan the work that has to be done in most small jobs like you will handle.

And you have a big opportunity if you prepare the design: you aren't going to be underbid by some cutthroat lo-ball operator from across town. If your customer likes your plan and your cost estimate, the job is yours at your price. That's an advantage most of the big engineering contractors can only dream about.

But don't kid yourself. Amateurs don't get many remodeling jobs based on their design. You have to know your way around the basic principles that distinguish good design from disaster. Construction costs too much and takes too long to start over again after the job is finished. You'd better be prepared to make some good suggestions and come up with an efficient and innovative room plan when your client asks for your help.

That's what this chapter is about: creating an innovative and efficient plan that reflects your skill and craftsmanship. You don't need a design staff or even a drafting board to do it. You really don't even need to draw up a formal plan—though you probably will for most jobs. You just have to create the concept of what your customer needs and wants. If that seems easy, read on.

Space Requirements

The American Indian had a good thing going. He had all the space he needed outside so he only required a little space inside. Now we have little space outside and need all we can get inside. Maybe this isn't progress. But progress or not, we're stuck with it.

Let's take a moment to discuss house design. Most people want a home designed so that the kids don't have to run through the master bedroom to get to the bathroom. And visitors shouldn't have to walk through the center of the dining room to get from the living room to the kitchen. Planning a house or addition with these goals in mind will create a well-designed, comfortable, pleasing, and well-equipped home where everyone can live and move about with privacy and minimum disturbance to others.

General Space Requirements

A house is a unit for living. Each unit must have sufficient space for living, sleeping, cooking, and eating. There must be space for sanitary facilities, laundry, and storage. The space has to be planned to permit the placement of furniture and essential equipment.

Figure 3-1 shows specific space requirements for a house with separate rooms. Space requirements for a house with combined rooms are given in Figure 3-2.

	Minimum Area for Living Unit				
Name of space	With 1 bedroom	With 2 bedrooms	With 3 bedrooms	With 4 bedrooms	Least Dimension
Living room	160	160	170	180	11'0"
Dining room	100	100	110	120	8'4"
Bedroom-primary*	120	120	120	120	9'4"
Bedroom-secondary	N/A	80	80	80	8'0"
Total area of bedrooms	120	200	280	380	---

*Primary bedrooms shall have at least one wall space of at least 10' uninterruped by openings less than 44" above the floor.

Minimum Room Area for Living Units with Separate Rooms
Figure 3-1

	Minimum Area for Living Unit				
Name of space	With 1 bedroom	With 2 bedrooms	With 3 bedrooms	With 4 bedrooms	Least Dimension
Living room/dining area	N/A	210	210	230	250
Living room/dining area/sleeping area	250	N/A	N/A	N/A	N/A
Living room/dining area/ kitchen	N/A	270	270	300	330
Living room/sleeping area	210	N/A	N/A	N/A	N/A
Kitchen/dining area	100	120	120	140	160

Minimum Room Area for Living Units with Combined Spaces
Figure 3-2

All other habitable rooms (including finished basement rooms) should have at least 80 square feet, with the minimum dimension no less than 8 feet.

Stairs or closet space cannot be included when determining the required room space.

Minimum Space Requirements for Bathrooms
Three fixtures are essential for a full bath: (1) commode, (2) lavatory, and (3) tub or shower. Each should be arranged for easy and comfortable use. One shouldn't have to stick his feet into the tub when using the commode, or have to climb over the commode to get into the shower. The door should open to at least a 90-degree swing. If not, a sliding door should be installed. Figure 3-3 provides an example of minimum spacing of bathroom fixtures for comfortable use.

Minimum Requirements for Other Areas
Halls: A hall has to be at least 3 feet wide to accommodate the moving of furniture.

Laundry area: The laundry room or space should be large enough for a washer and dryer. These appliances have to be repaired occasionally, so space is needed for maintenance purposes.

Bedroom closets: Every bedroom needs at least one closet. It should be at least 2 feet deep and 3 feet wide. A 5-foot clear hanging area is required over the required width. The lower shelf should not be over 74 inches above the room floor. At least one shelf and rod is essential, and there should be at least 8 inches of clear space over the shelf. The floor of the closet should be finished and at least half of the floor depth should be flat. Figures 3-4 and 3-5 reflect these dimensions. Keep this information handy. Few builders can remember closet figures.

Linen closet: A linen closet should be located near the bedrooms. The closet should be at least 14 inches deep, 18 inches wide, and have shelves spaced a foot apart. No shelf should be higher than 74 inches from the floor nor deeper than 2 feet. Nine

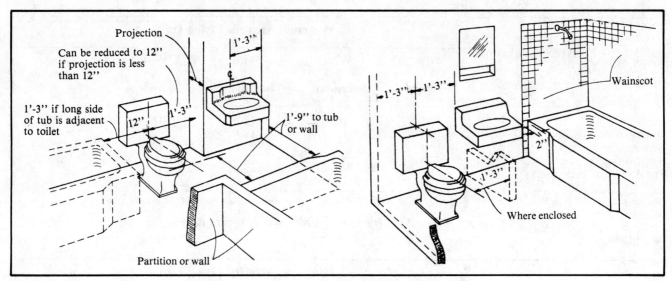

Recommended Dimensions For Fixture Spacing
Figure 3-3

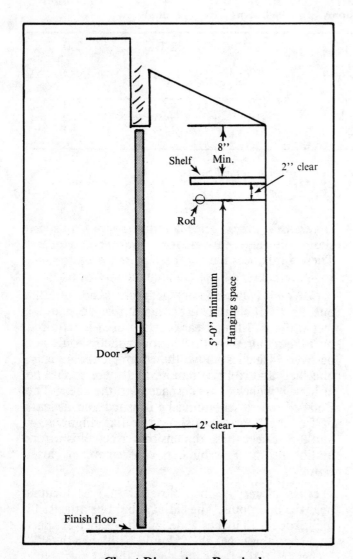

Closet Dimensions Required
Figure 3-4

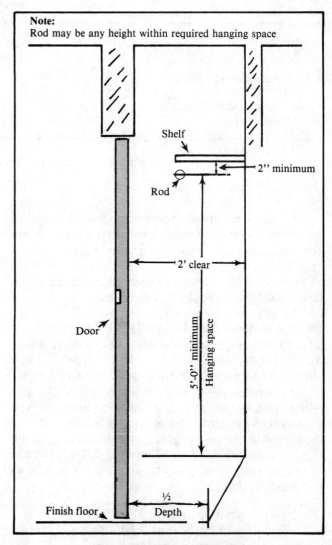

Closet Dimensions Required
Figure 3-5

square feet of shelf space is recommended for a 2-bedroom house. For 3- and 4-bedroom homes at least 12 square feet of shelving is needed.

Coat closet: A coat closet with the same minimum size and equipment as a bedroom closet should be near the living area, as close to the front door as possible.

Kitchen storage: The lack of storage and counter space can turn an excellent cook into a grouch. We're going to examine kitchens in greater detail when we remodel Brown's old kitchen in a later chapter. But for now there are a few minimum requirements you should follow:

Base and wall cabinets—50 square feet of shelving with not less than 20 square feet in either one. Drawers or shelves in cooking ranges may be included in the minimum figures.

Drawer area—11 square feet.

Countertop area—11 square feet. Sink and cooking tops are not counted in this total.

If a range is not included, add at least a 40-inch space for a range. If a 39-inch range or 40-inch range space is provided, it can be counted toward the required areas as 4 square feet of base cabinet shelving and 2 square feet of countertop.

Wall shelving above 74 inches and countertops higher than 38 or lower than 30 inches above the floor are not included in the required area.

The space between countertop and wall cabinets should be at least 24 inches over range and sink cabinets, and 15 inches over base cabinets.

Storage: Few American homes ever have sufficient storage space. The minimum storage space requirement is, including both exterior and interior, 200 cubic feet plus 75 cubic feet for each bedroom. Even this won't be enough for many families.

At least 25 percent of the space should be located within the house and should be suitable for storage of those things normally stored inside. Such space can be attic space, basement space, utility rooms, or separate closets. Storage space in attics should be accessible by a permanent or disappearing stairway.

Of the required total space, 50 percent should be located for easy storage of items and equipment used outside, such as garden tools and water hose. This space may be located within the house if it is accessible from the exterior. Garage, storage shed and carport storage area can be used in computing the minimum space requirements.

Ceiling height: In habitable rooms you can't drop the ceiling below 7 feet 6 inches. In attic rooms with a sloping ceiling, at least half of the room must have a ceiling at least 7 feet 6 inches

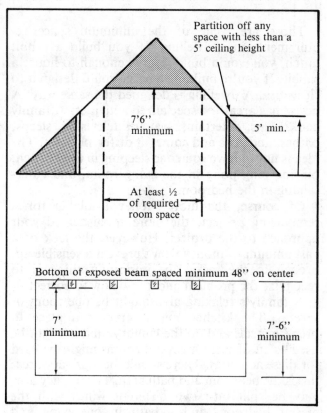

Ceiling Height Requirements
Figure 3-6

high. In rooms with exposed beams or girders spaced at least 48 inches on center, the bottom of the beam or girder should be at least 7 feet from the floor. Don't plan any floor area in attics with less than a 5-foot ceiling height; partition off such space.

A ceiling at 6 feet 8 inches is the minimum in bathrooms, toilet compartments, and utility rooms. Halls also fall into this category. Figure 3-6 illustrates some of these points.

Basement space: Where a basement is used as a habitable room (a recreation room, bathroom, or utility room are not considered habitable) for year-round occupancy, the basement floor can be no more than 48 inches below the exterior ground level. Underground houses come under a different rule, but there aren't enough of those around to worry about.

Later you'll be concerned with remodeling a basement into an addition. We'll cover habitable basement construction in a separate chapter. Lucky for us, Mr. Brown's house is on high ground. But we'll discuss all the problems inherent with basements located in low-lying areas as well.

Those are some of the minimum space requirements in a dwelling. If you build a rabbit hutch, you would build it large enough to house a rabbit. If you're building a barn, you'd design it to fit its use. A dwelling is designed the same way. A house has areas for specialized activities. A family cooks, eats, entertains, watches television, sleeps, bathes, and goes and comes at different times. The idea is not to have someone sleeping in the kitchen, playing ping pong in the living room, and entertaining in the bedroom.

Of course, the more money available for a remodeling project, the more sophisticated your approach to the project. However, the lack of a vast amount of money won't prevent a sensible approach to the task. The important thing is to recognize the problem and know how to correct it.

A family's relaxing area might be one room or several. The kitchen might serve as a place to prepare meals and do the laundry, and it might be used as an office. Or several rooms might be used for these activities. There should be a privacy area; this is the bedroom and bath area. The privacy area may be split into two different wings with the master bedroom and a bath in one wing and bedrooms and bath in another wing. There can be several privacy areas. A bathroom opening out into a walled-up private garden is a privacy area.

There are many different arrangements. A small house with only the minimum space requirements is usually modest in its surroundings. No need here for elaborate grounds and a guest parking lot. This doesn't mean that the owner is happy with a view of a littered back alley from his bedroom window. But some things are beyond your control.

Some owners might want facilities for outside activities such as a barbecue grill on a patio overlooking a fenced-in yard. They might prefer that the family or living room be on the back of the house and the kitchen on the front. Every room needs the appropriate access: from garages or carports to the kitchen, from the main entrance to the drive, parking area, or street. Any arrangement is possible as long as everything is compatible. Groceries are not usually carried through the front door and guests are not usually routed through the kitchen.

Traffic Flow

The movement of people into and through the various rooms is much like the movement of vehicles on streets and roads. There is a proper way to do it. The more cross traffic, the more the ag-

gravation. The interstate highway system eliminates cross traffic and expedites movement. The system isn't perfect by any means—neither is the traffic flow in most houses.

The purpose is to strive for balance—no traffic through the middle of rooms, such as when doors are located in the center of walls (Figure 3-7) or at opposite corners (Figure 3-8).

The *straight-shot* method in both the above plans would eliminate the awkward movement through the rooms and save steps as well. (See Figure 3-9.)

The straight-shot arrangement allows for easier placement of furniture by eliminating short or broken walls which are created when doors are centered in the walls. Relocating doors is expen-

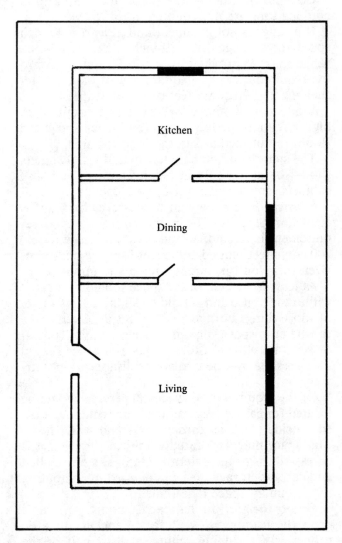

Center-of-Wall Layout
Figure 3-7

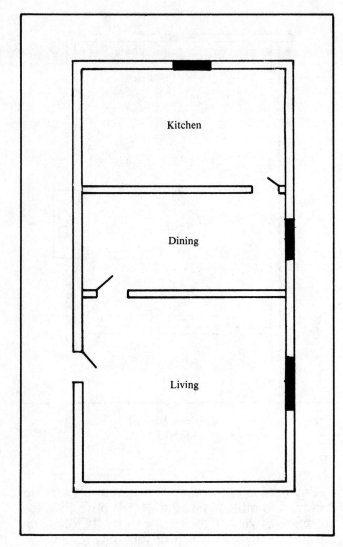

Opposing Corner Layout
Figure 3-8

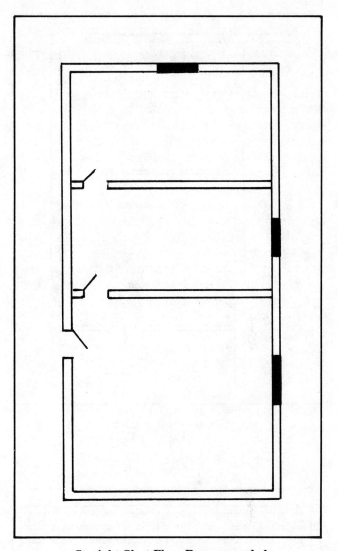

Straight-Shot Flow Recommended
Traffic Flow
Figure 3-9

Access To:	From	Thru
Habitable room	Habitable room	Bedroom
Bathroom	Habitable room	Bedroom
Bathroom	Bedroom	Habitable room
Bathroom	Bedroom	Another room
Habitable room	Habitable room	Bathroom

Unacceptable Access Patterns
Figure 3-10

sive. But it does allow better use of space without actually increasing space.

Brown's house has doors centered in walls.

Houses used to be built with larger rooms than are common today. Sufficient wall space for placement of furniture was apparently not a problem. And, in many old houses, access to one room is through the middle of two others. Figure 3-10 shows how *not* to design room arrangements.

Relocating and Removing Partitions

Brown is tired of the two small rooms near the kitchen. He wants the partitions between the two small rooms taken out and the kitchen wall partition removed. He wants the existing kitchen and the two small rooms remodeled into a modern kitchen-family room with a fireplace. Figure 3-11 is the existing layout. Figure 3-12 is the desired layout.

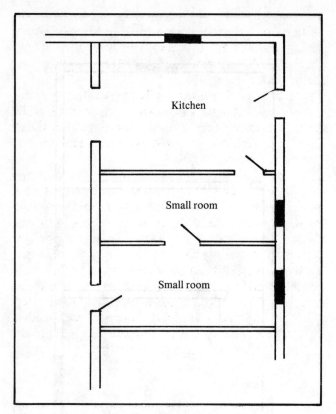

Existing Layout
Figure 3-11

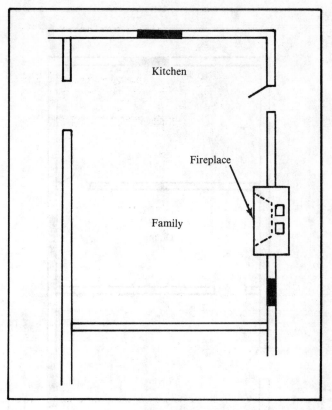

Desired Layout
Figure 3-12

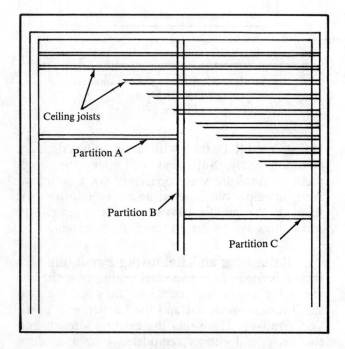

Load Bearing Partition - B
Non-Load Bearing Partitions - A & C
Figure 3-13

You need to know whether the two walls are load bearing or non-load bearing. Here's how. Find the direction of run of the ceiling joists. If joists run parallel to the wall it is probably only a non-load bearing partition. (See Figure 3-13.) Where a second story is involved, check to determine what support, if any, the wall provides the second floor. A load bearing wall can be removed, but a beam will have to take over the load. Temporary support is required during installation of the beam.

The removal of non-load bearing partitions will require repair to ceiling, floor, and walls where the partition was removed. If wires or plumbing are in the wall, these will have to be removed.

The removal of a non-load bearing partition itself is a simple task. It's the repair of the area where the partition was located that's ticklish. Unless the entire ceiling, walls, and floor of the rooms are going to be refinished, you'll have to think about matching things up. That can be one of those times that makes you wonder if selling magazine subscriptions wouldn't be a better way of making a living. In Brown's house the partition removal won't be a problem since the walls, ceil-

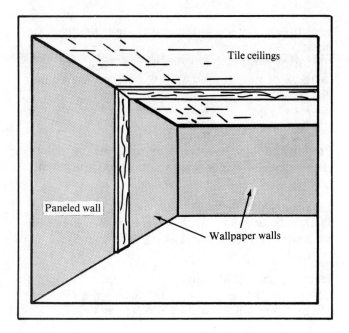

Shallow Beam and Post
Figure 3-14

ings, and floors are all going to be redone. In a job where the budget is slim and the customer wants the space without a whole new wall and ceiling, plan a shallow beam and post effect to cover the gaps. (See Figure 3-14.)

On drywall or plaster walls the solution is fairly simple: use the same materials to patch and feather out. But on meeting walls of non-matching finish and tiled ceilings, the shallow beam and post is your best choice.

Of course, there is still the floor. The best approach is to fill it in and cover it with carpet or a resilient floor covering. But a new floor covering might cost more than the customer wants to spend. That's something you should consider before knocking out a partition.

Sometimes your customer wants to grab a little extra space from one area and put it to use for another purpose. No space is gained, it's just used better. It's a trade-off. But if room space is needed more than a hallway, it's a good swap. Traffic will still move through the area but the room is bigger and better. (See Figure 3-15.)

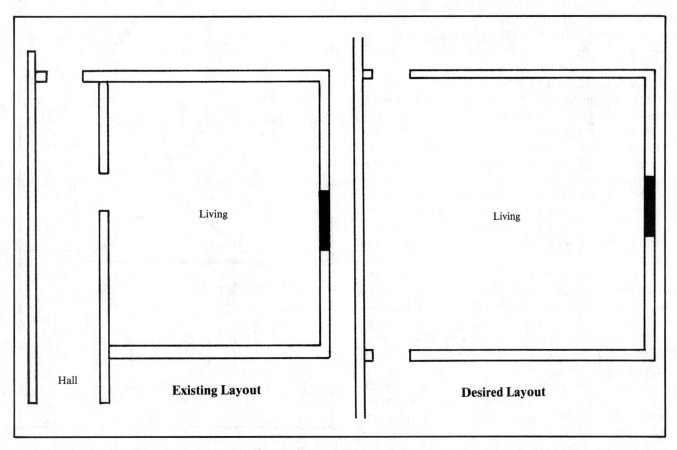

Gaining Better Utilization of Space
Figure 3-15

Window Arrangement

Window size and placement is important in the appearance of a house. It also affects the arrangement of furniture and influences the cost of heating and cooling. And few people enjoy spending time in a room without windows.

Placement of windows on the front of a house must be proportioned to maintain balance. Windows in two-story houses usually have the first and second story windows directly in line, one over the other. The windows on the front (and often on the sides, too) are placed with the left of center equal in distance to the windows right of center. (See Figure 3-16.)

Of course, there are many exceptions to this general rule. Some houses have small windows on one side of the front entrance and large windows on the other side—or perhaps none on one side. A particular design will call for different window sizes and placement. But balance is a consideration. Windows should be the same height as doors.

Another important factor in window placement is the direction faced—north, south, west, or east. The area of the country where the house is located will dictate the location and size of windows to a great extent. Windows facing the east and west receive a lot of sun. Those facing north battle the cool breezes. The south window permits the low-swinging winter sun to provide warmth.

Take advantage of various exposures and place windows intelligently, to the extent the design and location permits. But don't overdo it. If the only exterior wall of a bedroom is on the north side of the house, it still needs a window, cold winds notwithstanding.

The window (glass area) should be at least 10 percent of the room area according to current standards. Example: a 12 x 12 room equals 144 square feet. The recommended window area is 14.4 square feet.

It's nice to have a pleasant view. However, what's pleasant to one person might be unpleasant to another. Some people don't like to look out on a busy street and see cars zooming by. Others can sit by the window and watch cars fly by all day long. Be sure you know what your client wants and plan accordingly.

Windows are used for ventilating. The screened area should be not less than 4 percent of the floor space. In the warm areas of the country screens were dispensed with on many houses when central air condition became available. Now, with the high cost of energy, the screens are going back up. A cool breeze on a warm night is once again in demand.

Storm windows and insulated windows have their place and should be used where the climate dictates.

Doors Sizes

Prehung doors are a standard 6'8" in height. However, here are a few minimum measurements:

Exterior—Minimum

Main entrance: 3'0" wide x 6'6" high. Where double entrance doors are used, 2'6" wide doors can be used.

Service doors: 2'6" wide x 6'6" high.
Garage doors: 2 car - 15'0" wide x 6'4" clear height. 1 car - 8'0" wide x 6'4" clear height.

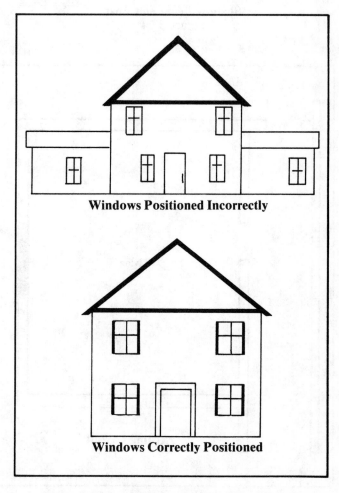

Windows Positioned Incorrectly

Windows Correctly Positioned

Proper Window Placement Maintains Balance
Figure 3-16

Interior—Minimum
Habitable rooms: 2'6'' wide x 6'6'' high.
Stair doors and basement: 2'6'' wide x 6'6'' high.
Bathrooms and clothes closets: 2'0'' wide x 6'6'' high.

Attic Access—Minimum
14 x 22 inches except where equipment such as a furnace is installed. But the opening must be large enough to allow placement of the equipment. A disappearing or permanently installed stairway is necessary.

Crawl Space Access—Minimum
18 x 24 inches except where equipment is installed. Again, the opening should be large enough to accommodate passage of the equipment.

Garages and Carports—Minimum
2-car garage: 18'4'' wide x 20'0'' long.
1-car garage: 10'0'' wide x 20'0'' long.
2-car carport: 18'4'' wide x 20'0'' long.
1-car carport: 10'0'' wide x 20'0'' long.

The carport length is measured at the opposite ends of the roof cover. The width of a detached carport is measured from inside face of posts to inside face of posts. (See Figure 3-17.)

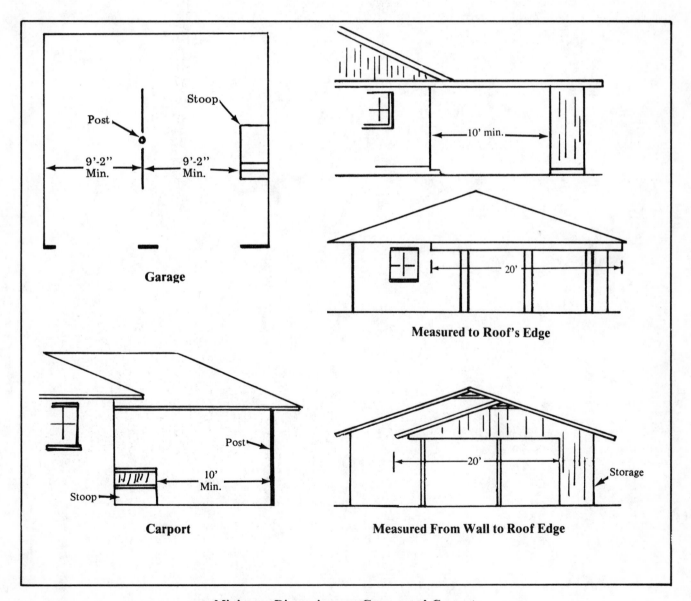

Minimum Dimensions — Garage and Carport
Figure 3-17

General Exterior Design

A house need not be elaborate or expensive to be attractive. While it is true that the location and grounds have a lot to do with it, the important factor, from the remodeler's view, is the house itself. A mixture of incompatible exterior finish materials, poorly designed roofs, disproportionate windows, and ill-designed entrance work and chimneys throws the whole thing out of symmetry. Simplicity is always a virtue and a good approach when remodeling.

Summary

We've covered some important ground in this chapter. But there's still more to learn. Your planning will always be geared to the one house under consideration. Ignoring the general requirements and guidelines in this chapter can result in a disaster. But there are many "right" ways to plan an addition or remodeling project. Work within the guidelines here and add your own good ideas. No one can say that his way is better than yours. Experience is still the best teacher. A builder who plans with common sense usually plans well.

4

Repairing Structural Defects

Any house can be repaired. The big question is when is it more economical to repair or abandon and rebuild? Some offer this rule: when repair costs exceed 2/3 of the replacement cost, repair is impractical. This is a good enough rule, but many considerations are involved. And the most important is what the owner wants. You can give advice. But what the owner wants is what really counts. Your job is to give him the best job possible for the money.

Step 1
Repositioning Girders and Sills
As you'll recall, Brown's home has unlevel floors and a dip in the roof. The cause of this was found under the house during our inspection. Two pillars near the center of the house had deteriorated to the point that one of the girders dropped down. Termite damage in one of the sills has rendered a few feet of the sill worthless, and the sill needs to be replaced. There isn't much sag in the wall over the termite-damaged sill, but enough to require jacking the wall back to its proper position.

The reason there is little sag in this wall is that 1 x 8-inch board sheathing was applied to the framing at a 45-degree angle. It supported the weight, preventing serious sag. (See Figure 4-1.) Back in the thirties this was a common sheathing method, and a good one.

We will handle the deteriorated pillars and sagging girder first. Let's jack the girder back into position. Since the interior walls will be redone, we

won't have to worry about cracking them. Had the walls been replastered with the girder sagging as it is, then raising the girder would damage the finish wall. Anytime you use a house jack, something will move, and when something moves that isn't designed for movement, things start groaning and cracking.

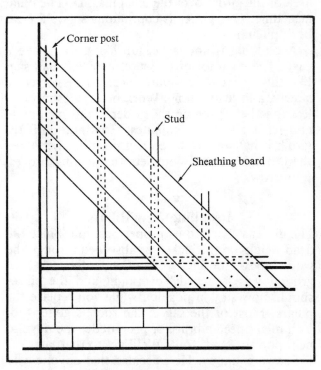

1" x 8" Sheathing on 45° Application
Figure 4-1

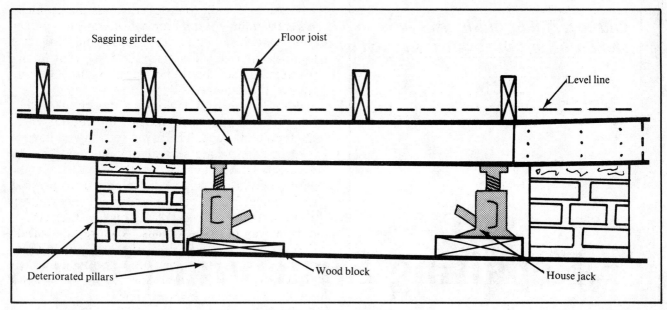

Repositioning Sagging Girder
Figure 4-2

Position the jacks as shown in Figure 4-2. Be sure the girder is continuous over the pillars. If there is a break in the girder over the pillar, position the jacks on each side of the pillar. Lift the area of the girder over the one pillar. Do one pillar at a time. Our girder is continuous so we'll clear both pillars.

A building is not jacked up like a car. Take it easy. There's a lot of pressure involved. Bad sags like this (3 inches or more) need to be corrected slowly, a little at a time. Work both jacks more or less together. Take up the girder a half inch or so and let it rest a few minutes. Continue until the girder is halfway to the correct position. Let it rest there overnight and complete the jacking process tomorrow.

Step 2
Installing New Pillars
The girder is back to its proper position. Since the dead weight on the jacks was absorbed through the wood block "footing" you put under the jacks, the ground was "packed" overnight and the girder should now stay in position while you replace the pillars. Most of the sag in the roof is gone, too. You might need a brace or prop under the ridge to help pop the "set" out. We'll get to that later.

Brown is happy. He's decided that anyone who can level his floor and take that ugly sag out of his roof so quickly is a real pro.

The brick used in the pillars appear in fair condition. It's the mortar that's bad. Tear the pillars out and build new ones with concrete block or new brick. Don't re-use the old brick. Strong mortar and weak brick is just as bad as weak mortar and strong brick.

After the pillars cure, put a termite shield or a vapor barrier (check your code) on top of the pillars and ease the pressure off the jacks. Then check that it's level.

Step 3
Replacing Sills
The sill can't be removed without some support under the joists to hold up the wall. The best method is to install an auxiliary sill about 3 feet back from the damaged sill and jack the joists back to the proper position. (See Figure 4-3.)

My practice has been to permanently install the auxiliary sill. After jacking the auxiliary sill into place, pillars are built on solid footings. When the pillars are cured, the jacks are removed. To ensure a firm rest for the auxiliary sill, place wedges between the pillar and sill. If a termite shield or vapor barrier is used, place the wedges on top of the shield or barrier and set the wedges as shown in Figure 4-4A. Always use two wedges for ease of set and a firm base for the sill. Space wedges where they will not make contact with each other when

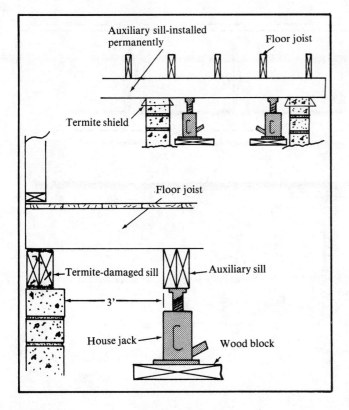

**Auxiliary Sill
Figure 4-3**

driven under the sill. (See Figure 4-4B.)

The auxiliary sill so installed solves the problem. Where economy is a major consideration, you can come out from under the house and leave the termite-damaged sill in place. If this is your intent, the auxiliary sill should be placed within a few inches of the damaged sill.

Since Brown's job is not on a tight-budget, let's remove the damaged sill. Locate the boundaries of the termite damage. This is the section you have to take out. There are several ways to do this. The hard way is with a chisel, which doesn't leave you with much in the way of a straight cut to join the replacement sill. Of course, you could always cover up the cut with a piece of scrap plywood, but this is not the way of professionals. A sabre saw will do the job, too. Just remember that on the other side of the sill there is sheathing and siding. The quickest and best method for cutting the sill is a small chain saw with a solid bar.

You might have to wrestle some with the sill after it's sawed to get it out. That's because termites always leave a little wood around the nails.

After the sill is out, break off the nails coming through the floor, wall sheathing, and joists and place the new sill in position. Toe nail through the sill into the joists.

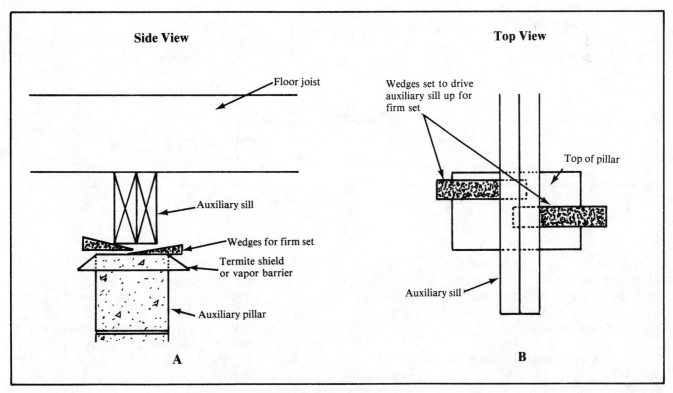

**Wedged Auxiliary Sill
Figure 4-4**

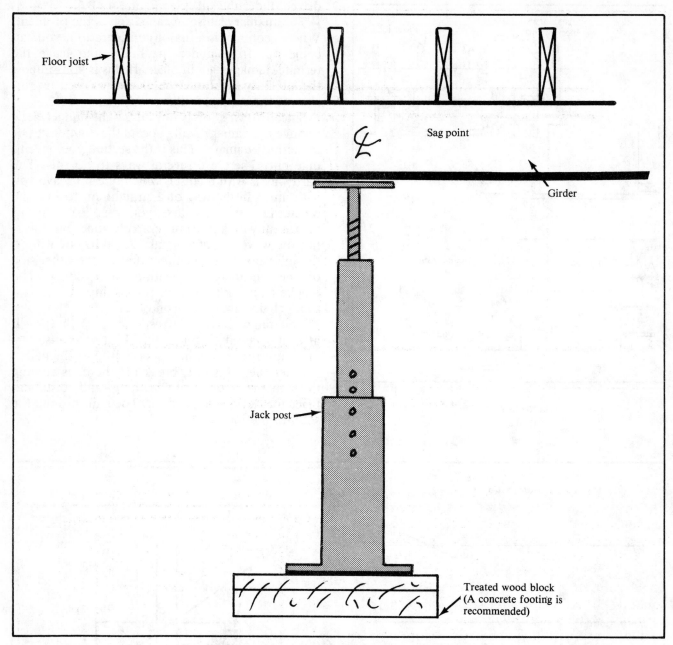

Repositioning Sagging Girder With Jack Post
Figure 4-5

The method used to replace the sill has not disturbed the floor, the sheathing, or the siding. Before you crawl out from under the house, be sure to remove all pieces of wood scrap so as not to invite the termites back.

Step 4
Correcting Sagging Girders

Jack posts can be used in crawl spaces and basements to correct girder sags. The jacks are left in place. In a crawl space, a footing is required for the jack. In a basement the concrete floor is enough for light loads. Heavy loads require a wider distribution of the load. A 3/8-inch steel plate under the jack can be used for this purpose. A jack post isn't enough for heavy jacking. A regular jack is used to lift the load and the jack post is then placed into position. Figures 4-5 and 4-6 show two uses of the jack post.

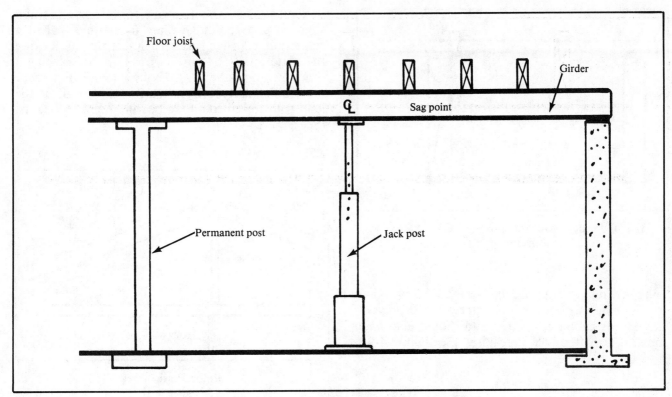

Basement Jack Post Supporting Sagging Girder
Figure 4-6

Step 5
Eliminating Floor Squeaks

The best thing to do for a minor squeak in machinery or a floor is oil it. A little mineral oil between the boards will often solve the problem. If the floor is covered with carpet or other floor covering, you'll have to tackle the problem from underneath. If it is a second story floor that is squeaking, there's little that can be done unless the carpet is taken up or the downstairs ceiling removed under the area of the squeak. Often, rearranging or shifting some heavy furniture over the spot will help.

Squeaks are caused by several things. A separation of the subfloor from the joists might be a cause. If this is the problem, screws can be angled through the floor joists into the subfloor and tightened, drawing the subfloor down against the joists as shown in Figure 4-7.

A quicker method is to drive wedges between the joist and subfloor. A light tap on the wedge for a snug fit is all that's necessary. Put wood glue on both top and bottom surfaces of the wedge to provide a "weld" that prevents the wedge from gradually working out. (See Figure 4-8.)

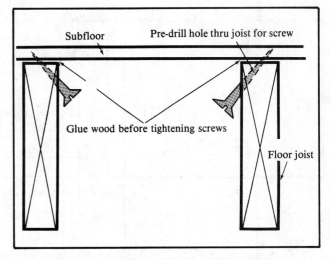

Eliminating Separation Causing Floor Squeaks
Figure 4-7

Undersized joists cause squeaks and a springy floor. The best bet here is to add a girder.

Wood flooring (board) is sometimes nailed down parallel to the joists. Such an installation is always a good candidate for squeaks. It doesn't provide

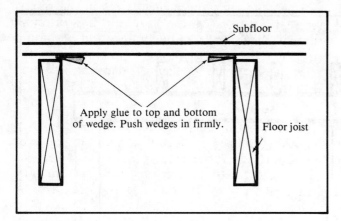

Use of Wedges to Eliminate Squeaking Floor
Figure 4-8

enough strength between the joists. Solid bridging or blocking at about 3-foot intervals will prevent the boards rubbing together and should eliminate most, if not all, of the squeaking. (See Figure 4-9.)

Step 6
Correcting Sagging Roofs
Restoring the girder to its proper position and replacing the termite-damaged sill corrected most

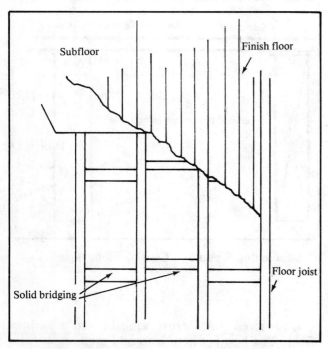

Solid Bridging Between Joists to Eliminate Squeaking in Parallel Laid Flooring
Figure 4-9

of the problem with Brown's roof. There is still a slight sag in the ridge plate (sometimes called a ridge pole or ridge rafter). Handle this by jacking the plate up to the correct position and placing a brace or prop under it. A 2 x 4 is enough. The brace must be supported by a load bearing wall or the load transferred to a load bearing point by beams. (See Figure 4-10.)

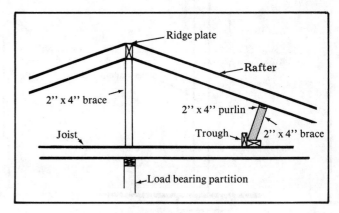

Ridge Plate Brace
Figure 4-10

Rafter sag along the span of the rafters can be eliminated by bracing. If there is no "trough" across the ceiling joists, add one to support the rafter braces. The hard part of this is getting the long 2 x 6 material for the trough up into the attic. An attic vent at the gable is your best bet. Remove the vent and run the material through to the attic. Once the trough is nailed down, braces can be positioned to remove the sag as shown in Figures 4-11A and 4-11B.

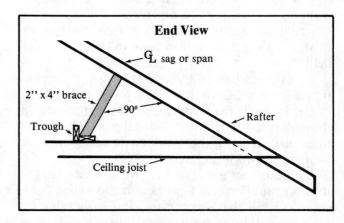

Trough Rafter/Joist Bracing
Figure 4-11A

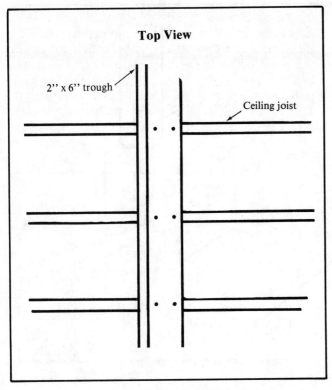

Top View

2" x 6" trough

Ceiling joist

Trough Rafter/Joist Bracing
Figure 4-11B

When the sag is permanently set in the rafter, locate the center of the sag, saw 2/3 or more through the rafter from the bottom and position it straight. Nail a short piece (4' minimum) of the same size material to the sawed rafter, centered with the saw cut as shown in Figure 4-12.

Step 7
Correcting Sagging Roof Sheathing

Sheathing can be boards or plywood. Either can be too thin for the rafter spacing and will sag between the rafters, resulting in a wavy roof surface. The best method for correcting this is to remove the shingles and install new sheathing over the old. The new sheathing (plywood) should be heavy enough to prevent sag. Plywood sheathing of at least 1/2-inch thickness is the minimum that most builders will use for sheathing where rafters are 16 inches on center. Nail the new sheathing at the rafters. Use a nail that will penetrate the rafter at least 1¼ inches. Plywood sheathing should always be applied across the framing—the length perpendicular to the rafters.

A flat garden spade is a good tool for removing shingles. Working from the top of the roof down, insert the spade under layers of shingles to remove large sections with comparative ease.

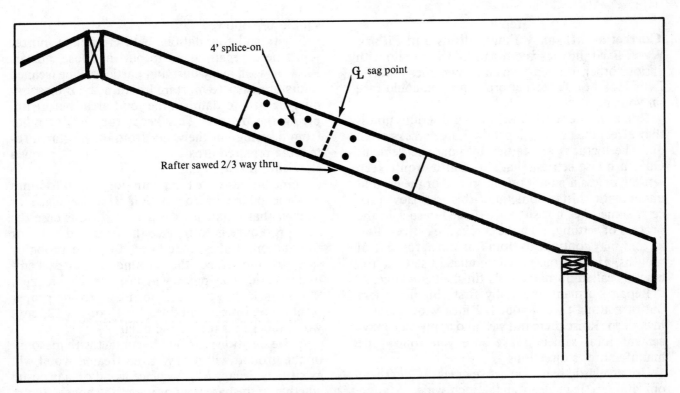

4' splice-on

C_L sag point

Rafter sawed 2/3 way thru

Correcting Sagging Rafter With Permanent 'Set'
Figure 4-12

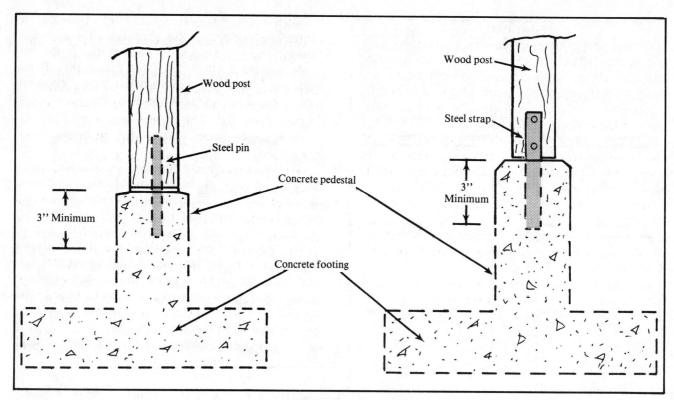

Pedestal Post Support
Figure 4-13

Step 8
Correcting Masonry Foundations and Pillars

We've fixed the sags and discussed what to do with various other problems which Brown doesn't have. Let's look at a few other problems you could face on your next job.

Cracks in foundation walls are not serious unless they affect the bearing capacity. Large cracks often do. These cracks are caused by house settlement. Once a house settles, that's it. But it won't keep settling unless it's built on marshland or the ground underneath shifts. Sometimes this is a slow process, sometimes not so slow. If the house is in the process of settling, then the cracks are active—that is, they may continue to "run" or widen for several months. If the house has completely settled, the crack is called dormant—it's finished cracking.

Repair dormant cracks by first chipping them out or routing them about 1/2 inch wide and 1/2 inch deep. Rinse them out well and apply an epoxy-cement joint sealer. Make sure you follow the manufacturer's directions.

Use an elastic sealer on active cracks. Chip them out about 1 inch deep and 1 inch wide. A good sealer will expand with the crack if and when it widens. Again, the manufacturer's directions

should be followed.

Pillars and foundations with crumbling mortar joints need repairing. Chip out the loose mortar. Blow or wash away dust and particles. The cleaned joints are then re-mortared with a thick masonry mix. Be sure to dampen the joint areas before applying the mortar. This keeps the wall or pillar from drawing out the water from the mortar. Protect the repaired area from the weather to allow a slow cure for strength.

Some houses were built with poor or no footings and the pillars or foundation is so damaged or sunken that repair isn't practical. In this case the best procedure is to replace the damaged pillar or foundation. Pillars that have sunk are probably safe to use since the ground has "packed" underneath. The pillar will need to be blocked. Slowly jack the structural members to the proper position and place suitable (2 x 10 or 2 x 12, etc.) wood blocks on top of the pillar.

As we mentioned, wood in contact with masonry or the ground will decay. Even treated wood will eventually decay. Always place wood on a pedestal whether in the basement or under the house. Figure 4-13 shows two methods of securing wood pillars and posts.

Summary

Your best tool is imagination. It's amusing to listen to two experienced builders discuss a construction problem. Often, what appears to be a major and costly repair project is reduced to a simple and inexpensive task. Know-how coupled with imagination is the greatest of all assets.

Don't be bound by the obvious or restricted by what is customary. If all the sills on one side of the house need replacing, don't call in a crane and 10 men. A couple of men, a couple of jacks, and materials for auxiliary pillars and sills will do the job.

Most of all, be sure you locate the problem. It's not always where you expect to find it. A story will illustrate this.

My helper and I were working under a pre-Civil War church in the country. It was damp and dark, and we were taking out some old, hand-hewn 6 x 8 sills that the termites had worked on. It was one of those jobs where a big outfit wanted to send out eight men and big equipment for $12,000.00. My price for this two-man job was about 10% of that, and I made money on it.

My helper couldn't do anything but look for snakes. He just knew that a rattler was snoozing behind every other pillar. Well, we finally finished and crawled out. The helper was very relieved. From there we went to the attic to chase down some leaks.

My helper was cheerful when we climbed up to the attic through the bell tower. He was finally out of snake country. Then he came eyeball to eyeball with a long, blue racer draped over some braces. He nearly tore the rafters down getting out of there. The snake was harmless, but that made no difference to him. The point is that sometimes you find what you're looking for in the least expected place. A sagging roof might be due to a fault under the house. A binding door could be due to a decaying wood post in the basement.

There's no substitute for intelligence and skill—and there never will be.

5

The Right Material for the Job

Nearly anything you buy from a reputable building supply house will comply with Federal specifications and the various industry standards. That doesn't mean that you won't get some trash in your order when it's delivered to the job. Crooked and knotty 2 x 4's will show up in your lumber pile. There will be several sheets of paneling that have been out on another job and returned to your dealer. It will look like it's been walked on, thrown about, and run into by a forklift. A roll of felt that has been stored too long will melt so it won't unroll without tearing.

Most sheetrock deliveries will include several broken or damaged sheets. But if the supply people ever try to unload damp sheetrock on you, don't take it. The paper will peel off in strips or rolls when you slide a sheet off the stack. Besides, the stuff is heavy enough even without several pounds of water. And if you manage to nail it up, it'll take forever to dry out.

It has become a practice for lumber yards to deliver lumber in dump trucks. They dump the lumber on the site as though it were sand. The steel bands, when used, either break or the stack tilts. Every piece will fall in a crooked pile when the band is cut. If you leave it crooked, every piece will turn and warp before you can nail it up. Ask for stack delivery if you don't know the yard's practice.

Some yards build up their orders so that what's on the top of your take-off sheet is on the bottom of the pile. Usually that means the lumber you need

first will be on the bottom of the stack, requiring sorting and restacking. Your list should specify how the lumber pile is arranged. Some dealers will hand stack the material on the site if you insist. This depends, of course, on their volume of business and how much they value your business.

Send all trash back. You pay for quality materials, accept no less. You have to absorb your own losses, let the supplier absorb his.

You can't do first-class work with shoddy materials. There's no way. Crooked studs, bowed joists, and separated plywood are nothing but trash, and whatever is built out of such junk will be no more than junk itself, no matter how good you are. And never use a bowed sheet of plywood for cabinet work. The bow is set and will remain set if used in cabinet doors or drawers.

Don't use 3/4-inch paneling boards that are still green. If you put them up tight and neat, they will separate when they cure. They may even separate completely from the tongue and groove.

Watch the prehung doors. Watch the windows. Watch the bathtubs, the commodes, the sinks. The imperfections, the careless workmanship and flaws all become your work once you accept them. It's a lot easier to reject flawed or shoddy material at the outset than later when you're ready to install it or, worse, after it is installed. Your customer won't like it one bit if you try to pass off shoddy goods on him. He won't permit it and shouldn't.

Most of the material and products you buy will be of acceptable quality. Just be certain that

Maximum Size of Coarse Aggregate	Approximate Cement (bag) per C.Y.	Approximate Water (gals) per Bag	Approximate proportions (by volume) per bag of cement		
			Cement	Fine Aggregate	Coarse Aggregate
¾"	6.0	5	1	2½	2¾
1"	5.8	5	1	2½	3
1½	5.5	5	1	2½	3½
2"*	5.2	5	1	2½	4

*Not for slabs or thin work

**Concrete Proportions
(Field Proportions)
Figure 5-1**

everything you accept is good standard quality at least. It'll save you man-hours, money, and your reputation.

Concrete Mix

In most areas concrete mix is available from a plant and is delivered to the job site by truck. Ready mix saves time and money when pouring footings, foundations, and slabs. For these jobs you'll need 5 bags of cement per cubic yard. The mixture contains a proportion of sand and gravel (or stone) to give a strength of at least 2000 psi. Figure 5-1 shows proportions of mix on the site where transit mix is not available or on-site mixing is preferred.

When figuring how many yards are required, a good rule of thumb is to multiply the length of the area in feet times the width in feet times the thickness in inches and then divide by 314.

Thus, a footing 100 feet long, 18 inches wide, and 8 inches thick would be:

$$\begin{array}{r} 100 \text{ ft. (length)} \\ \underline{\times 1.5 \text{ ft. (width)}} \\ 50.0 \\ \underline{100} \\ 150.0 \\ \underline{\times\ 8\ \text{ inches (thick)}} \\ 1200\ \text{ divided by 314 equals 3.82} \end{array}$$

The number of cubic yards required for the job is 3.82.

A slab 20 feet long, 12 feet wide and 4 inches thick will take 3 yards.

Framing Lumber

The lumber sold for framing has been graded by qualified lumber inspectors. This grading is done by visual inspection at the mill before it is shipped to your dealer. When you order material for framing, you will generally get what you order. Figure 5-2 shows grades of various species suitable for wall framing.

The softwood species listed in Figure 5-2 may be used in floor, ceiling, and roof framing. The spacing, span, and load determine the size and grades to be used. Figure 5-3 gives a few examples for live loads of 40 pounds (which is the total design load for stress and deflection composed of 30 psf live load plus 10 psf dead load with no more than 1/2 inch sag over a 15-foot span).

Species	Minimum Grade
Cypress	Number 2 Common
Douglas Fir	Standard Grade
Fir, Balsam	Number 1 Dimensions
Fir, White	Standard Grade
Hemlock, Eastern	No. 2 Com. Dimensions
Hemlock, Western	Standard Grade
Larch, Western	Number 2 Dimensions
Pine, Eastern White	Number 1 Dimensions
Pine, Red (Norway)	Number 1 Dimensions
Pine, Southern Yellow	No. 3, 2" Dimensions
Pine, Western White *	Number 2 Dimensions
Red Cedar	Standard Grade
Redwood, California	Const. Heart or Sap Common Dimensions
Spruce, Eastern	Number 1 Dimensions
Spruce, Englemann	Number 2 Dimensions
Spruce, Sitka	Standard Grade
Tamarack	Number 1 Dimensions

*Includes Ponderosa, Lodgepole, Sugar Pine, and Idaho

**Wall Framing Grades
Figure 5-2**

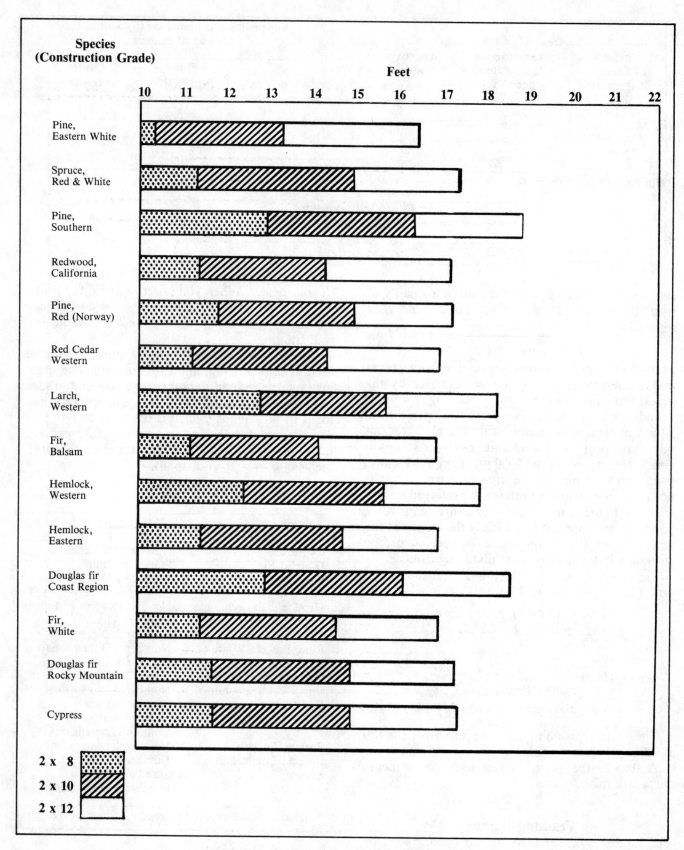

Floor Joist Span 16'' o.c. 40 Pound Live Load
Figure 5-3A

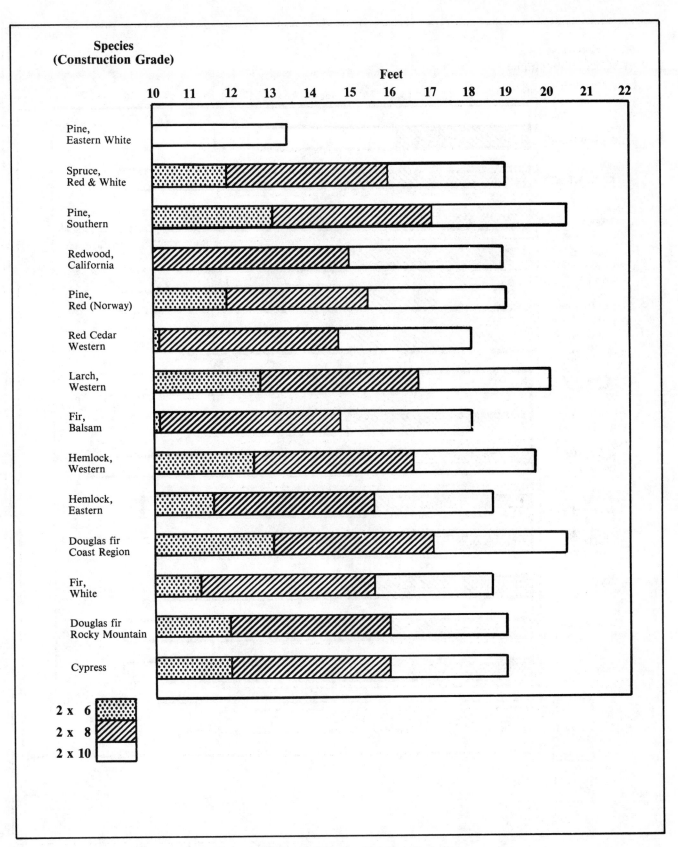

Ceiling Joist Span 16'' o.c. Limited Attic Storage
Figure 5-3B

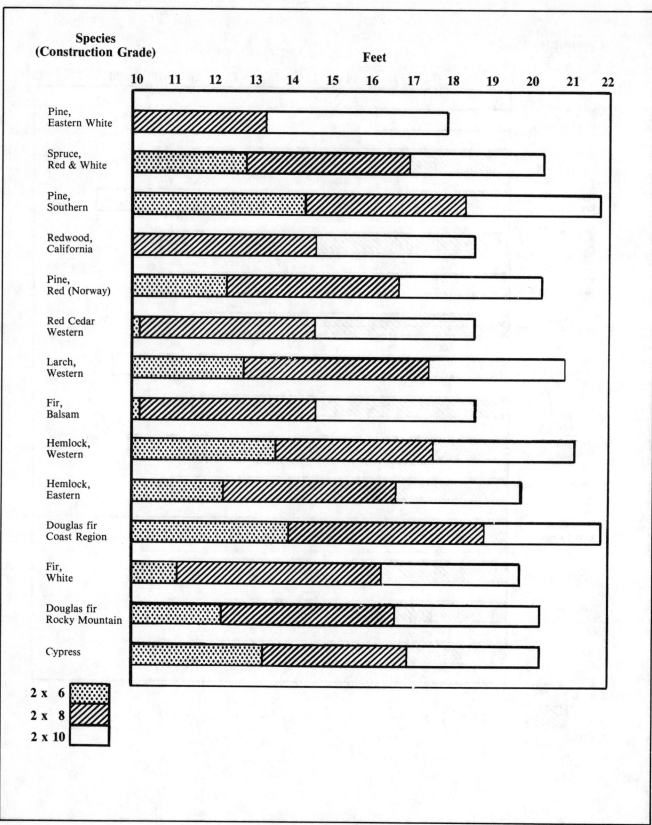

Rafter Span - Roof Pitch Over ³/₁₂
Heavy Roofing - 4 psf or Heavier
Figure 5-3C

Moisture Content

The moisture content of framing lumber should not be more than 19 percent. In those areas of the United States where the annual precipitation averages 15 inches or less, the moisture content should not exceed 15 percent. Figure 5-4 shows moisture content areas of the United States.

The moisture content of finish lumber should be as in Figure 5-5.

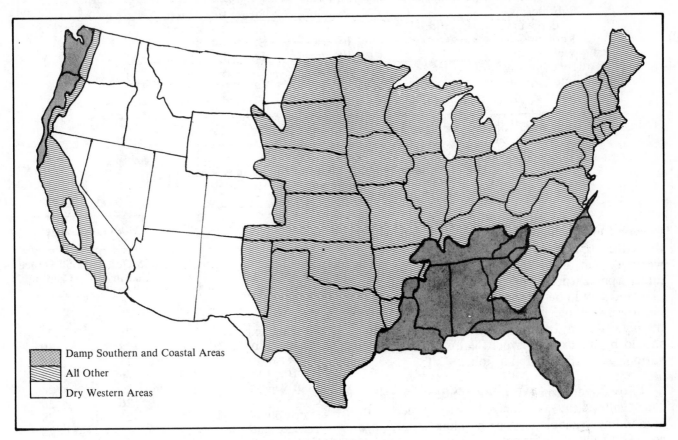

Damp Southern and Coastal Areas

All Other

Dry Western Areas

Moisture Content Areas
Figure 5-4

Use of Lumber	Moisture Content (Percent)					
	Dry Western Areas Average	Ind. Pcs.	**Damp Southern and Coastal Areas** Average	Ind. Pcs.	**Remainder of U.S. *** Average	Ind. Pcs
Interior trim, woodwork, and softwood flooring	5- 7	4- 9	10-12	8-13	7-9	5-10
Hardwood flooring	6- 7	5- 8	10-11	9-12	7-8	6- 9
Exterior trim, siding and millwork	8-12	7-12	11-13	9-14	11-13	9-14

* Local areas may require other moisture content limitations due to special climatic conditions.

Finish Lumber Moisture Content
Figure 5-5

Location	Exposed to Weather	Not Exposed to Weather
Porch or carport ceilings	Exterior	---
Soffits	Exterior	---
Truss gussets	Exterior	Exterior*
Built-up structural members	Exterior	Exterior*
Subflooring	Exterior	Struc. Int. or Ext.
Underlayment over subfloor	---	Struc. Int. or Ext.
Roof sheathing	Exterior	Struc. Int. or Ext.
Wall sheathing	Exterior	Struc. Int. or Ext.
Exterior wall finish	Exterior	---
Interior wall finish	---	Int. Struc.-Int. or Ext.
Interior ceiling finish	---	Int., Struc.-Int. or Ext.

*Structural-interior type having exterior glue may be used.

Plywood
Use of Various Types of Plywood for Various Areas
Figure 5-6

Plywood

Plywood grade should be appropriate for use. If the edge or surface of plywood is exposed to weather such as under roof overhangs, ceilings under open carports and porches, and soffits, the plywood should be exterior type. Plywood walls in bathrooms should have exterior glue. Subflooring of plywood in bathrooms and laundry rooms should be the exterior type. Figure 5-6 designates various grades for different jobs.

Galvanized and Aluminum Sheet Metal

The valley tin (galvanized sheet metal or aluminum) available at your building supply yard comes in rolls of different widths and can be used for many different purposes. Figures 5-7 and 5-8 suggest some of the ways valley tin can be used.

Glass

Use safety glass (tempered) whenever possible. If I had my way, only safety glass would be used in residential construction and remodeling. In storms and high winds flying glass is deadly. Falling against a window or glass door can cause a serious injury. Codes govern the use of glass in construction and should be checked when remodeling. Figure 5-9 shows glass thickness and sizes recommended for low, medium, and high wind areas. Figure 5-10 is a wind velocity map showing areas of low, medium, and high winds.

Nails

Once in a while you'll come across a run of 2 x 6's or 2 x 8's that won't take a toenail without splitting. In this case, nail rafters as shown in Figure

Use	Minimum Gauge*	
	1.25 Ounce Coating	1.50 Ounce Coating
Flashing, exposed	26	28
Flashing, concealed	28	30
Areaways	16	18
Roofing (over sheathing)	26	28
Roof valleys	26	28
Gutters	26	26
Downspouts	26	28
Termite shields	26	28
Ventilators	26	28
Gravel stops and fascias	26	28

* Thickness, including coating, is approximately 0.022 inch for 26 gauge and 0.019 inch for 28 gauge.

Valley Tin
(Galvanized Sheet Metal)
Uses and Gauge of Galvanized Sheet Metal
Figure 5-7

5-11. End nail the first rafter; then end nail the second rafter with the nail going in at an angle. This is appropriate when using 16d nails to secure the rafters to 2 x 6 or 2 x 8 ridge plate.

Figure 5-12 shows the nailing methods for various items in framing and sheathing. Figure 5-13 gives the sizes of common wire nails.

Uses	Min. Thickness (Inches)	Min. Tensile Strength
Flashing, exposed	.019	16,000
Flashing, concealed	.015	14,000
Head flashing	.024	16,000
Roofing, shingles	.019	16,000
Roofing, large sheets	.019	19,000
Roof valleys	.019	16,000
Gutters	.027	16,000
Downspouts	.020	14,000
Termite shields	.024	16,000
Ventilators	.027	16,000
Gravel stops and fascias	.024	16,000

Valley Tin (Aluminum)
Uses and Thickness of Aluminum Sheet Metal
Figure 5-8

Zone	S.S.	Wind D.S.	Maximum Glass Area in Square Feet* Nominal Glass Thickness (Inches)					
			$\frac{3}{16}$"	$\frac{13}{16}$"	$\frac{7}{32}$"	$\frac{1}{4}$"	$\frac{5}{16}$"	$\frac{3}{8}$"
Low	10.7	19.5	40.0	48.0	60.0	75.0	90.0	120.0
Medium	7.3	13.2	27.0	32.0	41.0	51.0	62.0	79.0
High	4.8	8.7	18.0	21.0	27.0	34.0	41.0	52.0

*Applicable to regular plate or sheet glass only.
S.S. Single Strength
D.S. Double Strength

Installation of Glass in Wind Zones
Figure 5-9

Fastening Wood to Concrete

Every remodeler has to fasten wood to concrete. The stud driver or "ramset" is a gun with an explosive cartridge similar to a .22 or .32 caliber blank. The driver uses a specially designed pin fastener which will "nail" wood to concrete, steel to steel, steel to concrete, and so forth. It is a handy tool when furring masonry walls or fastening wood plates to concrete floors. The gun fires the pin fastener into the work like a bullet. (See Figure 5-14.)

I have found the Remington "Power Mate," model 455A, stud driver a good tool for many jobs. I'm sure there are other brands that will do the job as well. The stud driver will pay for itself in short order. Like the chain saw, it demands a lot of respect. In some areas you might need a permit from the authorities to buy or use the driver.

Termite Protection

In those areas where termites are a problem, be sure to protect against them. Soil treatment and pressure-treated wood with termite resistant chemicals are two methods of protection. Metal shields on top of pillars and foundation walls are a physical barrier against infestation and should be used in conjunction with chemical treatment in heavy infestation areas. The best approach is to have a professional outfit take care of the soil treatment. They know how to protect the building for the best results. Figure 5-15 shows those areas of termite infestation.

Summary

As suggested at the beginning of this chapter, you'll have to keep a close eye on materials delivered to the job site. Check both the quantity and quality against what was ordered. Few suppliers will deliberately cheat you. But anyone can make a mistake.

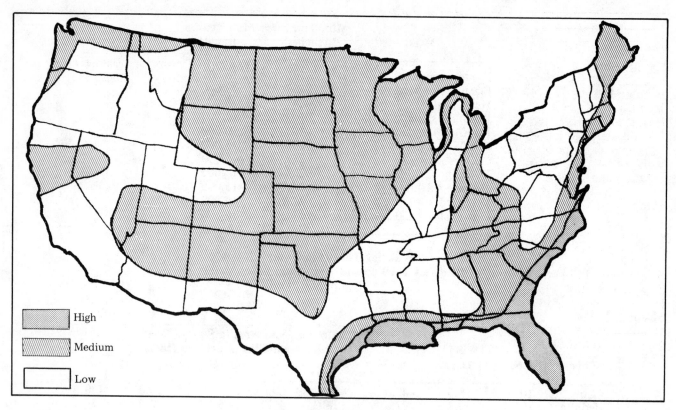

Zone Wind Velocities
Figure 5-10

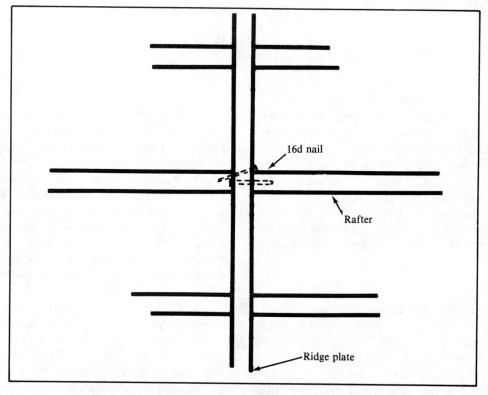

Double End Nailing Rafter to Ridge Plate with 16d Nails
Figure 5-11

Joining	Nailing Method	Nails		
		Number	Size	Placement
Header to joist	End nail	3	16d	--
Joist to sill or girder	Toe nail	2	10d	--
Header and stringer joist to sill	Toe nail	--	10d	--
Cross bridging to joist using 3'' stock	Face nail	--	6d	--
Ledger to girder or beam (2'' stock)	Face nail	3	16d	@ ea joist
Subfloor boards:				
1 x 6'' and smaller	Face nail	2	8d	@ ea joist
1 x 8''	Face nail	3	8d	@ ea joist
Subfloor, plywood:				
At edges	Face nail	--	8d	6'' O.C.
On each joist	Face nail	--	8d	8'' O.C.
Subfloor (2 x 6'', T&G) to joist or girder	Blind nail (casing) & face	2	16d	--
Soleplate to stud horizontal assembly	End nail	2	16d	@ ea stud
Top plate to stud	End nail	2	16d	@ ea stud
Stud to soleplate upright assembly	Toenail	4	8d	@ ea stud
Soleplate to joist or blocking	Face nail	--	16d	16'' O.C.
Double studs	Face nail	--	10d	16'' O.C. stagger
End studs, partition "T"	Face nail	--	16d	16'' O.C.
Upper top plate to lower top plate	Face nail	--	16d	16'' O.C.
Upper top plate, laps and intersections	Face nail	2	16d	--
Continuous header, two pc	Face nail	--	12d	12'' O.C. stagger
Ceiling joist to top wall plate	Toe nail	3	8d	--
Ceiling joist laps at partition	Face nail	4	16d	--
Rafter to top plate	Toe nail	2	8d	--
Rafter to ceiling joist	Face nail	5	10d	--
Rafter to valley or hip rafter	Toe nail	3	10d	--
Ridge plate to rafter	End nail	2	16d	--
Collar beam to rafter:				
2'' stock	Face nail	2	16d	--
1'' stock	Face nail	2	8d	--
1'' diagonal let-in brace to each stud and plate	Face nail	3	8d	4 nails @ top & 3 @ each stud
Built-up corner studs:				
Studs to blocking	Face nail	2	10d	Each side
Intersecting stud to corner studs	Face nail	--	16d	12'' O.C.
Built-up girders and beams, three or more members	Face nail	--	20d	32'' O.C. each side, stagger
Wall sheathing:				
1 x 8, or less, horizontal	Face nail	2	8d	@ each stud
1 x 6, or greater, diagonal	Face nail	3	8d	@ each stud
Wall sheathing, vertically applied plywood:				
3/8'' and less thick	Face nail	--	6d	6'' @ edge and
1/2'' and over thick	Face nail	--	8d	12'' each stud
Wall sheathing, vertically applied fiberboard:				
1/2'' thick	Face nail	--	1½'' RN*	3'' edge and 6''
25/32'' thick	Face nail	--	1¾'' RN*	@ each stud
Roof sheathing, board, 4-, 6-, 8'' wide	Face nail	2	8d	@ each rafter
Roof sheathing, plywood:				
3/8'' thick	Face nail	--	6d	6'' edge, 12''
1/2'' and over thick	Face nail	--	8d	each rafter

*RN: Roofing nail

Recommended Procedure for Nailing the Framing and Sheathing
Figure 5-12

Nail	Length (Inch)	Gauge	No. of Nails Per Pound (Approx.)
60d	6	2	10
50d	5½	3	13
40d	5	4	17
30d	4½	5	20
20d	4	6	30
16d	3½	8	46
12d	3¼	9	61
10d	3	9	65
9d	2¾	10¼	85
8d	2½	10¼	100
6d	2¼	11½	150
4d	1⅞	12½	275

Size and Number Per Pound of Common Wire Nails
Figure 5-13

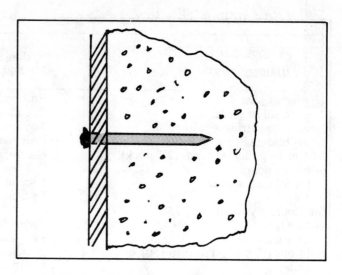

Stud Driver Pin Fastener - Wood to Concrete
Figure 5-14

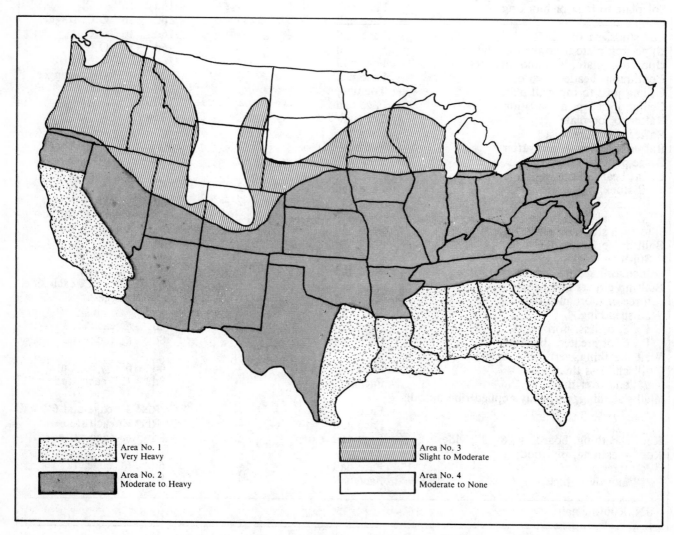

Approximate Areas and Density of Termite Infestation
Figure 5-15

6

Bathrooms

Each year thousands of bathrooms are remodeled or added. It's big business; a field in itself. But what about you? Can a builder with one or two men in his crew handle bathrooms? Maybe it's just you and a helper. Can you do it?

You can do it as well as the next man! Bathrooms aren't that complicated. It just takes a little know-how. Browse through a few magazines which feature articles on bathrooms. Look over the brochures put out by the companies that manufacture bathroom fixtures. You'll find that most of the planning and designing has already been done. Such items as wall finishes, colors, floors, lighting, design and layout have already been figured out.

Collect a notebook of attractive bathroom pictures from magazines and sales literature. Make up another notebook with color charts, fixture descriptions, color brochures, various floor plans showing fixture placement for different size bathrooms, lighting arrangements, and floor and wall finishes. Show these notebooks to potential customers so they get an idea of what is possible.

Use your imagination. These days just about anything goes where bathrooms are concerned. Walls can be paneled with barn planks. The room can have a sliding glass door facing a rock or flower garden. Bathrooms don't have to be confined to a small, drab cubicle. Flowers, plants, pictures, even music can enliven what was once strictly a functional necessity.

The bathroom can be a showplace. It can be one of the most beautiful rooms in the house. Who's to say it shouldn't be?

It used to be that planning the bathroom was time to start looking for ways to cut corners. Everyone thought all the fixtures had to be lined up against a common wall to save on pipe. If a second bathroom was built, it had to be back-to-back with the first bathroom. Well, it's true that less pipe might be required, but low cost plastic pipe makes this less important. Also, the second story bathroom does *not* have to be located over the first floor bath. In budget jobs the location of a bathroom might be a major consideration. Just keep an open mind and don't be controlled by previous practice.

Let's discuss subcontracting here. Even if you're one of those builders who is capable of doing plumbing, wiring and almost everything else, you probably can't because of code and licensing restrictions. You might be a top-notch carpenter but licensed as an electrician, or maybe a licensed plumber who is also a skilled carpenter. There are few builders who are licensed *and* qualified to do it all. So, you rely on subcontractors to do what you can't do or are not permitted to do. This can be a problem in bathroom work because of the different trades involved. A slow or slipshod subcontractor can cause delays and customer irritation.

Don't use any but the best subs you can for bathroom work. Some subs are more careful and reliable than others. You don't want tradesmen standing in an expensive tub grinding sand into the tub's finish while doing their work. The tub and other areas of the bathroom have to be protected from such carelessness. Painters will often stand step ladders in the tub without first protecting the tub. Plumbers will throw wrenches around like they were working in a garage. If the tub or other fixtures are scratched, it's your fault. If you've done much bathroom work, you already know what I'm talking about. If you haven't, then take measures to protect the tub and fixtures when you have sub work done.

The Modern Bathroom

A modern bathroom can accommodate just about any design. It can resemble the sterile operating room of a hospital or a flower garden. A water fountain can even be part of a bathroom. So can a waterfall. Dressing rooms and powder rooms can also be part of the bathroom. That doesn't mean every bathroom you remodel should have a waterfall. But there are imaginative, creative ways to handle any bathroom job. Your task is to find the features your customer wants and needs.

Bathroom remodeling is expensive. Most people realize this. They know it will cost a good deal to redo a bathroom and are not shocked by any reasonable quote. But they also realize that there are limits. A small four-room house probably isn't a candidate for a second bathroom with a rock garden and spa.

The size of the family usually determines minimum bathroom requirements. Anything beyond the minimum depends on money, taste, and space.

You will notice, in looking at various house plans, that bathrooms are in good proportion to the size of the house. Small houses have 1 bathroom. Medium-size houses have 1½ baths. Large houses usually have 2 full baths. The larger houses have 2 or more baths. In addition, some large houses have mud rooms, powder rooms, spas and saunas.

Thus, a bathroom can be a combination of bathroom, dressing room, closet and lounge. It can extend into a garden or have skylights or walls of glass or mirrors. It can have its own telephone extension and music center. Of course, not everyone wants an elaborate bathroom. But many do, so be prepared. A large variety of bathroom equipment is available in all colors, shapes and sizes.

Before starting on Brown's bathroom, let's take a closer look at bathroom fixtures, item by item.

Tubs

The shower/tub combination is your best bet for the single bathroom house. A shower uses less water, is quicker and, according to those who claim to know such things, more sanitary because you don't sit in your own bath water. While the tub is often used as a receptor for the shower, it's still a tub and available for bathing.

A standard rectangular tub is 4 to 6 feet long and 30 inches wide, and ranges in height from 12 to 16 inches. Bathtubs can be cast iron, steel or fiberglass. Cast iron tubs hold heat longer, are more durable, and can weigh as much as 500 pounds. The porcelain finish on a quality tub is approximately 1/16 inch thick, and with proper care will last a long time.

Steel tubs are not as expensive as cast iron and weigh around 100 pounds. While not as rugged as cast iron, steel tubs offer excellent service. Figure 6-1 shows a method of framing for a tub. Figure 6-2 shows an alternative method for supporting the tub. Framing is important. A full cast iron tub will weigh about 1000 pounds!

A modular fiberglass shower/tub unit is a one-piece fixture that eliminates the problem of finishing the walls in the shower area. There are no ceramic joints to clean. That's a big advantage. Unitized construction eliminates seams and corners. There are models for most standard tub openings in apartment houses, mobile homes, motels and condominiums. If you plan to install a modular unit, don't frame the bathroom until the unit has been positioned. Few of these units will pass through a doorway. None of them will fit through a doorway opening into a hall.

Fiberglass units are not as durable as cast iron or steel, however. And, they are more easily scratched or marred than metal tubs. But with reasonable care in installation and use, they can have a long life. Figure 6-3 shows typical framing and installation of a modular unit.

Most American homes are equipped with rectangular metal tubs recessed into a tub compartment or niche. The 3 surrounding walls are finished in ceramic tile to form a shower stall. Also available is the metal tub with one finished end. It is used in a corner because it needs no wall to enclose the finished end. While it can be used as a shower receptor with an appropriate shower curtain rod, it isn't as popular as the "niched" unit.

Other shapes are available in metal tubs. There is the square tub and the small tub that fits into a corner like a triangle. The small tubs are not large

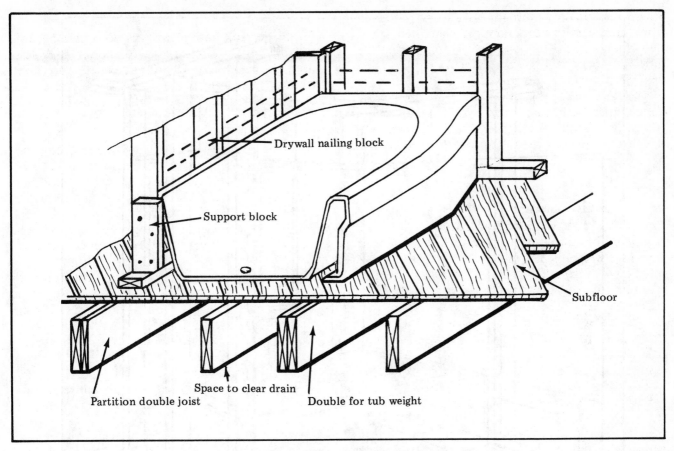

Framing For Bathtub (Block Support)
Figure 6-1

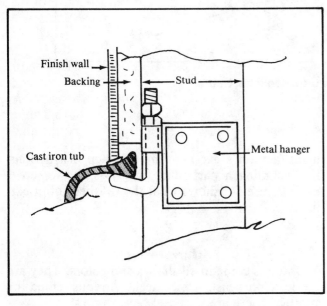

Tub Metal Hanger Support
Figure 6-2

enough for adults but are O.K. for children. The problem is that children grow up to be adults and the tubs don't.

The 60-inch tub is standard. Other lengths are available up to 72 inches. The tub bottom should have a non-skid surface. Most metal tubs are now built with a bottom non-skid surface. Grab bars and hand grips should be properly positioned and firmly secured.

Tub and shower glass enclosure doors should be safety glass. Shower heads can be 60 inches or higher. Every bathroom that doesn't have an openable window needs an exhaust fan, particularly bathrooms with showers. The fan pulls out excess moisture, protecting the walls and ceilings.

Shower Coves and Cabinets

Shower units constructed of polypropylene or fiberglass offer excellent service and are fairly easy to clean. The shower cabinet normally requires no framing and lends itself easily to budget jobs.

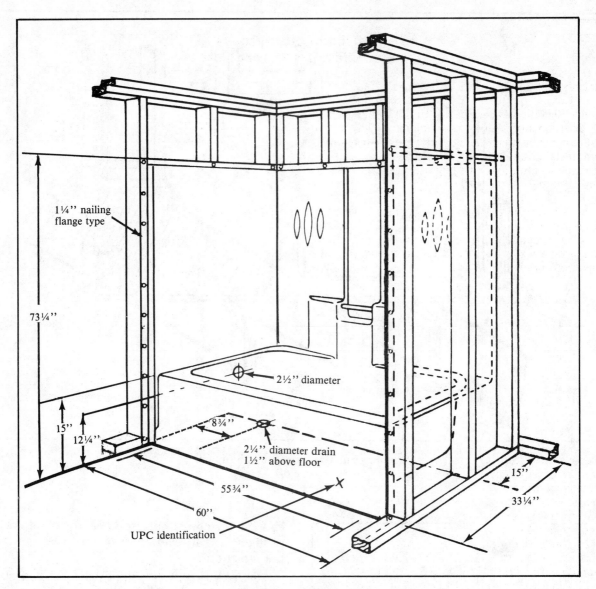

Framing and Installation of a Modular Unit
Figure 6-3

Fiberglass shower coves require special framing. The manufacturers usually include adequate installation instructions with their products.

Installation instructions provided by the manufacturers over the years have improved considerably. It used to be that the only instructions you received with most products were the packing instructions. Then, when installation instructions were introduced, they were usually written by someone who had never installed anything more complicated than a light bulb. You had ten pages of instructions where two would have done the job nicely. But all this has changed. Figure 6-4 is one manufacturer's excellent step-by-step instructions for installing a particular fiberglass shower cove unit. Figure 6-5 shows what the finished job looks like.

Lavatories
Lavatories come in all shapes and colors. They are made from cast iron, steel, vitreous china or plastic. There are six basic types:

1. Wall hung—The lavatory is attached to the wall. There is no cabinet. The pipes are exposed underneath.

Installation Instructions
For Fiberglass Shower Cove

Initial Instructions:

1. While the unit was designed to fit through most standard door openings, it is recommended that careful measurements be taken to determine if the unit must be within installation area before framing is complete. When moving unit, avoid flexing of sidewalls to prevent radius cracking.

2. In locations where plumbing is adjacent to masonry wall, provisions must be made for access to connections. Construct a separate wall a minimum of six inches from masonry wall. Install furring when back of unit is against masonry wall.

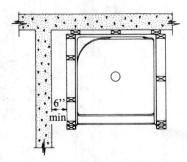

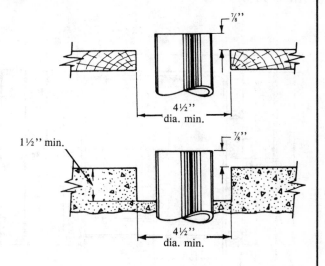

3. Being a one-piece shower cove, no shower pan or hot mopping is required. The rough plumbing for the shower cove drain must be located in accordance with dimensions specified on appropriate illustration. Note on the diagrams for wood and concrete the opening required in the subfloor to accommodate the drain fitting. The 2'' waste pipe must extend ⅞'' above the subfloor or slab. A pocket is required in slab construction to accommodate the drain fitting. No additional support under basin area is required when unit is securely nailed into access. *(Refer to Step 6 of installation instructions.)*

4. Before installation, arrange to provide access to connections. Consider these suggestions:

a. If units are installed on a slab at the corner of outside walls, connect at the accessible end or provide access to connections.

b. Units installed on a slab back-to-back against outside walls should be planned for access to connections. In wood partition construction at least one unit should have access through which other units may be reached. In masonry or fire rated partitions, both units should be provided with access.

c. Units installed on a slab in a cluster of four, back-to-back and end-to-end, should be planned with a plumbing wall which provides sufficient clearance for access through plumbing wall, or connection from the accessible end.

d. When units are installed on a floor covering the basement or in locations where drain connections are accessible from below, connections can be made from beneath.

When fire rated wall is specified, it should be in place before units are installed.

If unit is to be installed adjacent to a vertical duct or chase, it is suggested that fire rated gypsum drywall surround the unit.

5. If units must be stored out of doors, they must be protected from the elements and placed upside down.

6. Do not remove protective packaging until unit is ready for installation.

7. Shower floor protector should be in place during installation. Care should be taken to protect unit from possible damage after installation.

8. When exterior walls are to be insulated, they should be insulated before unit is installed.

Installation Instructions For Fiberglass Shower Cove
Figure 6-4

1. Construct framing as per illustration. Framing must be square and plumb. If grab bars are specified, provide 2 x 6 bridging for attachment at specified locations. *(Refer to step 10.)*

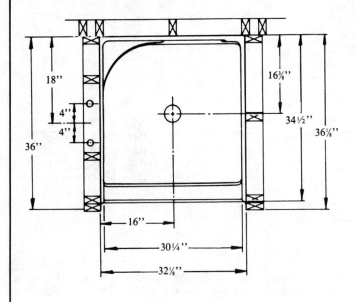

Fitting can be mounted on either side

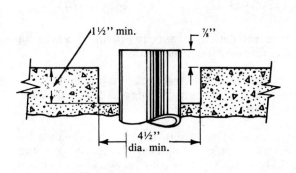

No additional wood bracing to support side walls.

No additional support under basin area, when unit is securely nailed into recess.

2. The rough plumbing for the shower cove drain must be located in accordance with dimensions on illustration. The 2" drain pipe must extend ⅞" above the subfloor or slab. A pocket is required in slab construction to accommodate the drain fitting.

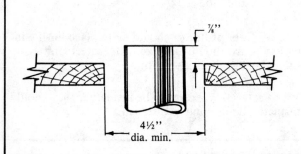

Installation Instructions For Fiberglass Shower Cove
Figure 6-4 (continued)

3. Rough plumbing for the supply may be positioned as required, but not strapped. Where access to supply fitting connections are limited, consider using soft copper tubing for risers to valve from supply line. Connect supply fitting and riser before positioning unit. The soft tubing permits positioning supply fitting and shower arm after unit is in place.

4. Install shower drain fitting in drain outlet of unit using a non-hardening mastic sealant to meet local code. On the outside of the unit lay out the location of the fitting holes in accordance with the illustration below. Using ¼" drill, make the pilot holes from the outside of the unit.

4a. For multiple installation where supply fittings are accurately located during rough plumbing, a template made of light weight plywood and containing all pilot holes can be used to locate holes on unit.

5. Working from the finished face of the unit, drill holes to required size using a hole cutting saw.

6. Clean recess thoroughly and set unit squarely into pocket in subfloor or slab and over drain pipe extension. Plumb front and side nailing-in flanges, and level rim of unit. Shim as required to level unit.

Unit must be plumb and level to insure proper shower enclosure installation. Some framing adjustments may be required because of variation in cove width and/or framing tolerances. *(Before nailing, check to see that nailing-in flanges are firmly against wall studs. This will help to prevent movement of the unit while nailing the unit into place.)* Using #6 large head galvanized nails, fasten the back wall nailing flange of unit to studs. Attach side wall nailing flanges in same way. The front vertical nailing flanges should be then nailed to studs on 8" centers; being careful to prevent possible damage.

Note: It is strongly suggested that the installer pre-drill the holes in flange before nailing to prevent buckling. In metal stud construction, holes are drilled through nailing-in flanges and metal studs and secured with sheet metal screws.

7. Caulk and lead the drain, plug and test.

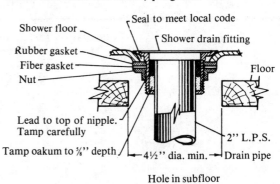

Installation Instructions For Fiberglass Shower Cove
Figure 6-4 (continued)

8. Apply furring strips ⅛'' thick to studs to insure a true and level wallboard installation. Apply water resistant sealer to nailing flange. Install wallboard horizontally with factory edge (paper bound) ¼'' maximum above horizontal finished surface of unit, extending at least to next stud beyond unit.

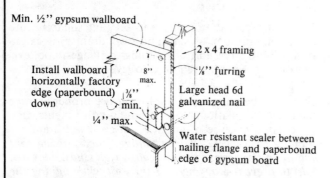

Vertical nailing flanges should be similarly furred and wallboard sealed to nailing-in flange. Mud, tape, and finish true and level. Seal area between paperbound edge of wallboard and unit with a water resistant sealer after wall finish has been applied.

9. Strap supply fittings. Caulk all openings around fittings and shower arm with water resistant sealer.

10. Install finish trim. Grab or towel bars can now be installed through the cove walls and fastened in place directly to the bridging previously installed. Screws, bolts, and backing should be selected to properly secure grab bars.

Grab bars can also be installed to cove, providing the proper fastening device is selected to securely attach bar to unit. Manufacturer's suggestions concerning grab bar installation should be carefully followed.

Shower rod should be installed above unit and positioned so shower curtain hangs inside dam. Follow manufacturer's installation instructions when shower enclosure is specified.

Clean-Up After Installation

When cleaning up after installation **do not use abrasive cleaners,** they may scratch and dull the unit surface. Use warm water and one of the liquid detergents to clean the surface.

Stubborn stains, paint or tar can be removed with turpentine or paint thinner. Plaster can be removed by scraping with a wood edge. Do not use metal scrapers, wire brushes or other metal tools. One of the powder type detergents may be used on a damp cloth to provide mild abrasive action to the residual plaster.

Dulled areas may be restored by rubbing with a liquid automotive type body cleaning compound followed with a light application of liquid wax.

Installation Instructions For Fiberglass Shower Cove
Figure 6-4 (continued)

Installed Cove Unit
Figure 6-5

2. Flush mount—The lavatory requires a metal ring or frame to hold it in place in the cabinet.

3. Self-rimming—The lavatory holds itself in place with an integral rim which rests on the counter top. (See Figure 6-6.)

Self-Rimming Lavatory
Figure 6-6

4. One-piece integral lavatory and counter top—The lavatory and top are formed as a unit from plastic or synthetic marble material.

5. Under counter—The lavatory is mounted beneath an opening in the marble of a simulated marble counter top. The fittings are mounted through the counter top.

6. Pedestal—The lavatory rests on a pedestal. The fixture offers a different decorative effect. (See Figure 6-7.) The fixture should be installed according to manufacturer's instructions, as shown in Figure 6-8.

The oval-shaped lavatory (19¼ x 16¼ inches in size) is available for installation in narrow counter tops. Figure 6-9 shows a self-rimming enamel cast iron lavatory of this design. Use the recommended adhesive on the counter top to ensure a good seal.

The lavatory cabinet or vanity comes in different heights. The 32-inch height is popular. A 30-inch high wall-hung lavatory is also popular.

In family-size bathrooms two lavatories go a long way toward providing harmony during those periods when all the kids have to brush their teeth at the same time (which is most of the time).

Where only one lavatory is installed in the counter top, locate it off-center in the vanity. This leaves counter room for those larger bathroom items such as hairdryers and make-up kits. A large counter area can be used for plant arrangements. Small mirrors are centered over the lavatory and not the vanity. Large mirrors the width of the vanity are centered over the vanity. Other mirror styles and placements are available. Two mirrors can be centered over each lavatory. Medicine cabinets might be placed at either side of a mirror with their doors opening toward the center mirror. This provides a three-sided view.

The nice thing about having a vanity in a bathroom is that it's multi-purpose. It hides the pipes, provides storage space, and can be decorative with ornamental trim and hardware.

Many decorative faucet controls are available. Taste and cost will determine the style to use. However, the controls should be easy to reach for operation, cleaning and maintenance. Cartridge-type controls offer good service. The faucet should be the mixing type.

Pedestal Offers New Styling
Figure 6-7

Initial Information

These instructions are for installing to frame type construction. For other installations, supply suitable bracing and fastening devices of sufficient strength and size.

Instructions

The following instructions assume that the wall and floor are square and plumb.

1. Rough-in the supplies and drain piping according to roughing-in dimensions. Observe plumbing codes.

2. Install sufficient backing behind finished wall for hangers and anchoring devices.

3. Install hangers according to roughing-in dimensions.

Note: Hanger may have to be repositioned. See Step 8.

4. Before installing lavatory, attach supply and drain fittings.

5. Trap should be installed so the inlet is 7⅛" from the finished wall. Slip the 1¼" O.D. drain tailpiece into trap.

6. Locate pedestal so mounting hole is approximately 8⁹⁄₁₆" from the finished wall.

7. Apply three of the self-adhesive gaskets provided, on to the top edge of the pedestal as illustrated.

8. Carefully place lavatory over hanger and pedestal, moving pedestal slightly as required to assure proper fit. If the bottom of the lavatory makes contact with any part of the china on the top edge of the pedestal, use the additional gaskets supplied to shim as required. **Lavatory must be supported by both the hanger and the pedestal.** Check for proper drain alignment and level of lavatory.

Note: Because of variations in china, hanger may have to be raised or lowered slightly so that lavatory contacts wall evenly and rests on pedestal.

9. Mark the centers of the anchoring holes for the lavatory and the pedestal.

10. Remove lavatory and pedestal. Drill holes for ¼" lag bolts and washers for securing lavatory to backing. Other fastening devices of sufficient size and strength may be used. Drill hole for attachment of pedestal to floor in similar manner.

11. Place pedestal and lavatory back into mounting position. Install ¼" lag bolts and washers and secure lavatory to wall. Attach pedestal to floor in similar manner. *Carefully tighten bolts leaving them approximately ½ turn loose.*

12. Connect hot and cold supplies to fitting. Connect previously installed 1¼" O.D. tailpiece to drain body. Secure trap connections.

Gaskets

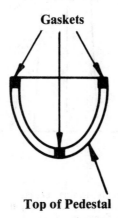

Top of Pedestal

Installation Instructions for Lavatory and Pedestal
Figure 6-8

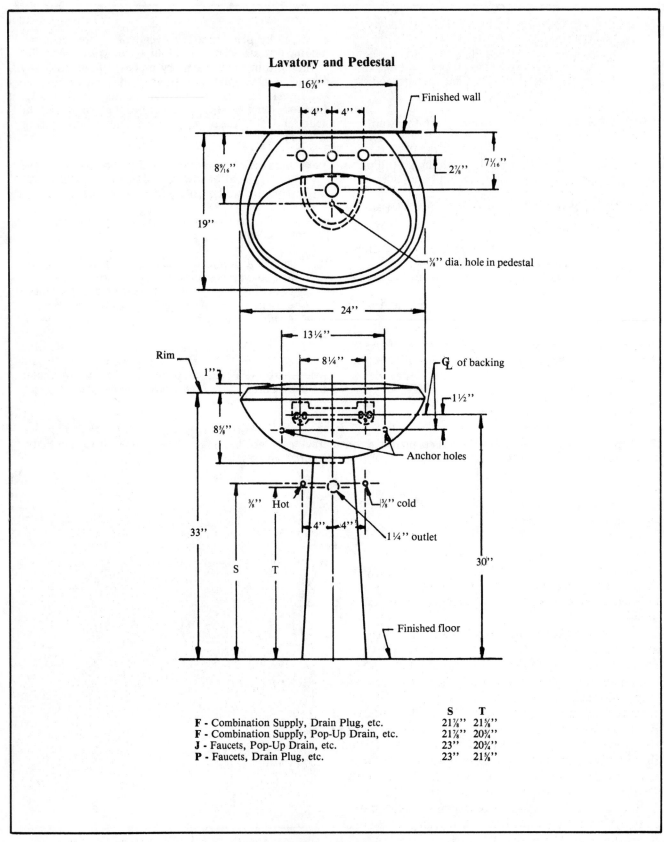

Installation Instructions for Lavatory and Pedestal
Figure 6-8 (Continued)

Narrow Counter Top Installation
Figure 6-9

Choose a lavatory and its hardware for its attractiveness as well as serviceability. Attention to these things can make the difference between a sale and no sale.

Commode

A commode is a toilet or a water closet. A good commode is expensive but the money is usually well spent. An inferior commode is likely to overflow, leak, or drip. The tank holds water for each flush and this water is needed almost instantly. The house water lines aren't large enough to provide the volume of water required for flushing.

So select a good quality commode when you're redoing or adding a bath. The two recommended commode bowl types are the "reverse trap" and the "siphon jet." The washdown bowl is not widely used anymore and is not acceptable in many areas.

The reverse trap bowl discharges into a trapway at the rear of the bowl. The trapway is generally round. This ensures proper flushing action.

The siphon jet operates similarly to the reverse trap, discharging to a rear trapway. Both the reverse trap and siphon jet bowls have enough water to cover most of the exposed surface within the bowl, thereby eliminating the fouling of unprotected dry areas within the bowl. Either type is available in round or elongated styles. The current preference is toward the elongated style. It's comfortable and for most homeowners has more "style" than the round bowl. But today's "style" is not necessarily tomorrow's.

Wall-mounted commodes have the advantage of being easy to clean under since they do not touch the floor. They are more expensive than the floor-mounted models and must be specially mounted. Thus, some rearranging of water lines and drain pipes will be necessary to change from floor to wall mount.

Vintage pull-chain commodes are also available for that bathroom decor of yesterday. (See Figure 6-10.)

Plastic toilet seats are longer lasting than wood seats. The paint on wood seats has a way of chipping off and becoming unsightly. Select sturdy seats with strong hinges.

Vintage Pull-Chain For Yesterday's Style
Figure 6-10

**Sunken Installation Method
Figure 6-11**

Bidets

The bidet is the latest addition to the modern bathroom. It's gaining in popularity. Eventually no bathroom in America will be considered modern without one. It is used for cleaning the perineum after using the commode. It is installed near the commode and needs hot and cold water hook-ups and a drain. The user sits astride the bowl facing the controls and adjusts the water volume and temperature. (A bidet is shown near the commode in Figure 6-12.)

Whirlpool Baths

A whirlpool bath is generally considered a luxury item, but many homes are having them installed. Figure 6-11 shows a sunken installation method of an enameled cast iron whirlpool tub. Various size tubs are available. Figure 6-12 shows another installation method.

Figure 6-13 gives framing instructions for corner, peninsular, recessed and sunken installations.

Spas

Spas are another luxury item gaining popularity. The spa is not considered generally a part of the bathroom since it is usually located outdoors. However, it can be placed indoors in an area adjacent to or as part of the bathroom area, if desired. Figure 6-14 is a body-contoured lounge designed for two people. It has six directional whirlpool jets, an injection system, underwater lighting and air-activated controls. It can accommodate four to six

One Method of Whirlpool Installation
Figure 6-12

persons. It measures approximately 86 inches on each side and is 40 inches deep. Other shapes and sizes are available.

Figure 6-15 shows typical piping for an electrical water heater. Figure 6-16 shows typical piping for a gas water heater. Figure 6-17 offers general instructions for outdoor and indoor installations. Because of the variety of installation techniques possible with a spa, framing and installation procedures other than those described may be required.

Local and state building, plumbing and electrical codes must be observed. Electrical connections should be made by a licensed electrician.

Installation Instructions

1. Carefully remove bath from shipping package. Do not handle bath by the piping, pump or power panel.

2. Install the drain fitting to bath.

3. Position the bath, referring to the framing instructions below. Connect the drain fitting to the trap.

Note 1: The piping or pump cannot be used for structural support nor should the bath be positioned by using any of the installed piping.

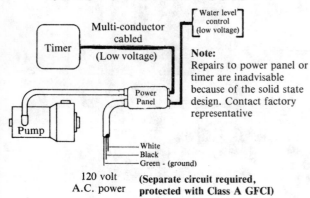

Timer

Multi-conductor cabled (Low voltage)

Water level control (low voltage)

Note: Repairs to power panel or timer are inadvisable because of the solid state design. Contact factory representative

Power Panel

Pump

White
Black
Green - (ground)

120 volt A.C. power

(Separate circuit required, protected with Class A GFCI)

Note 2: Construct suitable protection to cover the bath so it will not be damaged during construction.

4. The whirlpool system has been pre-wired at the factory. A routine electrical service connection should be made to the 3 foot length of ½'' water-tite conduit containing black, white and green (ground) wires, provided. **A separate equipment grounding conductor must be provided for whirlpool ground wire (green). Ground must not be connected to any current carrying conductor except at main service breaker box (per National Electrical Code, Article 250-61-b).**

Note: All installations shall have a Class A ground fault circuit protector which will provide additional protection against line-to-ground shock hazard. A 120 V., 15 amp., 60 hz. separate circuit is required.

5. Install the timer and connect to the power panel. *(See Timer Instructions.)* **Caution: Electrical power to whirlpool must be off before connecting to power panel.**

6. Provide access panels.

Framing Instructions

Corner Installation
With Ceramic Tile Walls and Tub Facing

1. Apply vinyl bead on sides to be built-in.

2. Install wallboard (2). Bevel bottom of wallboard to meet contour of vinyl bead.

3. Install wall tile and grout.

4. Frame out under tub rim (5). Cover with wallboard or equivalent (4). Finished face of tile must be ¹⁄₃₂'' under tub rim. Tile should not project beyond rim.

5. Install tile (3) on tub sides and apply grout. Seal under tub rim with water resistant sealer.

Note: Support under tub legs must be provided.

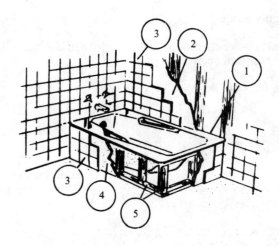

Installation and Framing Instructions
Figure 6-13

Peninsular Installation
With Ceramic Tile Walls and Tub Facing

1. Follow steps 1, 2, and 3 for corner installation.

2. Construct frame. Flange can be any width desired. Allow for thickness of tile and wallboard between header and underside of tub rim. (Tub weight should rest on tub legs.)

3. To top of header and vertical studs apply ⅜" wallboard (2). (Wallboard on top of header (4) may be omitted and tile applied directly to wood.)

4. When laying horizontal tile, apply mastic on bottom of tile which will be under tub rim.

5. Slide horizontal tiles into place, sliding only as necessary.

6. Apply vertical tile (3). Apply grout between tile and between tub rim and horizontal tile. Seal under tub rim with water resistant sealer.

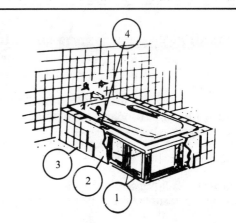

Recess Installation with Walls and
Tub Facing of Wood Paneling

1. See instructions for installation of vinyl tile bead (1) on sides to be built-in.

2. Apply finished paneling (2) with face ⅞" from studs. Bevel bottom edge of panels to meet contour of vinyl bead.

3. Frame out under tub rim (3). Face of paneling to be 1/32" under tub rim.

4. Apply panels.

5. Seal joints between tub and wall, and under rim with water resistant sealer.

Note: Support under tub legs must be provided.

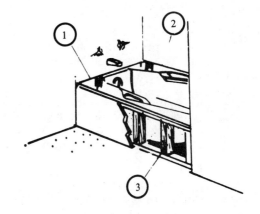

Sunken Installation

1. In sunken installations adequate framing to support tub legs must be provided.

2. Follow instructions for peninsular installation for construction of header.

Installation and Framing Instructions
Figure 6-13 (continued)

Installation Instructions for Vinyl Bead

For Recessed, Corner, and Peninsular Installations

1. The vinyl bead takes the place of the tiling in bead on a conventional tub and is necessary to obtain a permanent seal. The installation kit in the tub crate contains a 12' vinyl tile bead, a 12' strip of pressure-sensitive sealing tape and instructions.

2. For recess installations (wall on 3 sides) use entire vinyl bead and pressure-sensitive sealing tape. For corner installations (wall on long side and end) use peninsular installations (one short wall or end) use 3' vinyl strip and equal amount of sealing tape. Both bead and tape must be one piece, no matter what type of installation is used.

3. This tub is reversible. Allow ample working area close to installation position. A recessed installation is described below. It also applies to corner and peninsular installations.

4. Clean all dust, grease and foreign material from tub rim.

5. Position bead for proper installation (as per Instruction Number 7) and clip excess material at corners to form miter joint.

6. Starting ¾'' in from end of rim and using bottom edge of rim as a guide, apply predetermined length of pressure-sensitive sealing tape to tub. Clip excess material at corners to form miter joint. Remove paper backing.

7. Start vinyl strip ¾'' in from end of rim. Position bead properly on tub as shown and press firmly against tape to effect seal. Continue to do same while working towards opposite end.

8. When extrusion is in place, start at the area first applied, rub the gasket firmly in place with heel of hand to ensure continuous seal.

9. Tub is ready for installation, leveling, framing and application of wall material.

Caution: Do not apply vinyl tile bead or pressure-sensitive sealing tape in temperatures below 40°F.

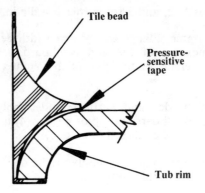

Installation and Framing Instructions
Figure 6-13 (Continued)

Outdoor Spa
Figure 6-14

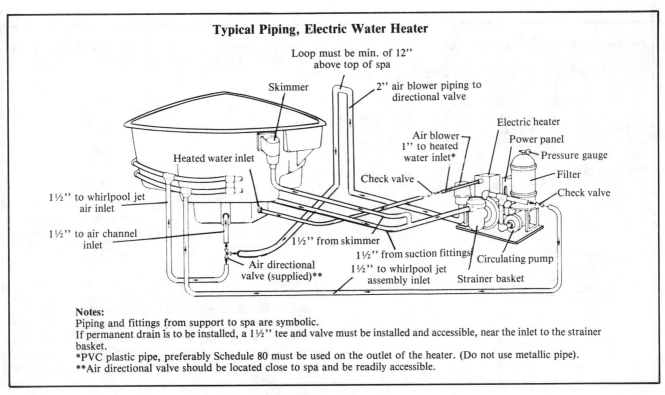

Typical Piping, Electric Water Heater

Loop must be min. of 12''
above top of spa

Skimmer

2'' air blower piping to
directional valve

Electric heater

Power panel

Air blower
1'' to heated
water inlet*

Pressure gauge

Filter

Heated water inlet

Check valve

Check valve

1½'' to whirlpool jet
air inlet

1½'' to air channel
inlet

1½'' from skimmer

1½'' from suction fittings

Circulating pump

Air directional
valve (supplied)**

1½'' to whirlpool jet
assembly inlet

Strainer basket

Notes:
Piping and fittings from support to spa are symbolic.
If permanent drain is to be installed, a 1½'' tee and valve must be installed and accessible, near the inlet to the strainer basket.
*PVC plastic pipe, preferably Schedule 80 must be used on the outlet of the heater. (Do not use metallic pipe).
**Air directional valve should be located close to spa and be readily accessible.

Typical Piping, Electric Water Heater Installation
Figure 6-15

Typical Piping, Gas Water Heater

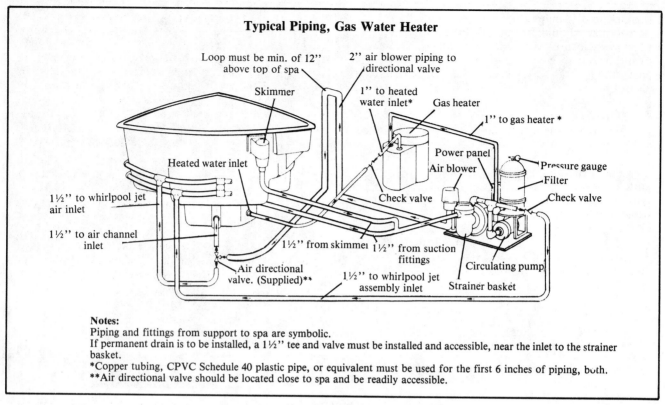

Notes:
Piping and fittings from support to spa are symbolic.
If permanent drain is to be installed, a 1 1/2" tee and valve must be installed and accessible, near the inlet to the strainer basket.
*Copper tubing, CPVC Schedule 40 plastic pipe, or equivalent must be used for the first 6 inches of piping, both.
**Air directional valve should be located close to spa and be readily accessible.

Typical Piping, Gas Water Heater Installation
Figure 6-16

Outdoor Installations - Above Grade

1. A 4" thick minimum concrete foundation, reinforced with 1/2" steel rods in a 6" cross hatch pattern, is recommended. Steel wire must secure each rod intersection and rods must be bonded to the bonding terminal on the power panel with minimum Number 8 solid copper wire, and approved bonding clamps. This foundation must be built on solid, undisturbed ground. Footings may be required around and in the center of the foundation. **Check with local building codes.**

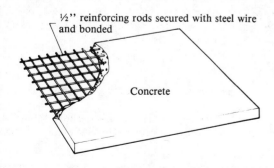

2. Provide proper drainage for water that may splash over rim of spa, or water that may collect at the spa base.

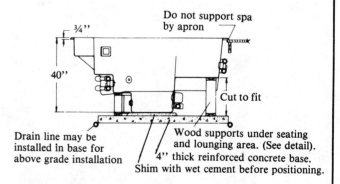

3. Locate the concrete slab for the support system so it is a maximum of 20 feet from the spa edge. It is recommended that the support system be installed below the water level of the spa, although not exceeding seven feet below. Locating the system below water level will eliminate manual priming of the pumps. If the support system is to be installed above the water level of the spa

Typical Spa Installation Instructions
Figure 6-17

(maximum 4'') manual priming will be required. Support system should be housed in an adequately ventilated enclosure that can be locked to prevent unauthorized access to the controls. Spa chemicals and combustible materials should not be stored in the enclosure.

4. A separate 240V., 60 amp circuit in rigid conduit is required from the main service to the spa support system power panel.

5. Metal structures including fencing, stairs, railings, gutters, down spouts, etc. within 5 feet of the spa edge must be bonded to the power panel bonding terminal with minimum #8 solid copper wire using approved bonding clamps.

6. Electrical receptacles must not be installed closer than 10 feet from the spa. Receptacles must have ground fault circuit protection. (G. F. C. I.).

7. Electrical light fixtures should **never** be installed directly over the spa. If lights are installed, they must be at least 5 feet horizontally away from the spa rim, and at least 5 feet above the ground. G.F.C.I. must be provided if a light fixture is located within 10 feet of the spa edge.

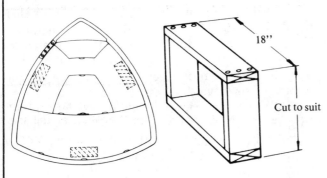

8. Install the spa so that support is provided under the seating areas. Supply reinforcing under the apron edge, but do not support the weight of the spa by the apron. Use 2'' x 6'' redwood or a treated wood which is resistant to deterioration, for the support.

9. Check with local codes and/or ordinances for regulations of gates, fences, covers, etc. for the purpose of excluding uninvited children or animals.

10. If a permanent drain is to be installed, a 1½'' tee and valve can be installed, and be accessible, near the inlet to the strainer basket. The outlet of the valve should be connected to the house drain, or drain field, in accordance with local codes.

11. There is a risk in using fuel burning appliances such as gas water heaters in rooms, garages or other areas where gasoline, other flammable liquids or engine driven equipment or vehicles are stored, operated or repaired. Flammable vapors are heavy and travel along the floor and may be ignited by the heater's pilot or main burner flames causing fire or explosion. Some local codes permit operation of gas appliances if installed 18 inches or more above the floor. This may reduce the risk if location in such an area cannot be avoided.

12. With the optional gas heater, conditions which cause back drafts must not exist. Failure to correct back drafts may cause air contamination and unsafe conditions. When exhaust fans are installed in the same room with the heater or in a connected room, sufficient air must be provided. **Insufficient make up air will cause outside air to be drawn into the room through the chimney, causing poor combustion and a risk of asphyxiation.**

Outdoor Installation - Below Grade

1. For below grade installation, follow same electrical and drainage requirements as for above grade.

2. A hole must be dug so that it is 12'' larger than the spa. The earth around the hole should be stable and be left undisturbed to provide solid support.

3 Trenching for the piping, air switches, and electrical light conduit must be provided. Piping from the spa drain must have a slight pitch from connection to the equipment assembly, and have as few bends and fittings as possible. Spa Support System should be installed so that pumps are located at a level below the water level in the spa to provide priming of the pumps. *(See Above Grade, Step 3.)*

4. After all electrical and plumbing connections are installed and tested, and supports under basin and seating areas are installed, back filling of sand or gravel around the spa, especially under the seating area, is required. Retaining walls may have to be constructed to withhold the back filling. A concrete wall may be constructed, enclosing the spa. With this type of construction, back filling is not required.

Indoor Installation

1. For indoor installation, the existing floor must be structurally able to support a minimum of 5,500 pounds.

Typical Spa Installation Instructions
Figure 6-17 (continued)

2. Follow the same electrical, drainage, and support requirements as specified for above grade installation.

3. A drain pan under the spa and spa support system is recommended to protect against possible water damage.

4. Materials specified for the room in which the spa is to be installed, must be selected to withstand high humidity. Provide adequate ventilation for the room, either naturally or by a properly sized exhaust fan or dehumidifier, to provide comfort and prevent possible damage to the structure.

5. Conditions which cause back drafts must not exist.

Failure to correct back drafts may cause air contamination and unsafe conditions. When exhaust fans are installed in the same room with the heater or in a connected room, sufficient air must be provided. **Insufficient make up air will cause air to be drawn into the room through the chimney, causing poor combustion and a risk of asphyxiation.**

6. Some of the chemicals used to maintain the correct chemical balance of the spa water can cause corrosion to certain metals found in various equipment, appliances and hardware. Therefore, only non-corrosive metals should be used in the spa area.

Typical Spa Installation Instructions
Figure 6-17 (continued)

Storage

Brown's old bathroom has no storage space. He wants plenty of storage space in his remodeled bathroom and also in the new bathroom you're going to add. When you plan the bathrooms, keep that in mind. You can build a cabinet above the commode. The vanity will have a storage space in the form of shelves and drawers. If the room is large enough, there can be a closet at the end of the tub. Such a closet can provide an access panel to the plumbing of the tub and shower.

It is hard to provide too much storage space in a bathroom. All sorts of gadgets are made for the bathrooms: hair dryers, electric tooth brushes, shavers, hot combs, waterpics, hot water bottles and heating pads. Also, there are the usual items for the medicine cabinet. Towels, wash cloths, soap and shampoos also need a place, as do dirty clothes. Think about these things when you're planning to build or remodel a bath. Remember, the more of these items and equipment that can be properly stored, the less cluttered the bathroom will be.

Lighting and Electrical Outlets

A bathroom without enough light and electrical outlets is as bad as one without adequate storage space. Most baths are short in both areas. A bathroom needs special lighting in specific areas as well as general lighting for the entire room. Your electrical fixtures dealer has hundreds of lighting fixtures that can give the needed light as well as the decorative effect desired.

Different fixtures have been designed for dif-

ferent areas. There are waterproof lights for over showers and tubs. The lights at the vanity area can be theatrical, hanging, spot, etc. You can even go for a luminous ceiling or indirect lighting from bulkheads.

A word of caution. Don't place switches or electrical outlets within reach of the shower or tub.

And put in enough outlets. Many accessories need electricity and will be of little use if there's no place to plug them in. Keep outlets above the vanity or lavatory level for easy access.

While you're thinking about outlets, don't forget sun lamps and heat lights. The heat lights, installed in the ceiling near the tub or shower, offer a quick warm-up while drying off.

Vanities and Cabinets

The vanity has probably done more for the bathroom than any other fixture. The vanity treatment—that is, the color, trim, doors, hardwood, etc.,—is virtually unlimited. You can paint it any color you wish. Just remember to keep the size of the vanity in proportion to the room.

Vanities come in as many styles as kitchen cabinets: early American, modern Mediterranean, etc. Your best sources of information on this subject are the many trade magazines that feature bathrooms. Such magazines offer many helpful suggestions for vanity treatment and design.

Bathroom Fittings and Hardware

Your sales pitch should include emphasis on faucets and hardware such as towel rings and rods, vanity pulls and hinges. Select good quality items in these areas and you can't go wrong. Lower-priced products will cheapen an otherwise excellent job. People notice these things.

Selection of water valve controls is a matter of preference. The various types are:

• Two valve system—2 handles and 2 valves. One for hot, one for cold.

• Single valve—1 lever or knob. Move right to left for temperature control. In or out, backwards or forwards for volume.

• Pressure valve—Preset for volume. The user sets the temperature, and the valve automatically maintains it.

• Thermostatic valve—A heat-sensing device automatically adjusts the hot and cold water flow according to a preselected temperature. The user controls the volume as well as temperature.

Walls and Floor Covering

The walls of a bathroom can range from plaster to wood paneling, from wallpaper to mirrors. You name it and it's being done. Even brick and stone are used for bathroom walls. It's the way things mesh that counts. Here you're not limited in your approach. Expand your thinking. Imagination is what brought the bathroom to showroom status in the first place. It was what brought the toilet from the outhouse to inside. Just because you're a small builder doesn't mean you're stuck with all the old methods.

Ceramic tile used to be the most used material around the tub and shower. The 4½ by 4½-inch square tile was a favorite wall size. The tile comes in other sizes, but this was the size most often used. Today many different sizes and colors of tiles are used. When blended with colored fixtures, a most beautiful bathroom emerges. A clash in colors is always possible, so colors must be identified early in the planning stage. But recognize that it's very easy to plan a bathroom without ceramic tile if your customer wants it that way.

Then there is marble. It can be either synthetic or natural. Plastic laminates and plastic-finished hardboards are very popular where cost is the prime consideration.

Whatever you use, blend the various bathroom components into a unified whole. A pink tub against a green wall won't do much for your customer's early morning shave.

Mottled or crystalline glazed tile is better for floors than highly glazed wall tile. Shapes of floor tile can be small squares, rectangles, hexagons or octagons. Quarry tile is often used on floors and makes a durable and attractive choice.

Sheet vinyl is an excellent floor covering. The six-foot-wide sheets permit a seamless floor in most baths. Carpet is another good bet, especially where sloppy kids have grown up to be sloppy adults.

Poured floors are built up in layers with polyurethane chips in a polyurethane glaze. These seamless floors are attractive, durable and easy to clean. Just be sure the chips are thoroughly covered. Otherwise they are rough on the bare feet. Resilient tile is not the best floor covering for a bathroom since water can eventually seep through the joints and loosen the tile.

Greenery and Bric-A-Brac

Have you ever wondered why plants and little bric-a-brac were placed in a bathroom? A green plant, live or artificial, is decorative. A fancy soap dish is also, as are fluffy colored towels.

Just remember not to overdo it. Don't have flower pots hanging everywhere. Just because a couple of plants look good, don't turn the place into a jungle. Again, check magazines or books for ideas.

Plans

Don't get the idea that the ideal bathroom has to be 12 by 16 feet and have a whirlpool tub and bidet. Bathrooms can be large or small. You should be prepared to handle anything from the small, basic bath (Figures 6-18A and 6-18B) to the larger ones.

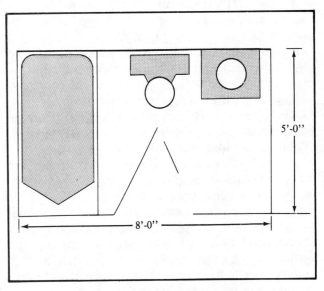

**Basic Bath-Common Wall Plumbing
Commode and Lavatory Positions Can Be Switched
Figure 6-18A**

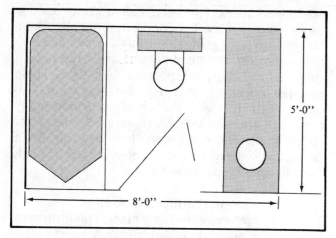

Fixtures Against Two Walls
Lavatory Pipe Runs Under Vanity
Figure 6-18B

Refer to Figure 3-3 for minimum space requirements for a basic bathroom. Figure 6-19A shows some additional guidelines. Keep these figures in mind when you're designing a bathroom. But these are minimums. Take a look at Figure 6-19B.

On the other end of the scale is the convenience or "half" bath. Located near the work and living areas, they save steps and help take the heavy traffic away from the main bath. (See Figure 6-19C.)

Brown's old bathroom is large. He wants it redone so that two or three people can use it at the same time with some privacy. Figure 6-20 shows the old bath and how it can be changed to accommodate two or three in privacy.

Brown also wants a bathroom added adjacent to the master bedroom. Since the bedroom is large, 14 by 16 feet (Figure 6-21A), one end of the room can be converted into a full bath as shown in Figure 6-21B. However, Brown prefers to keep his large bedroom as is, so Figure 6-22 is an alternative. We'll just enclose part of the back porch.

In old houses such as Brown's, bathrooms are frequently added in unused areas such as a large closet, pantry or a small back bedroom. But every unused space isn't a candidate for an additional bathroom. For example, if a pantry off the kitchen is converted to a bathroom, you may have to go through the kitchen to get to the bath. A half or convenience bath adjacent to the kitchen and family room can be an excellent alternative.

The minimum size for a bathroom with tub is 5 by 7 feet, as shown in Figure 6-23. Such a bath can often be added in a 1½ story house in the area under a shed dormer. Just don't forget that the plumbing has to run through a wall on the first floor.

Bathroom Checklist

Whether you are planning to remodel or add a bathroom, use a checklist to get your planning off to a good start. Indicate on the checklist what the customer wants. Remember, your design should reflect both a modern approach and what your customer wants and needs. You know what is being done with basic baths and what is appropriate for the larger and more elaborate baths. Your notebook has plans and information on various types of basic bath, half bath, bath and a half, and fancy jobs.

A basic bath remodeling job can cost anywhere from $3000 to $10,000, depending on the treatment. When the homeowner wants a ballpark figure, give him a range. Don't be specific in your quotes until the specifics are written down and added up. Figure 6-24 is a bathroom checklist that will help you and your customer get down to specifics. Make copies of it and use them when you go to see a potential customer.

Bathroom Plans and Specifications

The bath plan and specification sheet lists the materials included in the job and the cost. The cost is that amount charged to the contract, not your actual cost.

You're going to need accurate current costs for the cost estimate. Several trades are usually involved in bathroom construction, depending on how much of the actual work you will be doing yourself. Having prepared bids during periods of both stable and unstable prices, I have found it wise to obtain the latest prices for materials and subcontract work before figuring a job. It's the only way to guard against a loss. If the work is to be done 3 months from the date the job is figured, it's wise to put in a price escalation clause in the contract to protect yourself. Three months from now the tile sub might argue that his price is $100 above what he previously quoted you. If you're not protected, the loss is yours.

No matter what your approach will be or how you compute job costs, the result must reflect the total cost for doing the job. Don't forget to include your overhead, taxes, and insurance (more about these later on). It's better to spend your time fishing than working for nothing. Figure 6-25 is a specification sheet you might want to use. It should become a part of the contract with the customer.

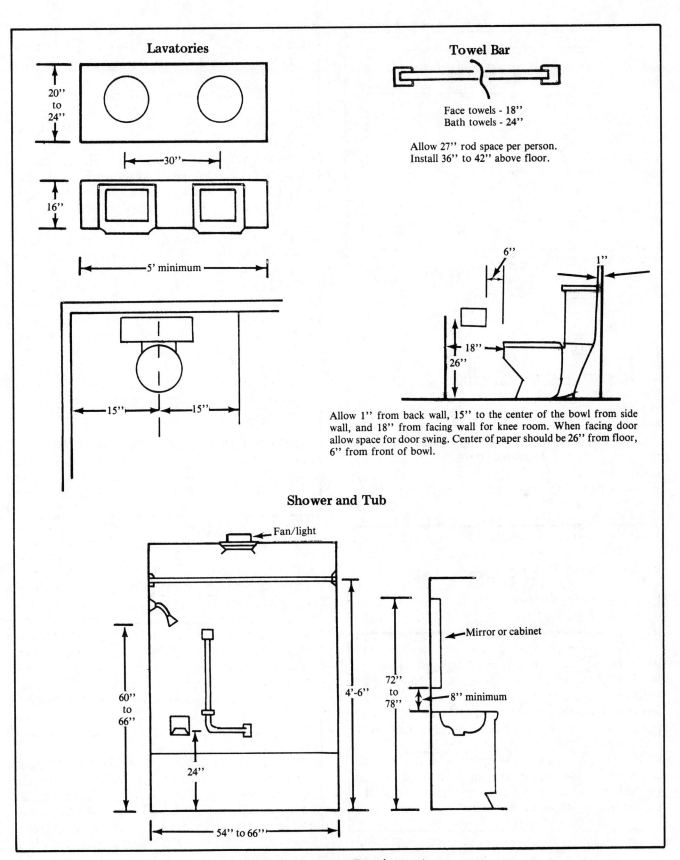

Lavatories

20" to 24"

30"

16"

5' minimum

15"   15"

Towel Bar

Face towels - 18"
Bath towels - 24"

Allow 27" rod space per person.
Install 36" to 42" above floor.

6"

1"

18"

26"

Allow 1" from back wall, 15" to the center of the bowl from side wall, and 18" from facing wall for knee room. When facing door allow space for door swing. Center of paper should be 26" from floor, 6" from front of bowl.

Shower and Tub

Fan/light

60" to 66"

24"

4'-6"

54" to 66"

72" to 78"

Mirror or cabinet

8" minimum

Minimum Space Requirements
Figure 6-19A

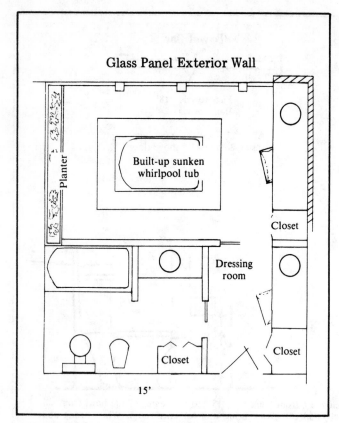

Modern Luxury Bath With Whirlpool Bath, Bidet and Dressing Room
Figure 6-19B

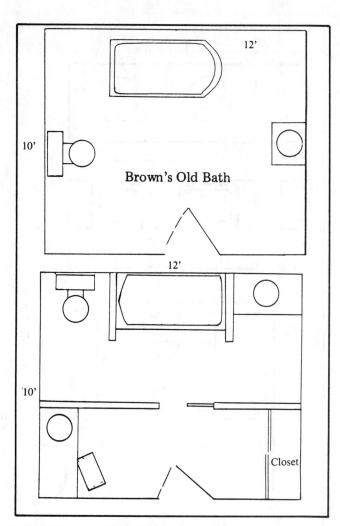

Brown's Remodeled Bath
Figure 6-20

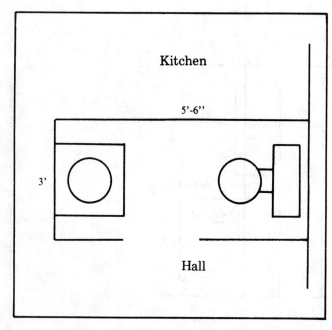

Utilizing Small Area For Good Advantage
Figure 6-19C

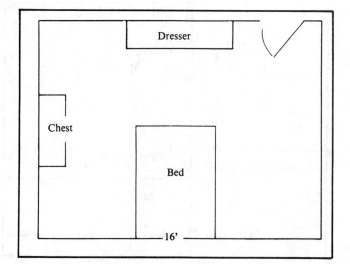

Large Bedroom Permitting Bath Addition
Figure 6-21A

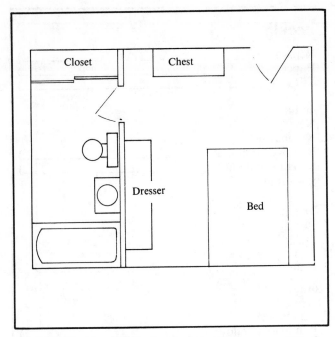

Conversion of Large Bedroom into Bedroom and Bath.
Figure 6-21B

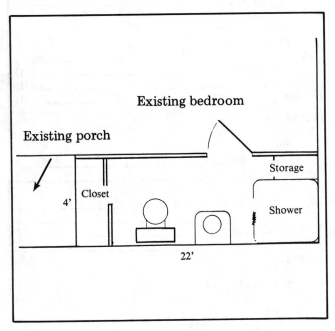

Converting Part of Back Porch Into a Bathroom
Figure 6-22

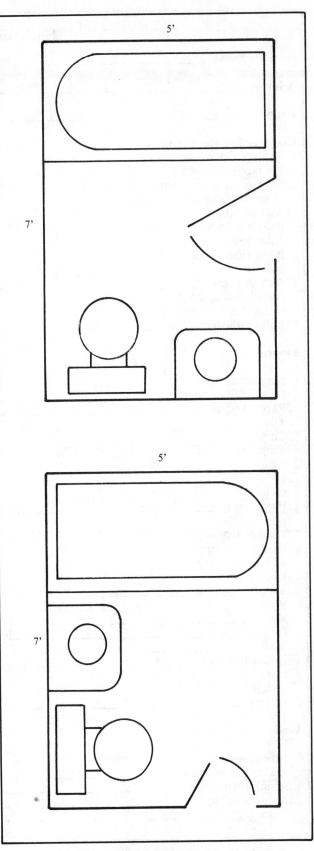

Minimum Size Bathroom
Two-Wall Plumbing and Common Wall Plumbing
Figure 6-23

Bathroom Checklist

Owner's Name _____

Address _____

Phone _____

Construction Requirements
☐ Replace floor joists
☐ Replace subfloor
☐ Replace underlayment
☐ Remove partition
☐ Add partition
☐ Relocate door
☐ Add closet
☐ Raise floor level
☐ New window
☐ Add shutters
☐ Install sliding door
☐ Install folding door
☐ Bulkheads (locations _____)

Floor
☐ New
☐ No change
☐ Remove (manhours _____)
☐ Repair (manhours _____)
☐ Tile
☐ Carpet
☐ Vinyl
☐ Other
☐ Style
☐ Color
☐ Floor size (_____)

Plumbing Requirements
☐ Remove fixtures
☐ Remove radiator
☐ Rough-in:
    Supply _____ feet
    Waste _____ feet
    Vent _____ feet
    Gas _____ feet
    Steam _____ feet
☐ Install fixtures
☐ Install spa
☐ Install bidet
☐ Other (_____)

Walls
☐ New
☐ No change
☐ Remove (manhours _____)
☐ Repair (manhours _____)
☐ At tub area (sq ft _____)
☐ Color (_____)

☐ Repaint (_____)
☐ Paper (_____)
☐ Tile (_____)

Ceilings
☐ New
☐ No change
☐ Remove (manhour _____)
☐ Repair (manhour _____)
☐ Sheetrock
☐ Luminous
☐ Skylight
☐ Beam
☐ Other
☐ Paint
☐ Color

Storage
☐ New
☐ No change
☐ Remove (manhours _____)
☐ Closets (type _____)
☐ Linen
☐ Laundry
☐ Washer/Dryer
☐ Expand shelving
☐ Electrical appliances
☐ Cleaning supplies
☐ Scales
☐ Panel (_____ sq ft)
☐ Glass (_____ sq ft)
☐ Mirrors (_____ sq ft)
☐ Other (_____ sq ft)

Enclosures
☐ New
☐ No change
☐ Remove (manhours _____)
☐ Repair (manhours _____)
☐ Shower stall
☐ Shower/tub modular (size _____)
☐ Tub doors (size _____)
☐ Shower doors (size _____)
☐ Style of doors _____
☐ Shower rod

Heating/Cooling
☐ New
☐ No change
☐ Relocate unit/registers
☐ New duct
☐ Close duct
☐ Baseboard
☐ Radiant
☐ Gas/space
☐ Electric/space

Bathroom Checklist
Figure 6-24

Electrical
- [ ] New
- [ ] No change
- [ ] Type wiring
- [ ] Ft. wiring
- [ ] Heater switch
- [ ] Infrared ht. lamp switch
- [ ] Fan switch
- [ ] Wall outlets
- [ ] Other

Ventilating
- [ ] New
- [ ] No change
- [ ] Ceiling fan
- [ ] Wall fan
- [ ] Vent
- [ ] Fan, lamp, heater unit

Lighting
- [ ] New
- [ ] No change
- [ ] Ceiling (type _____)
- [ ] Wall
- [ ] Bulkhead
- [ ] Indirect
- [ ] Waterproof
- [ ] Medicine cabinet
- [ ] Drop
- [ ] Other

Accessories
- [ ] New
- [ ] Remove (manhours _____)
- [ ] Towel ring
- [ ] Towel bars
- [ ] Paper holder
- [ ] Mirror (size _____)
- [ ] Soap dish
- [ ] Tumbler. TB holder
- [ ] Match tile type
- [ ] Other finish type
- [ ] Other

Medicine Cabinet
- [ ] New
- [ ] No change
- [ ] Recessed
- [ ] Surface
- [ ] Type
- [ ] Shape
- [ ] Size
- [ ] Mirror

Shower Cabinet or Cove
- [ ] New
- [ ] No change

- [ ] None
- [ ] Fiberglass
- [ ] Polypropylene
- [ ] Size
- [ ] Color
- [ ] Door

Lavatory
- [ ] New
- [ ] No change
- [ ] Self-rim
- [ ] One-piece
- [ ] Wall hung
- [ ] Flush mount
- [ ] Undermount
- [ ] Rectangular
- [ ] Oval
- [ ] Size
- [ ] Color
- [ ] Other

Tub
- [ ] New
- [ ] No change
- [ ] None
- [ ] Fiberglass tub/shower modular
- [ ] Steel
- [ ] Cast iron
- [ ] Sunken
- [ ] Whirlpool
- [ ] Size
- [ ] Style
- [ ] Color

Commode
- [ ] New
- [ ] No change
- [ ] Reset
- [ ] Floor mount
- [ ] Wall mount
- [ ] Style
- [ ] Color

Vanity
- [ ] New
- [ ] Existing
- [ ] None
- [ ] Replace
- [ ] Remove
- [ ] Finish (_____)
- [ ] Style
- [ ] Color
- [ ] Size
- [ ] Doors
- [ ] Pulls (style_____)
- [ ] Drawers
- [ ] Height

Bathroom Checklist
Figure 6-24 (continued)

☐ Knee space
☐ Hamper space
☐
Bidet
☐ New
☐ None
☐ Style
☐ Color

Vanity Top
☐ New
☐ No change
☐ Material
☐ Color
☐ Splash height
☐ Edging
☐ Molded splash
☐ Separate splash
☐ Number of cutouts
☐ Other

Fittings
☐ New

☐ No change
☐ Type valve (_____)
☐ Diverter type (_____)
☐ Massage
☐ Style
☐ Color
☐ Quality (_____)
☐ Other

Salvage Value
☐ Tub
☐ Shower
☐ Lavatory
☐ Commode
☐ Fittings
☐ Pipe
☐ Lumber
☐ Mirrors/glass
☐ Accessories
☐ Radiator
☐ Heating/cooling unit

Bathroom Checklist
Figure 6-24 (continued)

Job Specifications

Contractor's Name _____
Address _____
City _____ State _____ Zip _____
Phone _____
Prepared by _____

Contractor proposes to provide the building permit, labor, materials and equipment necessary to complete installation of the following:

Construction Requirements: Description Cost
Removal _____ $_____
Replacement _____ $_____
Addition _____ $_____
Relocate _____ $_____
Floor _____ $_____
Wall _____ $_____
Ceiling _____ $_____
Doors/windows _____ $_____
Total $_____

Plumbing Requirements Description Cost
Removal _____ $_____
Supply _____ $_____

Owner's Name _____
Address _____
City _____ State _____ Zip _____
Job Address _____
Phone _____
Date _____ Job No. _____

Waste _____ $_____
Vent _____ $_____
Gas _____ $_____
Steam _____ $_____
Total $_____

Floor
Removal _____
Underlayment _____
Cove _____
Finish _____
Sill _____
Other _____
Replacement _____
Total area _____
Type _____
Color _____
Total cost $_____

Job Specifications
Figure 6-25

Walls
Removal _____
Tub area _____
Finish _____
Other areas _____
Finish _____
Wainscot _____
Finish _____
Other _____
Replacement _____
Size _____
Color _____
Size _____
Color _____
Size _____
Color _____
Total Cost $_____

Ceilings
Removal _____
Description _____
Finish _____
Other _____
Replacement _____
Size _____
Color _____
Total Cost $_____

Ventilating
Fan _____
Venting _____
Switch _____
Other _____
Type _____
Timer _____
Humidstat _____
Other _____
Total Cost $_____

Accessories	Finish	Number	Description	Cost
Matched tile				
Tub trim				
Grab bars				
Bar soap dish				
Soap dishes				
Towel bars				
Tumbler holder				
Paper holder				
Tissue dispenser				
Mirrors				
Hooks				
Decorative items				
Folding stools				

Total Cost $_____

Medicine Cabinet(s)
Quantity _____
Mount _____
Color _____
Lights _____
Manufacturer's No. _____
Style _____
Mirror size _____
Type _____
Total Cost $_____

Fixtures & Fittings Color Description Cost $_____
Tub _____ $_____
Whirlpool _____ $_____
Shower cabinet _____ $_____
Shower cove _____ $_____
Commode & seat _____ $_____
Lavatory _____ $_____
Bidet _____ $_____
Lavatory faucets _____ $_____
Lavatory fittings _____ $_____
Lavatory valve _____ $_____
Bath valve _____ $_____
Shower head _____ $_____
Diverter _____ $_____
Tub fittings & overflow ____ $_____
Total Cost $_____

Vanity No. 1   **Vanity No. 2**
Cabinet style _____
Cabinet color _____
Manufacturer _____
Knob and pull No. _____
Hinge No. _____
Back plate No. _____
Doors _____
Shelves _____
Total Cost $_____   Total Cost $_____

Enclosures
Description _____
Color _____
Door size _____
Rod length _____
Replacement _____
Size _____
Glass Type _____
Total Cost $_____

Storage
Type _____
Doors _____
Shelves _____
Finish _____
Size _____
Drawer _____

Job Specifications
Figure 6-25 (continued)

Number _____
Hardware _____
                    Total Cost $_____

Heating/Cooling
Heating _____
Size _____
Manufacturer_____
Type _____
Cooling _____
Size _____
Manufacturer_____
Type _____
Duct _____
Registers_____
Size _____
Size/Color _____
                    Total Cost $_____

Electrical & Lighting
Removal _____
Service entrance_____
Wire outlets _____
Electrical heater_____
Wire lighting _____
Switches _____
Lighting _____
Replacement _____
New service _____
Wire switches_____
Electrical cooling _____
Wire fan _____
Outlets _____
Heat lamps_____
                    Total Cost $_____

Tops (   ) As per drawing attached
Material_____
Style _____
Color _____
Number of cutouts _____
Edge treatment _____
Splash height _____
Splash type_____
                    Total Cost $_____

Lavatories
Quantity _____
Mount_____
Style _____
Manufacturer & No. _____
Color _____
Material _____
                    Total Cost $_____

Contractor will do the following demolition and dispose of items removed:
☐ Vanity ☐ Top ☐ Lavatory ☐ Tub ☐ Commode
☐ Shower enclosure ☐ Radiator ☐ Medicine Cabinet
☐ Bath fittings ☐ Deteriorated pipe ☐ Flooring
(_____ sq. ft.) ☐ Wall covering (_____ sq. ft.)
☐ Ceiling cover (_____ sq. ft.) ☐ Partition_____ ☐ Doors ☐ Windows ☐ Ducting
☐ Heating equipment ☐ Cooling equipment ☐ Electric
                    Total Cost $_____

Contractor will make the following repairs:

Item	Description	Cost

                    Total Cost $_____

Total Costs above $_____Tax $_____Total $_____

Owner will furnish labor and material as follows:

Item	Description

These are the total and complete specifications for this job. Only the items checked or for which a cost is indicated are included in this job.

Contractor_____ Owner_____

Date _____

Job Specifications
Figure 6-25 (continued)

Figure 6-26 is a sample contract form you can use for most remodeling jobs. Such items as the specification sheet and drawings of floor plans become part of the agreement by reference.

Do what you say you will do! The remodeling business is given a bad name by those who are out to fleece the homeowner. They are not professional builders. They are professional thieves. They give the rest of us a bad name. You're a pro. You do what you promise and charge a professional price.

Make your specifications sheet clear and detailed enough to eliminate any misunderstanding. Make certain that the subs perform professionally. Let them know in advance that you require first-class work, and anything less is unacceptable.

By the way, complete Brown's new bathroom before you tear out the fixtures in his old one. Otherwise, he'll have no place to go.

Summary

Bathroom remodeling is not all that specialized. Once you've learned a few basics, it's rather simple to design and build a modern and beautiful bathroom.

And the work is out there. Let the potential customers know you're available to do bathroom work by placing a small ad in the local newspaper or in the yellow pages of the phone book.

When you go to see a potential customer, do the same thing you did when you went to see Mr. Brown. Drive over in your pickup, dressed in your work clothes. You might do it during lunch break. You don't need to wear a coat and tie. You're not that kind of a builder. Merely explain to your prospect that you just came off the job. He'll be happy that you're actually a working contractor. Don't go overboard, though. There's a difference between a carpenter in overalls and a bum.

Most people prefer to deal with the person who will be supervising or doing the work. When you're going to have an operation, you want to talk to the doctor who's going to hold the knife, not the hospital administrator or chief billing clerk. Make it clear to your customers that this will be your job and you will be working on it daily. Ask them if they are sure your competition is not just a "paper contractor" who subs out everything but the sales pitch.

Be the kind of remodeler who's on the job, getting the work done, supervising his crew, and building a reputation for attention to detail and quality with every project. You don't need heavy volume—just a handful of good jobs each year. At year's end your bank balance will be more solidly in the black than your "manager" competitor whose overhead continues even when volume doesn't.

Proposal and Contract

Date_____19_____

To _____

Dear Sir:

We propose to furnish all materials and perform all labor necessary to complete the following:

_____

_____

_____

Job Location:

All of the above work to be completed in a substantial and workmanlike manner according to the floor plan, job specifications, and terms and conditions on the back of this form for the sum of

Dollars ($_____)

Payments to be made as the work progresses as follows:_____

_____

_____

the entire amount of the contract to be paid within_____days after substantial completion and acceptance by the owner. The price quoted is for immediate acceptance only. Delay in acceptance will require a verification of prevailing labor and material costs. This offer becomes a contract upon acceptance by contractor but shall be null and void if not executed within 5 days from the date above.

By_____

"YOU, THE BUYER, MAY CANCEL THIS TRANSACTION AT ANY TIME PRIOR TO MIDNIGHT OF THE THIRD BUSINESS DAY AFTER THE DATE OF THIS TRANSACTION. SEE THE ATTACHED NOTICE OF CANCELLATION FORM FOR AN EXPLANATION OF THIS RIGHT."

You are hereby authorized to furnish all materials and labor required to complete the work according to the plans, job specifications, and terms and conditions on the back of this proposal, for which we agree to pay the amounts itemized above.

Owner _____

Owner_____Date_____

Accepted by Contractor_____Date_____

Proposal and Contract
Figure 6-26

1. The Contractor agrees to commence work within (10) days after the last to occur of the following, (1) receipt of written notice from the Lien Holder, if any, to the effect that all documents required to be recorded prior to the commencement of construction have been properly recorded; (2) the materials required are available and on hand, and (3) a building permit has been issued. Contractor agrees to prosecute work thereafter to completion, and to complete the work within a reasonable time, subject to such delays as are permissible under this contract. If no first Lien Holder exists, all references to Lien Holder are to be disregarded.

2. Contractor shall pay all valid bills and charge for material and labor arising out of the construction of the structure and will hold Owner of the property free and harmless against all liens and claims of lien for labor and material filed against the property.

3. No payment under this contract shall be construed as an acceptance of any work done up to the time of such payment, except as to such items as are plainly evident to anyone not experienced in construction work, but the entire work is to be subject to the inspection and approval of the inspector for the Public Authority at the time when it shall be claimed by the Contractor that the work has been completed. At the completion of the work, acceptance by the Public Authority shall entitle Contractor to receive all progress payments according to the schedule set forth.

4. The plan and job specification are intended to supplement each other, so that any works exhibited in either and not mentioned in the other are to be executed the same as if they were mentioned and set forth in both. In the event that any conflict exists between any estimate of costs of construction and the terms of this Contract, this Contract shall be controlling. The Contractor may substitute materials that are equal in quality to those specified if the Contractor deems it advisable to do so. All dimensions and designations on the plan or job specification are subject to adjustment as required by job conditions.

5. Owner agrees to pay Contractor its normal selling price for all additions, alterations or deviations. No additional work shall be done without the prior written authorization of Owner. Any such authorization shall be on a change-order form, approved by both parties, which shall become a part of this Contract. Where such additional work is added to this Contract, it is agreed that all terms and conditions of this Contract shall apply equally to such additional work. Any change in specifications or construction necessary to conform to existing or future building codes, zoning laws, or regulations of inspecting Public Authorities shall be considered additional work to be paid for by Owner as additional work.

6. The Contractor shall not be responsible for any damage occasioned by the Owner or Owner's agent, Acts of God, earthquake, or other causes beyond the control of Contractor, unless otherwise provided or unless he is obligated to provide insurance against such hazards, Contractor shall not be liable for damages or defects resulting from work done by subcontractors. In the event Owner authorizes access through adjacent properties for Contractor's use during construction. Owner is required to obtain permission from the owner(s) of the adjacent properties for such. Owner agrees to be responsible and to hold Contractor harmless and accept any risks resulting from access through adjacent properties.

7. The time during which the Contractor is delayed in this work by (a) the acts of Owner or his agents or employees or those claiming under agreement with or grant from Owner, including any notice to the Lien Holder to withhold progress payments, or by (b) any acts or delays occasioned by the Lien Holder, or by (c) the Acts of God which Contractor could not have reasonably foreseen and provided against, or by (d) stormy or inclement weather which necessarily delays the work, or by (e) any strikes, boycotts or like obstructive actions by employees or labor organizations and which are beyond the control of Contractor and which he cannot reasonably overcome, or by (f) extra work requested by the Owner, or by (g) failure of Owner to promptly pay for any extra work as authorized, shall be added to the time for completion by a fair and reasonable allowance. Should work be stopped for more than 30 days by any or all of (a) through (g) above, the Contractor may terminate this Contract and collect for all work completed plus a reasonable profit.

8. Contractor shall at his own expense carry all workers' compensation insurance and public liability insurance necessary for the full protection of Contractor and Owner during the progress of the work. Certificates of insurance shall be filed with Owner and Lien Holder if Owner and Lien Holder require. Owner agrees to procure at his own expense, prior to the commencement of any work, fire insurance with Course of Construction. All Physical Loss and Vandalism and Malicious Mischief clauses attached in a sum equal to the total cost of the improvements. Such insurance shall be written to protect the Owner and Contractor, and Lien Holder, as their interests may appear. Should Owner fail so to do, Contractor may procure such insurance, as agent for Owner, but is not required to do so, and Owner agrees in demand to reimburse Contractor in cash for the cost thereof.

9. Where materials are to be matched, Contractor shall make every reasonable effort to do so using standard materials, but does not guarantee a perfect match.

10. Owner agrees to sign and file for record within five days after substantial completion and acceptance of work a notice of completion. Contractor agrees upon receipt of final payment to release the property from any and all claims that may have accrued by reason of the construction.

11. Any controversy or claim arising out of or relating to this contract shall be settled by arbitration in accordance with the Rules of the American Arbitration Association, and judgment upon the award rendered by the Arbitrator(s) may be entered in any Court having jurisdiction.

12. Should either party bring suit in court to enforce the terms of this agreement, any judgment awarded shall include court costs and reasonable attorney's fees to the successful party plus interest at the legal rate.

13. Unless otherwise specified, the contract price is based upon Owner's representation that there are no conditions preventing Contractor from proceeding with usual construction procedures and that all existing electrical and plumbing facilities are capable of carrying the extra load caused by the work to be performed by Contractor. Any electrical meter charges required by Public Authorities or utility companies are not included in the price of this Contract, unless included in the job specifications. If existing conditions are not as represented,

thereby necessitating additional plumbing, electrical, or other work, these shall be paid for by Owner as additional work.

14. The Owner is solely responsible for providing Contractor prior to the commencing of construction with any water, electricity and refuse removal service at the job site as may be required by Contractor to effect the improvement covered by this contract. Owner shall provide a toilet during the course of construction when required by law.

15. The Contractor shall not be responsible for damage to existing walks, curbs, driveways, cesspools, septic tanks, sewer lines, water or gas lines, arches, shrubs, lawn, trees, clotheslines, telephone and electric lines, etc., by the Contractor, subcontractor, or supplier incurred in the performance of work or in the delivery of materials for the job. Owner hereby warrants and represents that he shall be solely responsible for the condition of the building with respect to moisture, drainage, slippage and sinking or any other condition that may exist over which the Contractor has no control and subsequently results in damage to the building.

16. The Owner is solely responsible for the location of all lot lines and shall if requested, identify all corner posts of his lot for the Contractor. If any doubt exists as to the location of lot lines, the Owner shall at his own cost, order and pay for a survey. If the Owner wrongly identifies the location of the lot lines of the property, any changes required by the Contractor shall be at Owner's expense. This cost shall be paid by Owner to Contractor in cash prior to continuation of work.

17. Contractor has the right to subcontract any part, or all, of the work agreed to be performed.

18. Owner agrees to install and connect at Owner's expense, such utilities and make such improvements in addition to work covered by this Contract as may be required by Lien Holder or Public Authority prior to completion of work of Contractor. Correction of existing building code violations, damaged pipes, inadequate wiring, deteriorated structural parts, and the relocation or alteration of concealed obstructions will be an addition to this agreement and will be billed to Owner at Contractor's usual selling price.

19. Contractor shall not be responsible for any damages occasioned by plumbing leaks unless water service is connected to the plumbing facilities prior to the time of rough inspection.

20. Title to equipment and materials purchased shall pass to the Owner upon delivery to the job. The risk of loss of the said materials and equipment shall be borne by the Owner.

21. Owner hereby grants to Contractor the right to display signs and advertise at the job site.

22. Contractor shall have the right to stop work and keep the job idle if payments are not made to him when due. If any payments are not made to Contractor when due, Owner shall pay to Contractor an additional charge of 10% of the amount of such payment. If the work shall be stopped by the Owner for a period of sixty days, then the Contractor may, at Contractor's option, upon five days written notice, demand and receive payment for all work executed and materials ordered or supplied and any other loss sustained, including a profit of 10% of the contract price. In the event of work stoppage for any reason, Owner shall provide for protection of, and be responsible for any damage, warpage, racking, or loss of material on the premises.

23. Within ten days after execution of this Contract, Contractor shall have the right to cancel this Contract should it be determined that there is any uncertainty that all payments due under this Contract will be made when due or that any error has been made in computing the cost of completing the work.

24. This agreement constitutes the entire Contract and the parties are not bound by oral expression or representation by any party or agent of either party.

25. The price quoted for completion of the structure is subject to change to the extent of any difference in the cost of labor and material as of the date and the actual cost to Contractor at the time materials are purchased and work is done.

26. The Contractor is not responsible for labor or materials furnished by Owner or anyone working under the direction of the Owner and any loss or additional work that results therefrom shall be the responsibility of the Owner. Removal or use of equipment or materials not furnished by Contractor is at Owner's risk, and Contractor will not be responsible for the condition and operation of these items, or service for them.

27. No action arising from or related to the contract, or the performance thereof, shall be commenced by either party against the other more than two years after the completion or cessation of work under this contract. This limitation applies to all actions of any character, whether at law or in equity, and whether sounding in contract, tort, or otherwise. This limitation shall not be extended by any negligent misrepresentation or unintentional concealment, but shall be extended as provided by law for willful fraud, concealment, or misrepresentation.

28. All taxes and special assessments levied against the property shall be paid by the Owner.

29. Contractor agrees to complete the work in a substantial and workmanlike manner but is not responsible for failures or defects that result from work done by others prior, at the time of or subsequent to work done under this agreement.

30. Contractor makes no warranty, express or implied (including warranty of fitness for purpose and merchantability). Any warranty or limited warranty shall be as provided by the manufacturer of the products and materials used in construction.

31. Contractor agrees to perform this Contract in conformity with accepted industry practices and commercially accepted tolerances. Any claim for adjustment shall not be construed as reason to delay payment of the purchase price as shown on the payment schedule. The manufacturers' specifications are the final authority on questions about any factory produced item. Exposed interior surfaces, except factory finished items, will not be covered or finished unless otherwise specified herein. Any specially designed, custom built or special ordered item may not be changed or cancelled after five days from the acceptance of this Contract by Contractor.

Proposal and Contract
Figure 6-26 (Continued)

Notice To Customer Required By Federal Law

You have entered into a transaction on_____which may result in a lien, mortgage, or other security interest on your home. You have a legal right under federal law to cancel this transaction, if you desire to do so, without any penalty or obligation within three business days from the above date or any later date on which all material disclosures required under the Truth in Lending Act have been given to you. If you so cancel the transaction, any lien, mortgage, or other security interest on your home arising from this transaction is automatically void. You are also entitled to receive a refund of any down payment or other consideration if you cancel. If you decide to cancel this transaction, you may do so by notifying:

_____

(Name of Creditor)

at _____

(Address of Creditor's Place of Business)

by mail or telegram sent not later than midnight of_____. You may also use any other form of written notice identifying the transaction if it is delivered to the above address not later than that time. This notice may be used for the purpose by dating and signing below.

I hereby cancel this transaction.

_____

(Date)   (Customer's Signature)

   Effect of rescission. When a customer exercises his right to rescind under paragraph (a) of this section, he is not liable for any finance or other charge, and any security interest becomes void upon such a rescission. Within 10 days after receipt of a notice of rescission, the creditor shall return to the customer any money or property given as earnest money, downpayment, or otherwise, and shall take any action necessary or appropriate to reflect the termination of any security interest created under the transaction. If the creditor has delivered any property to the customer, the customer may retain possession of it. Upon the performance of the creditor's obligations under this section, the customer shall tender the property to the creditor, except that if return of the property in kind would be impracticable or inequitable, the customer shall tender its reasonable value. Tender shall be made at the location of the property or at the residence of the customer, at the option of the customer. If the creditor does not take possession of the property within 10 days after tender by the customer, ownership of the property vests in the customer without obligation on his part to pay for it.

Notice to Customer Required by Federal Law
Figure 6-26 (continued)

Kitchens

Any builder interested in profitable remodeling projects should be ready and able to handle kitchen jobs.

The Brown project is going well. You added the new bath and remodeled the old one. You found at least one sub you won't do business with any more. In time you'll know who's reliable and who isn't. There are competent people in all trades. Eventually, you'll get to know who they are.

The Browns' kitchen is outdated. Mrs. Brown wants an efficient, modern kitchen. Let's see what's wrong with hers.

Old Kitchens

Few older kitchens have an efficient layout. The refrigerator is in the far corner, across the room from the back door, and it's too small for today's needs.

Then there isn't much counter space. The short counter top at the sink is way too small to be efficient. There is a pantry out on the back porch, but it might as well be in the next block for all the good it does for kitchen efficiency.

As for lighting, Mrs. Brown has a 2-bulb globe type fixture in the center of the ceiling. Of course, the old electric stove has a light in the backsplash, but the only thing it illuminates is the backsplash. Anyway, the stove will probably be replaced.

The kitchen layout is awkward. You have to walk a mile to fry an egg. The traffic pattern is centered in front of the refrigerator and the stove.

There's no ventilation equipment. The window over the sink sticks so much that it's almost impossible to open.

The old plastered walls and ceilings are cracked and chipped. It's so dreary it must make even strawberry shortcake taste bad.

This kitchen is just right for the magic at your disposal. You're going to make Mrs. Brown's kitchen the most beautiful and efficient on the block—and get some very complimentary attention for yourself in the process.

The Present Concept

Figure 7-l shows the geometry of a modern kitchen arrangement. It's called the "Work Triangle." What that means is that you don't have to walk so far to fry an egg.

Let's talk about the work triangle. A kitchen has three main elements:

1. Refrigerator/storage center—At the center of this area is the refrigerator. The adjacent cabinet area, or pantry, stores the non-perishables. All the food is stored in one area. When a meal is being prepared, the cook doesn't have to chase all over the kitchen gathering up the various ingredients.

The fridge should be located near the kitchen entrance door (back door). The idea is to have a short route from the car to the storage area when bringing in the groceries. Counter space is required adjacent to the fridge. The best bet is to locate the fridge at the end of the cabinets. This gives balance, having the large appliance at the end of the kitchen area. The door handle of the

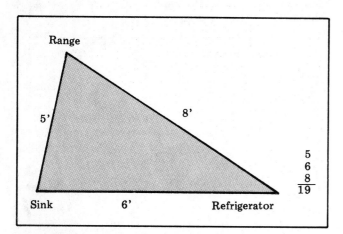

Distance to Stove, Sink and Refrigerator
Figure 7-1

refrigerator should be next to the counter top to provide direct access from the fridge to the counter top.

The cabinet space next to the fridge is used for storage of canned goods, cereals, and other non-perishables. If there is to be a walk-in or closet-type pantry, it should also be near the fridge.

2. Cleaning/preparation center—This area includes the sink and adjacent counter top area. Now nothing is more deficient than a sink without adequate counter top space on each side. Allow at least 30 inches on each side of the sink. A lot of work goes on here. A swing-out stool at the sink is an excellent accessory to this center. When it's not needed, it can be swung back under the sink.

Since this is the cleaning and preparation area, items such as cleansers, sponges, and scrub brushes should be close at hand. A garbage pail storage compartment should be near the sink. The bottom of the compartment can be finished with the same material as the counter top for easy cleanup of spills. It's a good idea to provide some method of ventilation for the compartment. A decorative screen or air holes can do much to help carry off food odors. Of course, a garbage disposal unit is desirable, too.

The dishwasher should be located near the sink so that dishes can be moved from the sink to the washer or vice versa without having to take more than one step. If a dishwasher is not to be installed now, it's a good idea to plan a compartment for installation at a later date. The compartment should have a minimum 24-inch-wide clearance for inserting the appliance. For the moment, temporary shelves can be installed in this space. Just remember that the washer has to sit on the floor.

The cabinet should be designed accordingly. It saves a lot of trouble later.

3. Cooking/serving center—In this area you have the range, oven, microwave oven and any other pots, pans, and plug-in gadgets designed for cooking. Storage space in this area is required for pots, pans, and all the other items required in the center. The microwave unit needs a separate storage area for those special dishes and items it requires. Some cooks might prefer to have some food items such as coffee, tea or soups kept in this area. That's O.K., too. Everyone has his own way of doing things.

The center is also a serving center. That is, food is served from this area. So toasters, waffle irons and such will need storage space here.

Now, the distances involved in connecting these three centers usually form a triangle. Some bright individual somewhere along the line decided that this was a "Work Triangle," and it's been called that ever since. Anyway, it beats having to go around in circles as some kitchen layouts require.

The work triangle is supposed to serve as a measure of the kitchen's efficiency. It works like this: The sum of the three sides of the triangle should be no more than 22 feet. The distance is measured from the front of the three major appliances: fridge to sink is 4 to 7 feet; sink to range is 4 to 6 feet; and range to fridge is 4 to 9 feet. Normal traffic lanes (the paths people, other than the cook, use to walk through or around in the kitchen) should not pass through the triangle. If they do, the efficiency of the triangle is reduced.

If the fridge is located on the back porch or in the hall, the triangle is bound to be more than 22 feet. If the microwave unit is across the room from the range, you'll be going around in circles.

Whether you agree with the triangle concept or not, it is popular, so you should be familiar with it. What we're after is efficiency. We builders will have to go along with it until something better comes along. Figure 7-2 shows how the triangle concept looks in a kitchen.

The U-Shaped Kitchen
Probably the best arrangement is the U-shaped kitchen. It is closed on three sides, eliminates through traffic, and eliminates wasted steps because the area of work is an actual triangle. Adequate cabinet space is available on three walls. One of the walls can be a pass-through serving counter into a dining or family room. The pass-through can be a decorative focal-point. U-shaped kitchens should be at least 10 feet wide at the base. If they are less,

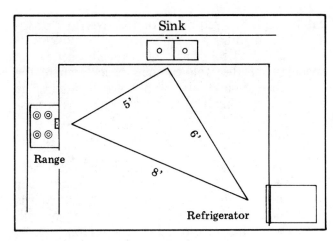

Work Triangle
Figure 7-2

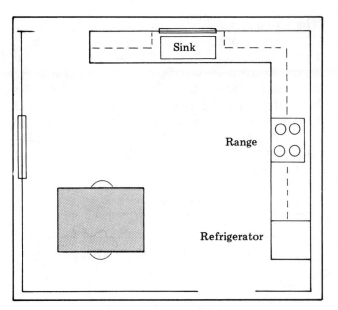

L-Shaped Kitchen
Figure 7-4

the work space at the sink becomes cramped and efficiency suffers. Figure 7-3 shows a U-shaped plan.

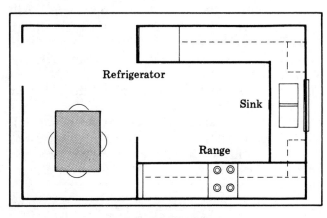

U-Shaped Kitchen
Figure 7-3

and is ideal for a long, narrow room because it takes up very little floor space. The many good points of this design are offset if traffic is directed through the kitchen. Thru traffic is no problem if there is only one way in and out. Cabinet space is generous in this plan. Be careful to avoid closing the room in too tightly. Allow 5 feet between the two cabinet walls (at floor level). A greater distance will reduce efficiency. Figure 7-5 shows a corridor kitchen.

The L-Shaped Kitchen

If the walls are not too long, the L is a good arrangement. Watch the location of the doors. If the wall space is broken by a door on each wall, you can have traffic interfering with triangle efficiency. Continuous counters and appliances on two adjoining walls permit comfortable space for eating in the kitchen. Not as much cabinet space is available in the L as in the U. Figure 7-4 shows an L-shaped kitchen.

The Corridor Kitchen

The corridor kitchen is similar to the U-shape but occupies only two walls. It is an excellent step saver

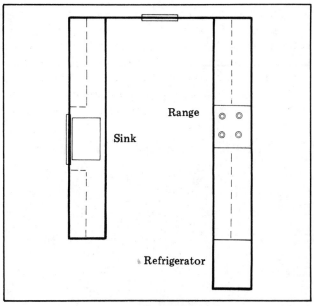

Corridor Kitchen
Figure 7-5

The Sidewall Kitchen

The sidewall or single wall kitchen is usually a part of another room. And while cabinet and storage space is obviously sacrificed in this plan, you can combine the functions of the kitchen with the functions of the entire room. The work centers are in a line against one wall. Efficiency is good but counter work space is limited. The sidewall is ideal for small houses and apartments. Figure 7-6 shows a sidewall plan.

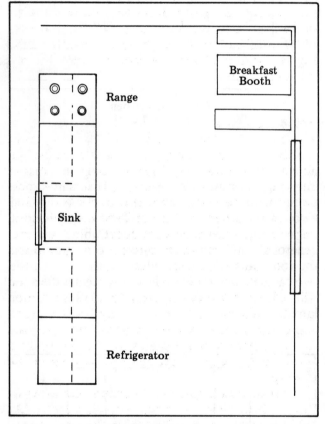

**Sidewall Kitchen
Figure 7-6**

Specifics on Storage and Counters

Keeping the triangle in mind, let's look at some requirements for counter tops and storage.

Here are the generally accepted standards: First, count the number of bedrooms. That's the best measure of the expected occupant load of the house. It's assumed that the master bedroom will contain two people and the other bedrooms one person. Each permanent resident of the house needs 6 square feet of storage space in wall cabinets. An additional 12 square feet of wall cabinet space is required for the overflow of in-laws, overnighters and guests.

It's assumed that base cabinets will occupy the space under the wall cabinets except where the stove, fridge, and dishwasher are located. The assumption is that if sufficient wall space is provided, the base cabinets will take care of themselves. Base cabinets are, by the way, measured by the linear foot.

I'm sure you know of 2-bedroom houses where three people occupy the master bedroom and four others occupy the other bedroom. The standards mentioned above don't fit all cases, so you will want to modify them to fit your situation.

So, in a two-bedroom house there would be 6 square feet of wall cabinet space for each person for a total of 18 square feet. Adding the 12 square feet for entertaining we reach a total of 30 square feet.

See how easy it is when you have some good assumptions to go on? Suppose there were 5 bedrooms. Well, it would be two in the master for 12 square feet, and 6 square feet for each of the four persons in the other 4 bedrooms, which will add 24 square feet to the 12 for a total of 36 square feet. That plus the 12 square feet allotted for guests and relatives makes a total of 48 square feet for wall cabinets.

Here are some more assumptions to aid your planning:

• The counter space between the stove and sink should be no less than 3 feet and no more than 4.

• The counter space between the refrigerator and sink should be no less than 4½ feet and no more than 5½.

When you redo Mr. Brown's kitchen, he might have some suggestions of his own. Take his suggestions even if they violate these rules, and improve on them if necessary. The important thing is that you know what goes into planning a kitchen. A modern kitchen can break any of these standards and still be beautiful and efficient. But know what the generally accepted rules are and know why you are not following them.

What do you do when the work triangle exceeds that magic 22 feet—the distance beyond where kitchen efficiency is supposed to decline? I wouldn't recommend relocating a wall to make the room smaller. The triangle concept is sound logic, but it's not part of the building code. A lot of cooks still favor a large, spacious kitchen even if the triangle is 30 feet or more.

It's all right if you want to extend a snack bar into the kitchen at the end of the cabinets. Dress it up, though. Round off the corners and add a shelf or two under the counter top. In a large kitchen you can even put a couple of short base cabinets back-to-back and cover them with a one-piece top to form a work island as shown in Figure 7-7. Look at the extra storage space you get.

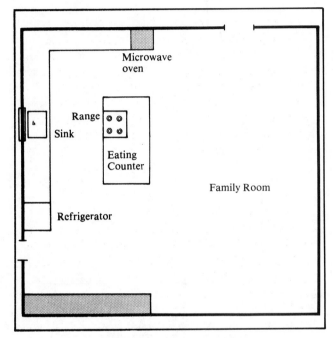

Work Island / Room Divider
Figure 7-7

Base cabinets are 36 inches from floor to top. This height corresponds with the height of stoves and sink cabinets. The space between the top of the counter and the bottom of the wall cabinet should be around 18 inches. The cabinet over the stove should be about 26 inches from the top of the stove.

It isn't essential, but most people working at the sink like to have a window to look out. If you're going to install a window, don't put it too close to the corner of the room. Stay back from the corner at least 15 inches so there will be space on each side of the window for a wall cabinet. The standard wall cabinet is about 13 inches deep.

And don't use a window that will come down too low. Allow space for the counter backsplash. A four-inch backsplash on the integral or molded counter top is a popular height. The window rough opening should be about 44 inches from the floor.

Doors should be hung so that they don't open against the front of an appliance. Keep the fridge

away from the stove. The cost of operating a refrigerator is expensive enough without having a hot stove heating it up. And if a built-in oven is installed, don't put it too close to the counter top range. Leave about 12 inches between the units. Sometimes it's necessary to get rid of a hot pan or pot in a hurry. Space on both sides of the range also permits removing hot containers off the stove without lifting them over other containers still on the stove.

A few paragraphs back we mentioned microwave ovens. If there's a shortage of space in the kitchen, microwave ovens are now made to fit in the space over the range where the hood is normally installed. Some even have a ventilating unit. Figures 7-8A and 7-8B show how to install various microwave units.

Kitchen Cabinets

If you have been around a while, you might remember the old safe type of cabinet. Some had screens on the doors to keep out the flies. Before the screen wire version there was a type that had tin sections in the doors. Small holes were punched through the tin to permit ventilation. Cakes, pies, and other goodies were kept in the safe. The safe, cupboard and orange crates were the forerunners of our modern day kitchen cabinets.

You probably won't run into a remodeling job that has a kitchen dating so far back. And not all the jobs will have cabinets that need replacing. In some remodeling jobs the cabinets are adequate and only need refinishing and new hardware. In a few projects the doors and drawers may have to be removed.

A lot of people just like to change the looks of things from time to time. You can make old cabinets sparkle by redoing the finish and replacing the old hinges and pulls. You can junk all the old doors and drawers and install doors that overlap the framing for a continuous paneled look. Concealed hinges are used and the doors are fitted flush. The doors are cut with an angle edge on all sides, or just on the bottom, which acts as a finger hold so no pulls are required. Of course, ornamental pulls can be added, if desired.

Cabinets should support the major decorative theme of the kitchen: from country kitchens to sleek, modern ones. Modern appliances can be used in just about any style kitchen. A kitchen does not have to switch from counter top range and built-in oven units to a wood-burning stove to qualify as country. The cabinets, floor, walls, ceil-

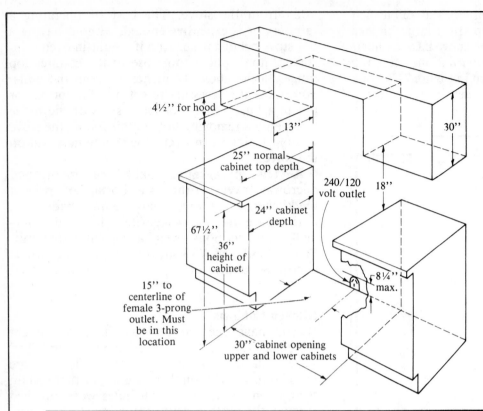

4½" for hood

13"

30"

25" normal cabinet top depth

240/120 volt outlet

18"

24" cabinet depth

67½"

36" height of cabinet

8¼" max.

15" to centerline of female 3-prong outlet. Must be in this location

30" cabinet opening upper and lower cabinets

Overall Range			Microwave or Electric Upper Oven Interior			Lower Oven Interior			Approx. Shipping Wt. (Lbs.)	Total Connected Load (KW)
Width	Height	Depth	Width	Height	Depth	Width	Height	Depth		
30"	67½"	26-5/8"	21"	12"	13"	22"	15"	16"	290	13.0KW at 120/240V 9.8KW at 120/240V

Overall Range			Microwave or Electric Upper Oven Interior		
Width	Height	Depth	Width	Height	Depth
30"	67½"	26⅝"	21"	12"	13"

Lower Oven Interior			Approx. Shipping Wt. (Lbs.)	Total Connected Load (KW)
Width	Height	Depth		
22"	15"	16"	290	13.0KW at 120/240V 9.8KW at 120/208V

Cabinet Specifications
Upper-Lower Cooking Center 30-Inch Width
Figure 7-8A

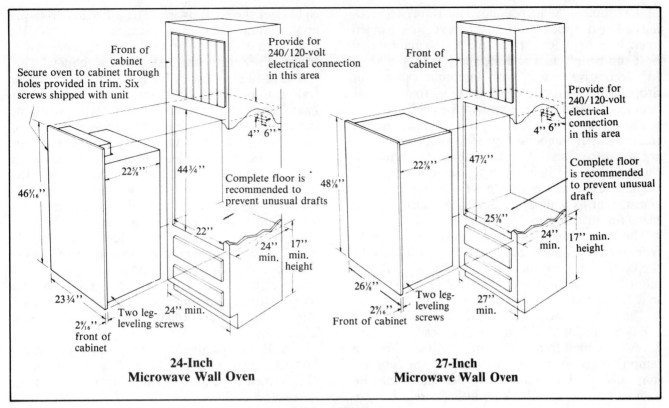

**Cabinet Cut-Out
Microwave Oven Installation
Figure 7-8B**

ing, and furnishings can do this without sacrificing modern conveniences.

Kitchen cabinets are more or less standard. The only exception is the wall cabinets designed for over-the-refrigerator installation. It can be anywhere from 12 inches in depth to 21 inches. Figure 7-9 shows the basic measurements.

Stock cabinets are available in most areas or can be ordered if not locally available. Most stock jobs are available in various widths from 12 inches to 48 inches; the increase is in 3-inch increments. They come assembled or knocked down. The knockdowns usually come with assembly instructions and sufficient filler strips.

Level floors and plumb walls are essential for proper installation. If you don't have a level floor and vertical wall, some additional work is needed before the cabinets are installed.

If the floor is unlevel, shim the cabinet so that it will rest level. If the wall is out of plumb, furring will be required. Filler strips will finish the wall cabinets to the wall. Quarter-round or shoe moldings will finish the cabinets to the floor.

Wall cabinets seldom go to the ceiling. The top

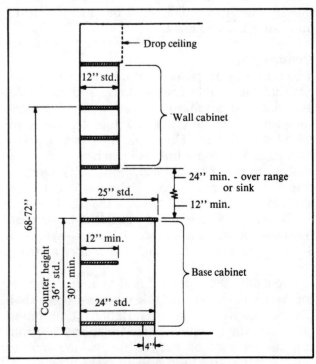

**Kitchen Cabinet Proportions
Figure 7-9**

of the cabinet is 84 inches above the floor. The ceiling can be dropped at this point. A soffit is framed to enclose the space between the ceiling and cabinet or it can be left as is and the space used as a shelf for decorative items. With luminous ceilings or drop ceilings with recessed lighting fixtures, the cabinets are usually extended to the ceiling.

The drop ceiling lends itself to various decorative treatments. If paneling is used in the kitchen, the soffit can be finished with the same paneling. Or, the soffit can be finished in the same materials as the cabinets. Another option is to use a contrasting finish such as wallpaper. Soffits are also a good place for soft, indirect lighting.

The base cabinets can be installed first, working from the corner outward. Here it is important to fit the corner properly since the counter top will not allow for a variance and still fit. Watch the openings for appliance-fitting requirements.

The wall cabinets are secured to the studs. The cabinets are built with a hanging brace or strip. Screws are used to make the installation. If you're tempted to use nails, don't. Someday the cabinets may have to be removed without damaging the cabinet or wall, and that's no little chore if they've been installed with lots of nails.

Brown wants a U-shaped kitchen with cabinets on three walls. He's going fancy with this kitchen. Unless you have special tools and skills, leave the building and installation of cabinets to the cabinet-maker. They can usually do it more economically and faster than you can.

Counter Tops

This is the day of specialization. Now there are outfits that make only counter tops. You can order any size and arrangement you want. The back splash is an integral part of the top. The edges are rounded or square. The whole thing is molded or formed as a complete unit. All you have to do is install it on top of the cabinet. Use screws and pre-drill the holes. Do it from underneath. Use care. You don't want to drill through the top.

The surface is laminated plastic—a hard, heat-resistant and durable surface. It's usually glued to 3/4-inch-thick particleboard and makes a strong top.

There are other approaches if you are doing the top yourself. Sheet and roll materials other than plastic laminate are available. Laminated polyester, vinyl, laminated melamine and linoleum can also be used on the job for making counter tops. Apply it to 3/4-inch-thick exterior plywood. Vinyl and linoleum are flexible and can be formed

right over the back splash. The other materials are applied flat. Metal trim is necessary to finish the joint between the top and splash. The edges can be trimmed with the same material or with metal trim.

The ill-fit between the back splash and an uneven wall can be filled by metal cove molding. On manufactured tops a dab of caulk will usually do the job if the walls are not too uneven.

Other materials sometimes used for tops are ceramic tile and marble. For my money, both are too hard and result in a lot of broken dishes. But the customer's wishes are the guiding factor. Marble has to be precut to size in the shop with special tools. The best thing about it is that it's self-edged.

A good tile man can do wonders with a counter top. Tile can also be used in other areas of the kitchen to complement the counter top.

Someone has to clean the tile joints on tile counter tops. Sometimes they have to be repaired. Careful placement of tile can save time and trouble later.

Mrs. Brown wants the manufactured top with a formed edge. (See Figure 7-10.) It's a good choice. The formed edge prevents spills from running off the cabinet and onto the floor or your feet.

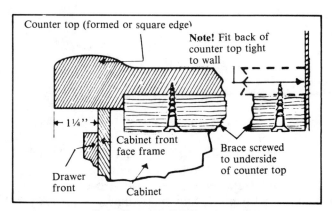

Counter and Cabinet Detail
Figure 7-10

Drop-In Range Installation

The 30-inch drop-in range is a favorite unit. It's built-in and makes a neat package in the kitchen. The unit is supported by the counter top and the cabinets. The counter top should extend over the edge of the cabinet 1¼ inches as shown in Figure 7-10. The gas or electrical hook-up is completed before the unit is installed. Slide the unit into place with the top trim of the range resting on the counter top. Figures 7-11A through 7-11I illustrate installation methods for various units.

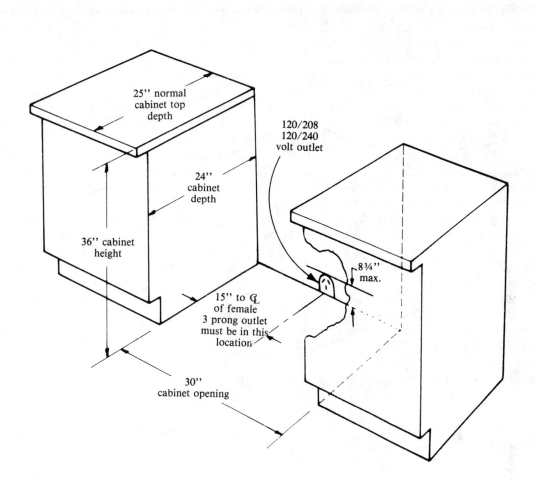

25" normal
cabinet top
depth

120/208
120/240
volt outlet

24"
cabinet
depth

36" cabinet
height

8¾"
max.

15" to ℄
of female
3 prong outlet
must be in this
location

30"
cabinet opening

Oven			Cutout Dimensions			Approx. Shipping Wt. (Lbs.)	Total Connected Load (KW)
Width	Height	Depth	Width	Height	Depth		
22"	15"	18"	30"	36"	24"	175	10.2KW at 120/240V 7.6KW at 120/208V

Electric Free-Standing Ranges
Figure 7-11A

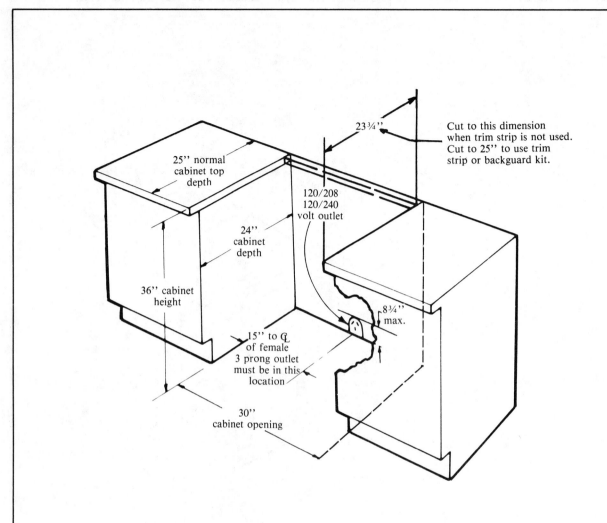

23¾"

Cut to this dimension
when trim strip is not used.
Cut to 25" to use trim
strip or backguard kit.

25" normal
cabinet top
depth

120/208
120/240
volt outlet

24"
cabinet
depth

36" cabinet
height

8¾"
max.

15" to ℄
of female
3 prong outlet
must be in this
location

30"
cabinet opening

	Oven			Cutout Dimensions		Approx. Shipping Wt. (Lbs.)	Total Connected Load (KW)
Width	Height	Depth	Width	Height	Depth		
22"	15"	18"	30"	36"	24"	190	10.2KW at 120/240V 7.6KW at 120/208V

Electric Slide-In Ranges
Figure 7-11B

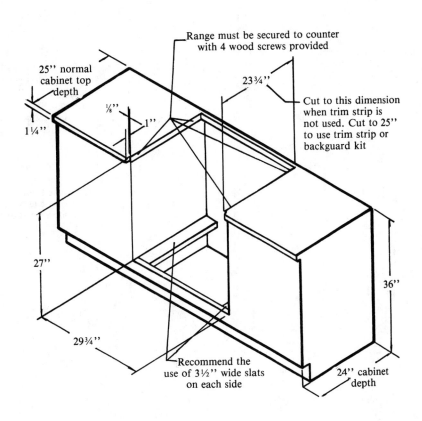

Range must be secured to counter
with 4 wood screws provided

25" normal
cabinet top
depth

23¾"

Cut to this dimension
when trim strip is
not used. Cut to 25"
to use trim strip or
backguard kit

1¼"

⅛"

1"

27"

36"

29¾"

Recommend the
use of 3½" wide slats
on each side

24" cabinet
depth

Oven			Cutout Dimensions			Approx. Shipping Wt. (Lbs.)	Total Connected Load (KW)
Width	Height	Depth	Width	Height	Depth		
22"	15"	18"	29¾"	27"	23¾"	170	10.2KW at 120/240V 7.6KW at 120/208V

Electric Drop-In Ranges
Figure 7-11C

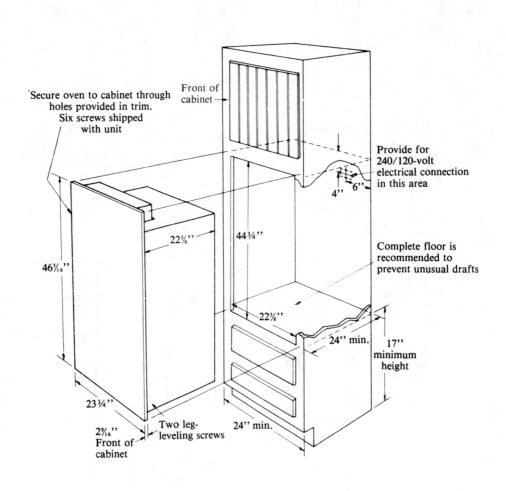

Secure oven to cabinet through holes provided in trim. Six screws shipped with unit

Front of cabinet

Provide for 240/120-volt electrical connection in this area

4" 6"

22⅝"

44¾"

Complete floor is recommended to prevent unusual drafts

46⁵⁄₁₆"

22⅝"

24" min.

17" minimum height

23¾"

2⁹⁄₁₆"
Front of cabinet

Two leg-leveling screws

24" min.

Upper Oven			Lower Oven			Cutout Dimensions			Approx. Shipping Wt. (Lbs.)	Total Connected Load (KW)
Width	Height	Depth	Width	Height	Depth	Width	Height	Depth		
18"	14"	19"	18"	12"	19"	22"	44¾"	24"	185	7.0KW at 120/240V 5.8KW at 120/208V

Electric Wall Ovens
Figure 7-11D

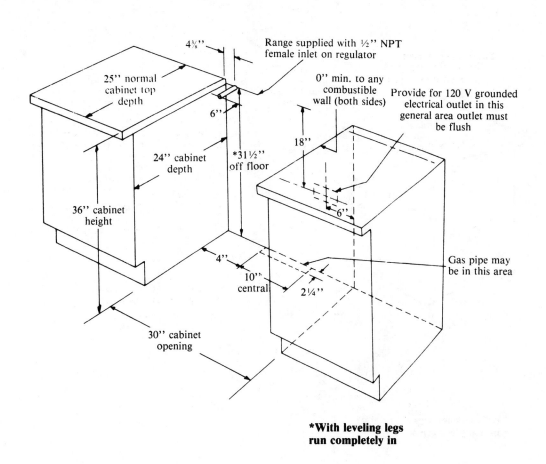

Gas Free-Standing Ranges
Figure 7-11E

Oven			Broiler			Cutout Dimensions			Approx. Shipping Wt. (Lbs.)
Width	Height	Depth	Width	Height	Depth	Width	Height	Depth	
22"	15"	18"	13½"	3½"	15"	30"	36"	24"	195

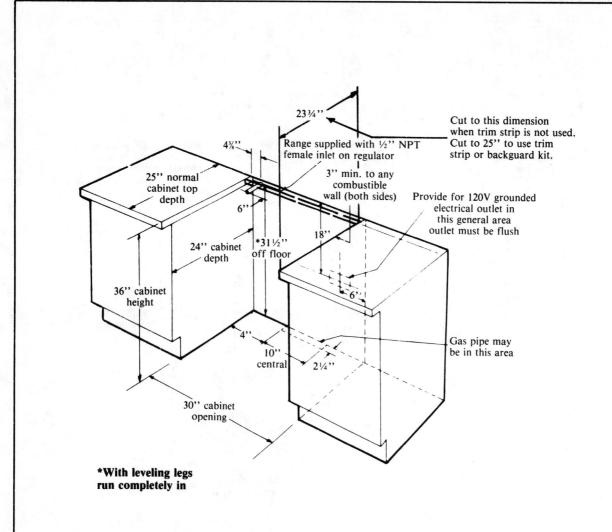

25" normal cabinet top depth

4⅛"

23¾"

Cut to this dimension when trim strip is not used. Cut to 25" to use trim strip or backguard kit.

Range supplied with ½" NPT female inlet on regulator

3" min. to any combustible wall (both sides)

Provide for 120V grounded electrical outlet in this general area outlet must be flush

6"

24" cabinet depth

*31½" off floor

18"

36" cabinet height

6"

4"

10" central

2¼"

Gas pipe may be in this area

30" cabinet opening

*With leveling legs run completely in

Oven			Broiler			Cutout Dimensions			Approx. Shipping Wt. (Lbs.)
Width	Height	Depth	Width	Height	Depth	Width	Height	Depth	
22"	15"	18"	13½"	3½"	15"	30"	36"	24"	195

Gas Slide-In Ranges
Figure 7-11F

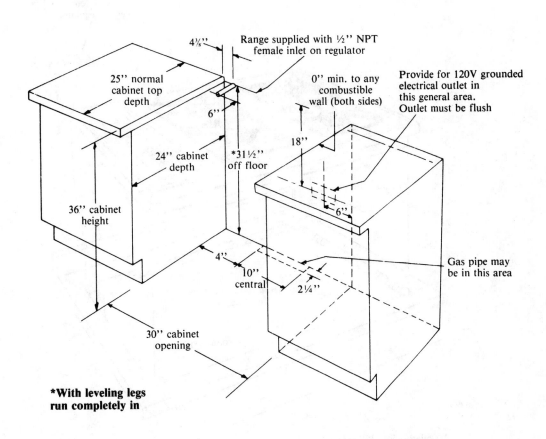

25" normal cabinet top depth

4⅜"

Range supplied with ½" NPT female inlet on regulator

0" min. to any combustible wall (both sides)

Provide for 120V grounded electrical outlet in this general area. Outlet must be flush

6"

24" cabinet depth

*31½" off floor

18"

36" cabinet height

6"

4"

10" central

2¼"

Gas pipe may be in this area

30" cabinet opening

*With leveling legs run completely in

	Oven			Broiler			Cutout Dimensions			Approx. Shipping
Width	Height	Depth	Width	Height	Depth	Width	Height	Depth	Wt. (Lbs.)
22"	15"	18"	13½"	3½"	15"	30"	36"	24"	195
22"	12½"	18"	13½"	3½"	15"	30"	36"	24"	220

Gas Free-Standing Ranges
Figure 7-11G

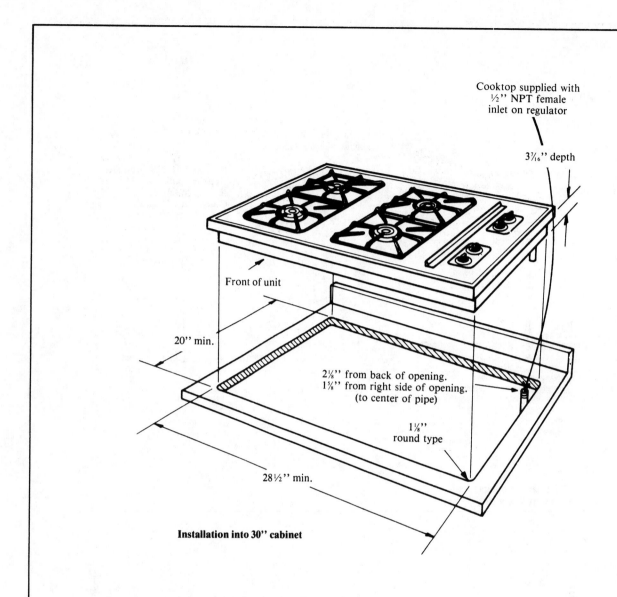

Cooktop supplied with
½'' NPT female
inlet on regulator

3⁷⁄₁₆'' depth

Front of unit

20'' min.

2⅛'' from back of opening.
1⅝'' from right side of opening.
(to center of pipe)

1⅛''
round type

28½'' min.

Installation into 30'' cabinet

Cooktop Width	Cooktop Depth	Cooktop Width	Cutout Depth	Approx. Shipping Wt. (Lbs.)
30''	21''	28½''	20''	50

**Gas Cooktops
Figure 7-11H**

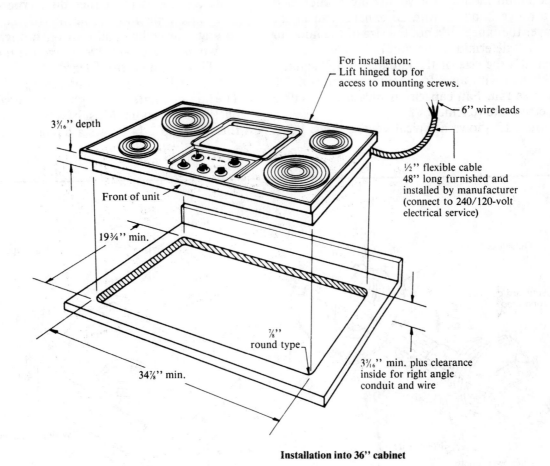

For installation:
Lift hinged top for
access to mounting screws.

6" wire leads

$3\frac{3}{16}$" depth

½" flexible cable
48" long furnished and
installed by manufacturer
(connect to 240/120-volt
electrical service)

Front of unit

$19\frac{3}{4}$" min.

$\frac{7}{8}$"
round type

$3\frac{3}{16}$" min. plus clearance
inside for right angle
conduit and wire

$34\frac{7}{8}$" min.

Installation into 36" cabinet

Cooktop Width	Cooktop Depth	Cutout Width	Cutout Depth	Approx. Shipping Wt. (Lbs.)	Total Connected Load (KW)
36"	21"	34-7/8"	$19\frac{3}{4}$"	50	7.6KW at 120/240V 5.7KW at 120/208V

Electric Cooktops
Figure 7-11I

Kitchen Ventilation

The common method for ventilating a kitchen is with a range hood or with an exhaust fan in the wall over the range. It's not the size of the hood or stove that determines the amount of ventilation needed; it's the size of the kitchen. A 160-square-foot kitchen with an 8-foot ceiling requires a 320 c.f.m. size fan. Fan capacity is measured in cubic feet per minute (c.f.m.)

Figure 7-12 shows a typical vented range hood and its various dimensions.

Range hoods are either duct type or ductless. The duct type moves the air to the outside up through the ceiling and roof or through the wall as shown in Figure 7-13. Don't end the duct in the attic. This can be a fire hazard.

The ductless hood is not vented to the outside. The air circulates through a charcoal filter and back into the room. Most have an aluminum grease filter. Figure 7-14 shows a ductless hood.

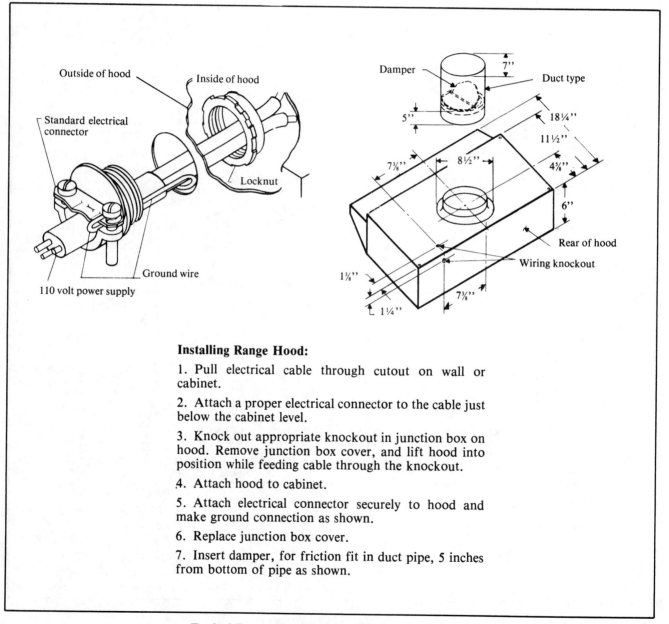

Installing Range Hood:

1. Pull electrical cable through cutout on wall or cabinet.

2. Attach a proper electrical connector to the cable just below the cabinet level.

3. Knock out appropriate knockout in junction box on hood. Remove junction box cover, and lift hood into position while feeding cable through the knockout.

4. Attach hood to cabinet.

5. Attach electrical connector securely to hood and make ground connection as shown.

6. Replace junction box cover.

7. Insert damper, for friction fit in duct pipe, 5 inches from bottom of pipe as shown.

Typical Range Hood and Installation Procedures
Figure 7-12

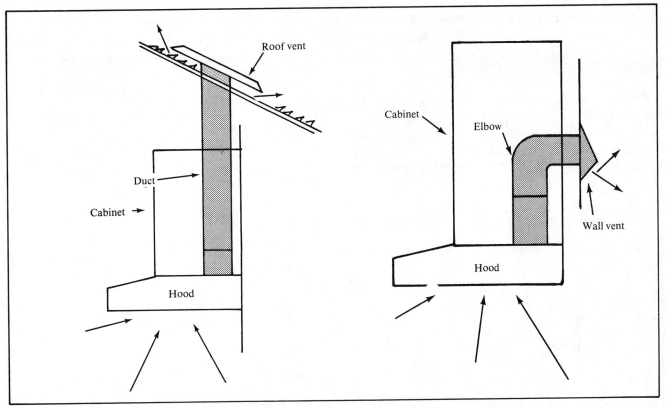

Range Hood Vented Thru Roof and Wall
Figure 7-13

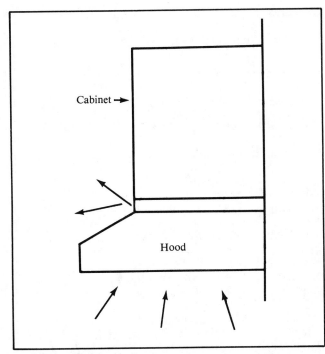

Ductless Range Hood Filters Air Back Into Room
Figure 7-14

Kitchen Planning

As previously mentioned, Brown wants a U-shaped kitchen with all the modern conveniences. Your planning begins with a budget cost and a good understanding of what your customer wants and doesn't like about the existing kitchen. Use the form in Figure 7-15 to help you understand what your customer needs and to help them decide on the kitchen they want.

Design it with proper lighting—lighting to work by, lighting to live by. Soft lighting is comfortable and relaxing. And with all sorts of dimmer switches and fixtures on the market, it's easy to be a lighting expert. Place lights over such areas as work islands. False beams with light strips are also available.

Keep in mind that some people tend to overreact. A person stuck with a place that has too few or too small windows will want huge windows placed everywhere when they remodel. If the rooms are too small, they'll want twice the existing size. It's up to you to keep everything in perspective, whether you're designing a kitchen, a bathroom or a bedroom.

Figure 7-16 shows the layout of an old kitchen, and Figure 7-17 shows the recommended layout for remodeling it.

Kitchen Preference Form

Name_____ Phone_____

Address_____ Date_____

To design a kitchen to an individual's personal preference, it is essential to know what that individual desires in a kitchen. In order that we can personalize your kitchen, according to your own personality, please check the following:

1. My personal kitchen should have:
☐A quiet corner where I can relax while alone or with a friend over a cup of coffee.
☐Soft lighting for a "meal for two" or a quiet moment.
☐A radio.
☐A television set.
☐An intercom system to the rest of the house.
☐A view of my garden or the outside.
☐A telephone extension.
☐Access to a convenience bath.
☐Open view to the family room.

_____

_____

2. My favorite colors are:
☐Bright tones   ☐Soft tones   ☐Medium tones
☐A mixture of colors for soft contrast.
☐A mixture of colors for bright contrast.
☐White____ Green____ Yellow____ Orange____
Red____ Blue____ Purple____
(Other Comments)_____

_____

3. I want my kitchen to have the following:
☐Refrigerator with freezer compartment in the:
Top____ Bottom____, Side by side____.
☐Refrigerator with Ice Maker____, Ice and water available on the outside____.
☐A double sink
☐Trash compactor
☐A single sink
☐Breakfast nook
☐Dishwasher
☐Breakfast bar
☐Garbage disposal unit
☐Ceiling fan
☐Built-in oven
☐Built-in mixer
☐Range top
☐Lazy Susan shelves
☐Range hood

☐Molded counter top
☐Cabinets
Type_____
Style_____
☐Air conditioning
☐Cabinets extended to ceiling
☐Lowered ceilings
☐Microwave oven
☐More wall outlets
☐Separate freezer
☐New floor covering
☐Chopping block
☐New wall covering
☐Built-in ironing board
☐Repainted
☐Broom closet
☐Multiple lighting for:
  ☐bright
  ☐soft
☐Pantry
☐Conventional lighting

4. I prefer a kitchen with:
☐More window space   ☐Less window space   ☐Same window space   ☐More floor space   ☐Less floor space   ☐Same floor space

5. Of the four basic kitchen plans I like the following best:
☐U-shaped   ☐L-shaped   ☐Corridor   ☐One-wall

6. The worst points about my present kitchen are:

_____

_____

7. My favorite color for appliances is:_____

_____

8. My favorite theme is:
☐Contemporary (clean lines, bold colors, natural wood
☐Early American (pewter, copper utensils, traditional woods)
☐Traditional (flexible style, simple, informal)
☐Spanish/Mediterranean (wrought iron, massive wood)

9. My choice of floor covering is:
☐Vinyl   ☐Carpet   ☐Inlaid style   ☐Tile
☐Resilient tile ☐Other_____

Kitchen Preference Form
Figure 7-15

10. My choice of walls is:
☐Sheetrock ☐Wallpaper ☐Paneling ☐Brick ☐Stone

11. My choice of ceiling is:
☐Luminous ☐Beamed ☐Textured ☐Drop ☐Paneling
☐Sheetrock

12. I will consider:
☐Changing locations of windows.
☐Changing location of doors.
☐Opening new doorway or closing existing one.
☐Opening new window or closing existing one.
☐Extending kitchen beyond house walls (addition).
☐Changing location of sink_____, range_____,
 refrigerator_____.
☐Adding new cabinets.
☐Replacing old cabinets with new.
☐Redoing existing cabinets.

13. I would like my kitchen to look like:
☐The attached picture.
☐My friend's whose address is_____.
☐My own creation.

14. If I am unable to completely redo my kitchen at this

time then the following may be put off until later:
☐New appliances ☐New cabinets ☐New counter tops
☐Flooring   ☐Walls   ☐Ceiling

15. How many in your family? Adults_____?
Children_____?Ages of children_____.

16. The height of the person who will use the kitchen the
most_____.

17. The kitchen will be used to prepare meals for:
☐Breakfast ☐Lunch ☐Dinner on a ☐regular
☐irregular basis.

18. I entertain adult guests ☐seldom ☐often about____
times a month.

19. I would like to have the following in the kitchen:
☐Laundry area ☐Eating area ☐Desk/art area ☐Bar
☐Children's play area

20. If you had to put a dollar limit on a personalized kitchen, what would that limit be? $_____.

21. The kitchen layout I would like would look
something like this:

Kitchen Preference Form
Figure 7-15 (continued)

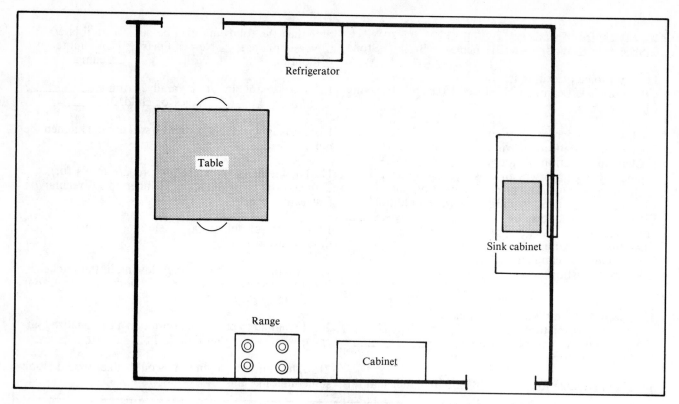

Old Kitchen
Figure 7-16

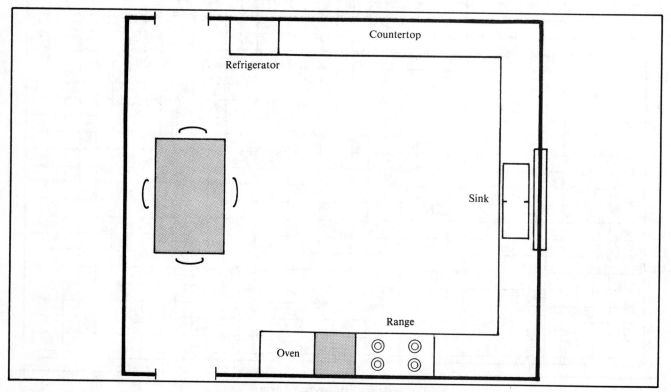

Recommended Layout of Old Kitchen
Figure 7-17

Now, try handling a little design problem of your own. Figure 7-18 is an old kitchen that a potential customer is thinking about remodeling. How would you do it? Use the graph scale in Figure 7-19. Each block represents 1 square foot.

Your Services and Their Value

Designing a kitchen takes thought and work. A lot of people get the remodeling bug and want someone to tell them what they should or should not do to have a modern, efficient, and beautiful kitchen. You shouldn't be doing this design work for nothing. If your customers want more than a quick sketch, tell them that you'll be glad to design a personalized kitchen—for a certain fee. Explain that if you do the job, the fee will be applied to the remodeling contract. If they are sincere about redoing the kitchen, the fee won't scare them off.

More than one builder has developed a good plan at no cost to his customers only to find that they hired some carpenters to do the actual work. That's all right if the builder was paid for his design. The fee depends on what you're going to produce. A floor plan, elevations, drawings, specifications, and a detailed estimate should cost several hundred dollars, at least.

Figure 7-20 is a *Memorandum of Agreement* that should be completed before you begin your planning work.

Planning Steps

Work your planning in steps. Once you have the Memorandum of Agreement signed, have the owner(s) complete the Kitchen Preference Form. Draw a floor plan of the existing kitchen with all measurements.

With every new customer you have to develop trust and confidence. You have to show them that you are a "pro" with lots of good ideas and the skill to turn those ideas into reality. Your planning service is the most important step in developing this confidence. You're not charging for this initial session. You're establishing yourself as a builder who knows what he's doing and, hopefully, convincing the potential customer of that fact.

But when you get the "All right, go ahead and work up a design (or bid) for me," it's time to bring out the Memorandum of Agreement.

How much "designing" you do depends on your method of operating. But you'll need to provide the customer with sufficient data for him to make up his mind. Pictures, floor plans, elevation drawings, specs, color charts, material samples and fixture brochures can be an effective part of the

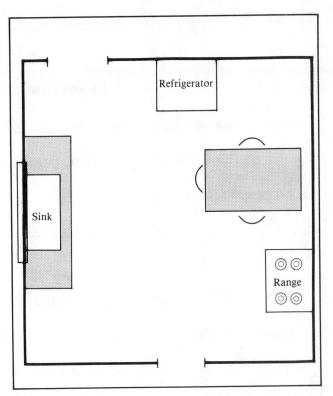

Old Kitchen You Will Re-Design
Figure 7-18

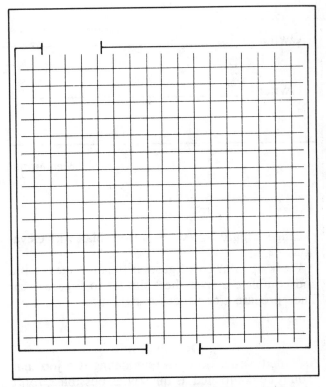

Your Recommended Layout of Figure 7-18
Figure 7-19

Memorandum of Agreement

The undersigned agree to have_____

prepare a floor plan, elevation view, construction cost estimate and outline specifications for

a proposed remodeling of the_____in the residence located at

_____

at a cost of_____.The sum of_____

is hereby acknowledged as received and the balance is due upon completion of the plans,

estimate and specifications. If_____

is authorized to complete all or a substantial portion of the work proposed, the fee for

preparing the plan, estimate and specifications will be credited in full against the cost of

construction.

Owner _____ Builder _____

Owner _____ Date _____

Memorandum of Agreement
Figure 7-20

package. Figure 7-21 shows a floor plan with elevations.

Work up your cost figures if the package is a bid and complete a spec sheet similar to Figure 6-25. Then prepare a contract for signature.

On the Job

As stated earlier, kitchen remodeling is a job that should be completed as quickly as possible. Don't go in and disconnect the stove and other appliances. With a little planning, you can work things so the household will continue to have use of the kitchen even if it's a mess. Get in, do the job, and get out.

There are some builders who like to have three or four jobs going at once. That is, they do enough on a job to "tie it down" and then run off and work on something else. This is jack rabbit contracting—hopping here and there trying to tie down several cabbage patches so no one else will get them. It's bad practice for a kitchen remodeler. Carrying several jobs at the same time is fine only if you have the crews to handle each.

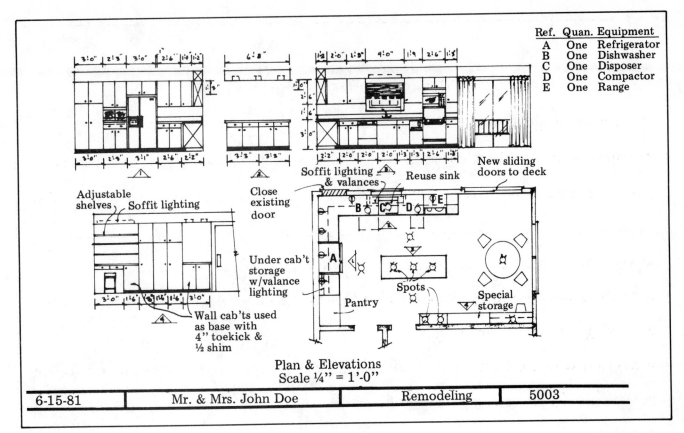

Ref.	Quan.	Equipment
A	One	Refrigerator
B	One	Dishwasher
C	One	Disposer
D	One	Compactor
E	One	Range

Plan & Elevations
Scale ¼" = 1'-0"

| 6-15-81 | Mr. & Mrs. John Doe | Remodeling | 5003 |

Floor Plan and Elevations
Figure 7-21

8

Flooring

An attractive floor can do more for a house than almost any other component. People marry ugly partners, wear ugly clothes, eat ugly food and speak ugly words. But when it comes to floors, everyone wants something pretty to walk on.

In this chapter we're going to talk about flooring material. We'll also discuss underlayment. After all, you have to have a smooth surface on which to place your floor.

The first thing you have to know is the measurements of the floor. (See Figure 8-1.) But before we get carried away with all those measurements, let's talk about the different floor coverings.

Wood

There are hardwoods and softwoods. Softwood flooring is made mostly from southern yellow pine. Douglas fir comes next. Western hemlock and larch are sometimes used. The flooring is available in several lengths and thicknesses. The most common is 2¼ inches wide and 25/32 inch thick. It is tongued and grooved and the underside is grooved to help reduce warping. Softwood flooring is hard to get in some areas and must be special ordered. It was once a popular flooring material in low-priced construction. Heavy traffic will wear it out, and it is almost impossible to keep such a floor "presentable."

Hardwood flooring was once the ultimate floor. It was used in most houses. You couldn't sell a house if it didn't have hardwood floors. Carpet and vinyl are now the most popular types of flooring. Hardwood, like softwood, is considered a special order item in many areas. Oak and maple are the two most popular hardwoods for flooring. Hickory, birch, beech, and a few other hardwoods are used also.

Like softwood, hardwood is grooved on the underside. It's available in thicknesses of 3/8, 1/2 and 25/32 inch and widths ranging from 1½ to 3¼ inches. The 2¼-inch-wide strip flooring, tongued and grooved, is the most commonly used. Simulated plank, pegged and random width styles are available. The boards or strips come in 2- to 16-foot lengths. Prefinished hardwood squares come in 9- or 12-inch sizes and are 5/16 or 1/2 inch thick. These squares form a parquet floor and can usually be obtained in oak, maple, cherry, teak and mahogany.

The strip flooring is delivered in bundles. The boards can be nailed down on underlayment or on an existing floor. It is recommended that building paper be used under the flooring. In new construction, lay the strips crosswise to the floor joists. When laying over an existing strip floor, lay crosswise to the old flooring.

"Cut" flooring nails are used for installing soft or hardwood strips. Use eightpenny nails on 25/32-inch flooring and sixpenny nails on 1/2-inch flooring. Fourpenny casing nails can be used on the 3/8-inch-thick strips. If you use a flooring hammer (it's easier and faster), specially designed nails are used.

Install strip flooring by starting the first strip at

117

Accurate Measurements are Essential
Figure 8-1

least 1/2 inch from the wall to allow for expansion. (See Figure 8-2.) First, face nail the strip at a place that will later be covered by the base board or shoe mold. Then nail through the tongue. All other courses are tongue nailed only. Tongue nail at about a 45-degree angle. Be careful not to beat the edge of the strip with your hammer. It'll dent the strip in an area that is hard to sand out. With the flooring hammer this is no problem since the rig is designed to fit along the strip and tongue and the nail driver is itself hit with a special hammer.

Stagger the strip butts so all the ends don't line up like little soldiers. Crooked strips should be discarded. They're more trouble than they're worth. The last course is face nailed like the first course. Leave expansion room of at least 1/2 inch between the strip and the wall. A rise in the moisture content of the flooring causes it to expand.

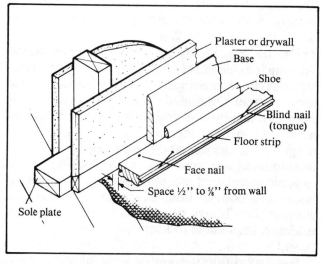

Installation of First Strip of Flooring
Figure 8-2

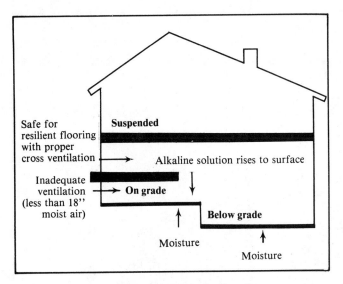

Safe for
resilient flooring
with proper
cross ventilation

Suspended

Alkaline solution rises to surface

Inadequate
ventilation
(less than 18"
moist air)

On grade

Below grade

Moisture

Moisture

Identification of Floor Levels
Figure 8-3

Square-edged strip flooring is face nailed. There is no tongue and groove. For this reason there must be a solid subfloor to lay it on. Even then it will squeak sometimes.

The various wood and wood-base tiles are installed with adhesives. A solid subfloor is required. If it is a concrete floor, make sure a proper vapor barrier is in place. The adhesive is spread on the underlayment or concrete slab with a notched trowel. Don't leave the adhesive too thick. Follow manufacturer's instructions. The tile are pressed onto the adhesive.

The wood block flooring is tongue and groove, with the tongue on two edges and the grooves on the other two sides. The block is nailed through the tongue in the same manner as when installing strip flooring.

Particle board tile is installed in the same manner as wood tile but without nailing. Don't use this tile on concrete. The tile is 3/8 inch thick and 9 by 9 inches in size. It's installed with adhesive but has tongue and groove edges for uniform fit.

Sheet Vinyl Flooring

Vinyl is probably one of the most widely used floor coverings today. Polyvinyl chloride (PVC) is the main ingredient. The vinyl composition also contains resin binders with mineral filters, plasticizers, stabilizers and pigments.

Vinyl can be filled or clear. The clear consists of a layer of opaque particles or pigments covered with a wearing surface of clear vinyl bonded to a vinyl or polymer-impregnated asbestos fiber or resin-saturated felt. It has a high resistance to wear. Filled vinyl is made of chips of vinyl having different colors and shapes immersed in a clear vinyl base. It is all bonded by heat and pressure.

Let's look at a typical installation procedure for covering an old floor.

Room Preparation

Remove all furniture and movable appliances from the area to be resurfaced. Make sure the floor area is dry. Resilient flooring is unsuitable for areas that are continually wet.

Remove any wax or floor finish on the existing floor, using a heavy duty cleanser or other wax remover. Floor adhesives will not bond readily to wax or floor polish.

Remove the baseboard, quarter-round or shoe molding, or vinyl wall base. A thin screwdriver and wide putty knife work well. The molding or wall base is replaced after the installation is complete to secure the edges of the vinyl floor covering.

Where possible, cut the bottom of doorway moldings so the flooring slips under easily. This can be done with a handsaw held flat on a cardboard scrap that is equal in thickness to the new flooring. (See Figure 8-4.)

Smooth and sweep the floor area, driving down all protruding nail heads. Unroll the new floor covering in another room.

Establishing Reference Points

1. Select the wall against which you'll put one of the factory edges of the new covering. (Factory edges run the entire length of the flooring, not the width of the roll.) In general, the factory edge should run along the longest straight wall in the room.

2. Measure out from this wall a given distance—to about the center of the room—at each end of the room. (See No. 2, Figure 8-4.)

3. Snap a chalk line through these two points.

4. Mark a second line at a 90-degree angle to the first line at a point where the second line can run the entire width of the room. It is extremely important that these two lines be exactly perpendicular. They now become your reference points by which you determine how to mark and cut the vinyl sheet. (See No. 4, Figure 8-4.)

5. On a sheet of paper (preferably graph paper), draw the two reference lines from the floor. Sketch the room roughly around the intersection of the two lines. Indicate all cabinets and closets.

6. Measure from the reference lines on the floor

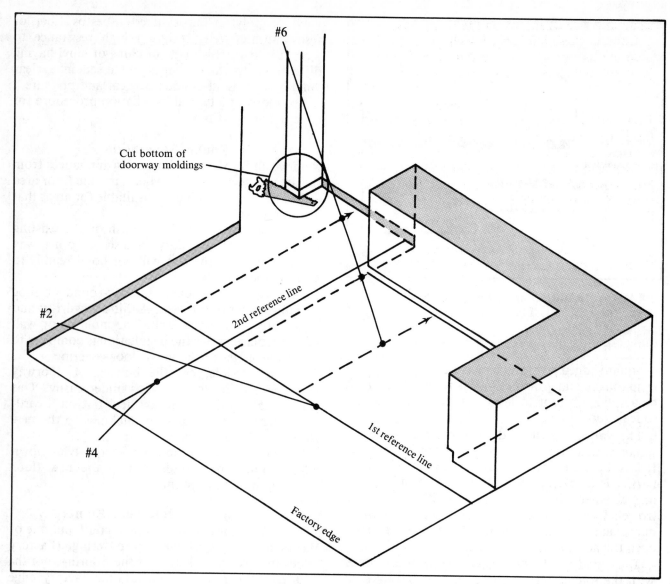

Establishing Reference Points
(#2, 4, 6 refer to paragraph number in text)
Figure 8-4

out to the walls or cabinet every 2 feet around the room, including at least two measurements for every offset. Record these measurements on your paper layout. Careful! Because walls are seldom exactly straight, measure every two feet. (See No. 6, Figure 8-4.)

Transferring Measurements

Position your floor-plan sketch on the new flooring material so that your first reference line is parallel to a factory edge of the flooring. Using the measurement recorded on your floor plan, measure in from the factory edge on the new flooring the distance from your first reference line to the wall. Mark this distance at both ends on the new flooring, and snap a chalk line through the points.

To establish a second reference line on the new flooring, measure along the first reference line from the edges to where the floor plan indicates your reference lines intersect. Add 1½ inches and snap a chalk line through the points at an exact 90-degree angle to the first line. This second line should run the entire width of the floor.

You now have reference lines on the new covering which correlate to those in the room. Measure out from these lines to all walls and cabinets at the same intervals as shown on the floor plan.

Connect these marks with a chalk line, being as accurate as possible. These outside lines become your cut lines. (See Figure 8-4.)

Cutting the Flooring

Don't make any cuts until all room measurements have been transferred and rechecked.

Using a straight blade utility knife and a metal straightedge as a guide, cut along all outside cut lines. (If you're cutting over a concrete floor, put a scrap of plywood or cardboard beneath the cut lines before cutting.)

One-Piece Installations

Before carrying the new covering into the room where it is to be installed, roll the material face in, so it rolls out into the room along the longest straight wall.

Don't force the covering under offsets or cabinets, but unroll enough to know that it is in the correct position.

Fold half of the covering back onto itself, being careful not to move the other half out of position.

You are now ready to spread the adhesive. Be sure to follow instructions on the container.

Spread the adhesive evenly, using a trowel with notches 1/16 inch wide, 1/16 inch deep and 3/32 inch apart.

Don't let the adhesive dry more than 15 minutes before placing the material onto it.

Fold back the other half of the covering and repeat the process.

Use a heavy roller to ensure a good bond and remove any air pockets. Roll toward the edges.

Cap all doorways and openings with a metal threshold strip.

Two-Piece Installations

Where a seam is needed because the room has a side larger than 12 feet, follow procedures used for one-piece installations with the following exceptions:

1. Before transferring measurements onto the new covering, overlap the two pieces at the seam area, making certain that the pattern is matched.

2. Place strips of masking tape across the overlap to hold pieces together.

3. Proceed to transfer measurements and cut the covering to the size and shape of the room. (Do not cut overlap seams.)

4. Untape the seams before moving the floor covering to the room where it is to be installed.

5. Move and position one piece at a time.

6. Fold back the top piece halfway and draw a pencil line on the subfloor along the edge of the second piece (at seam area).

7. Fold back the second piece and spread adhesive to within 12 inches of either side of the line. (See Figure 8-5)

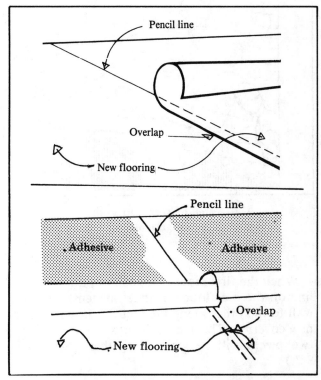

Procedures for Two-piece Installation
Figure 8-5

8. After placing the flooring in the adhesive, repeat the preceding two steps for the other half of the room.

9. Using a straight-blade utility knife and a metal straightedge as a guide, cut through both pieces where they overlap.

10. Fold back both seam edges, spread adhesive, and place the covering into position.

Seam sealing is the last step when installing most vinyl sheet flooring. However, there are some coverings in which seam sealing is one of the first steps. Determine this from the supplier and follow the manufacturer's instructions. Follow the seam sealing instructions on the container.

An Alternative Installation Method

Let's try another method. Unroll and position the first piece of the floor covering, allowing the excess to extend up the side of the walls and the front of the kitchen cabinets. (See Figure 8-6.) Be sure to allow enough material to cover the floor in every offset.

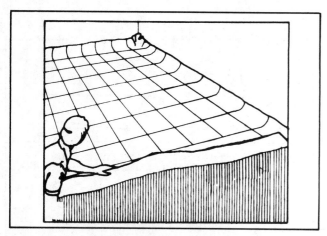

Extend Excess Up Walls
Figure 8-6

When the first piece is in position, draw a pencil line on the old floor (or underlayment) from one wall to the other along the factory-cut edge of the new covering. This marks the position at which the two pieces will be seamed together. (See Figure 8-7.)

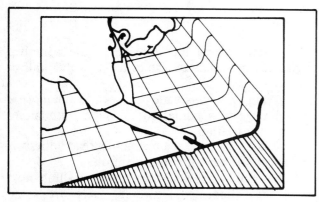

Mark The Old Floor
Figure 8-7

Place the second piece of flooring in position. Again, allow the excess to extend up the walls. Be careful not to move the first piece which has been positioned. (See Figure 8-8.)

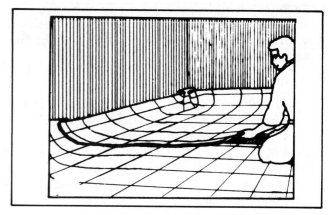

Place Second Piece Into Position
Figure 8-8

Overlap the two pieces of flooring where the seam is to be formed by matching up the pattern at the center (Figure 8-9), and cut both ends of overlay. When possible, use one of the simulated grout lines (a straight recessed line in the surface design of the flooring) as an overlap matching point.

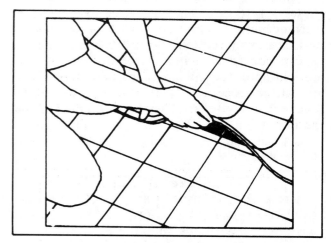

Match The Pattern At Overlap
Figure 8-9

Cut and remove a wide U-shaped piece of flooring from the excess extended up the wall at both ends of the seam. (See Figure 8-10.) The bottom of the U should form a perfect fit along the wall base. This partial fitting of the flooring allows both ends of the seam to lie flat on the floor instead of extending up the base of the walls. Be careful not to lose your pattern match at the seam while making the cut.

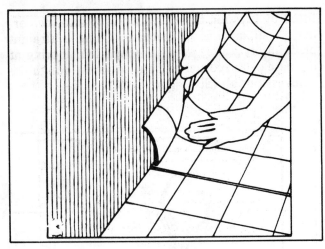

Remove Part of The Excess
Figure 8-10

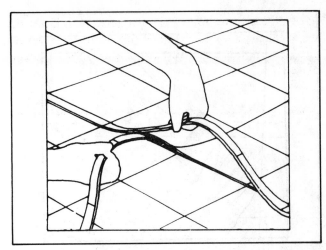

Remove Both Strips of Scrap
Figure 8-12

Using a metal straightedge and a straight-blade utility knife, cut along the overlapped grout lines (whenever possible) found in the flooring's design. Cut from one wall to the other, being sure to cut through both layers of flooring. (See Figure 8-11.)

Remove both strips of flooring scrap created by the pattern-matching cut. (See Figure 8-12.)

The two pieces of flooring are now pattern-matched and cut to fit each other. Be careful not to move the two pieces out of position.

Seam edges may also be straightedged and butted. Cut both seam edges using a sharp utility knife and straightedge. When the seam edges are butted together, the pattern should flow evenly and uniformly.

Gently fold back both edges of the seam and apply a 3-inch band of adhesive under the seam. (See Figure 8-13.) Lay one piece of material onto the adhesive, then slowly lay the second piece onto the adhesive, making sure that the seam is tight and that the pattern is matched.

Follow up by using a hand roller to firmly press the flooring onto the fresh adhesive. (See Figure 8-14.)

Apply seam sealer with the applicator bottle, moving the bottle from one end of the seam to the other as the adhesive is squeezed out. (See Figure 8-15.) Small amounts of excess sealer on the surface of the seam need not be removed, as they will gradually match the gloss level of the floor's sur-

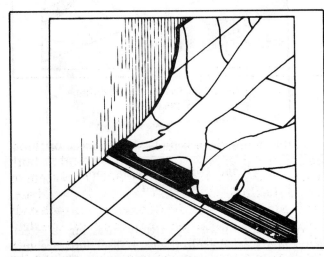

Cut The Overlap
Figure 8-11

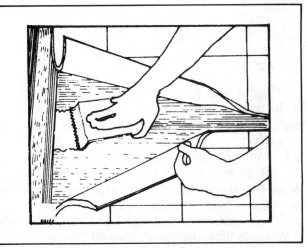

Apply 3-inch Band of Adhesive Under Seam
Figure 8-13

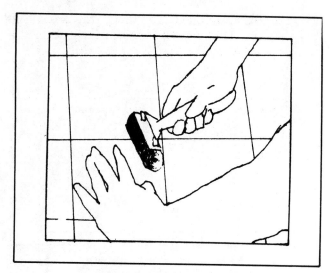

Press The Flooring Down
Figure 8-14

face. After the seam surface is sealed, begin the cutting and fitting sequence of the floor's perimeter.

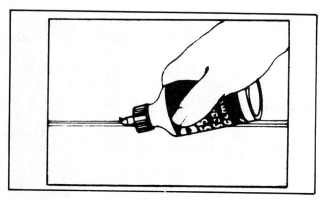

Apply Sealer
Figure 8-15

Cutting and Fitting One-Piece Installations

If the installation requires only one piece of flooring (no seam), then start the installation with the following:

Unroll and position the piece of flooring, allowing any excess to extend up the sides of the walls. Be sure to allow enough material to cover the floor in every offset. Whenever possible, line up the floor's pattern with the longest unobstructed wall.

When cutting the flooring to fit the room, cut and fit the corners first. Inside corners are quickly

fitted by pressing the excess flooring into the corner's wall and floor juncture. Using a straight blade utility knife, punch through the flooring at the base of the corner and cut upward through the excess material. (See Figure 8-16.)

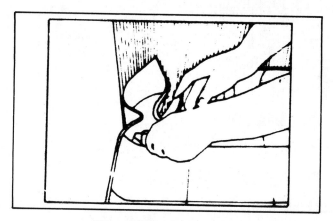

Fit and Trim for Corners
Figure 8-16

Outside corners are fitted by laying the flooring back at the base of the corner so the fold runs diagonally across it. (See Figure 8-17.) Hold the flooring against the corner, punch your knife through at the base, and cut upward through the excess flooring.

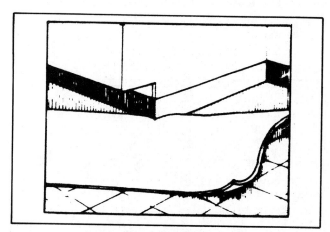

Fit and Trim for Outside Corner
Figure 8-17

Walls are fitted by firmly pressing a steel carpenter's square into the wall/floor juncture and cutting off the excess material with a utility knife held at a 45-degree angle. (See Figure 8-18.) When trimming flooring for a doorway, cut the edge so it will be directly under the closed door. Such exposed edges should be covered with a metal threshold molding.

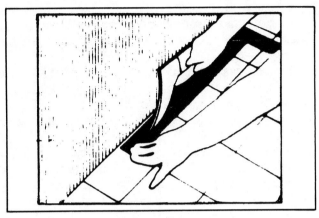

Trim at Walls For Fit
Figure 8-18

Adhering and Stapling

Now that the flooring has been trimmed to fit, it should be fastened down. The following steps should be completed within two hours after cutting and fitting the flooring. If you delay, the flooring will lose its elasticity, reducing its capacity to absorb any possible fitting mistakes. Minor mistakes in fitting can be overcome by stretching the flooring's edge and fastening it down.

Depending on the type of subflooring, the perimeter of the flooring may be fastened with adhesive, staples, or a combination of the two.

If the subfloor is concrete or some other hard surface, your installation will require adhesive. Some floor coverings require staples, others a combination of staples and adhesives. There are some areas such as the toe-kick overhang on the front of cabinets where you can't reach with a staple gun.

To fasten the perimeter of the floor with adhesive, apply a 3-inch band of adhesive around the perimeter and all fixtures. Don't get the adhesive on the face of the material and don't apply more than the recommended amount. A slight picture-frame effect may occur after installation, but this will disappear within approximately 24 hours.

Fold back the flooring along one wall. Starting one foot past the corner, spread a 3-inch-wide band of adhesive along the edge of the floor with a

trowel to within to one foot of the next corner. (See Figure 8-19.) This area left unspread will allow the flooring to be folded back from the next wall.

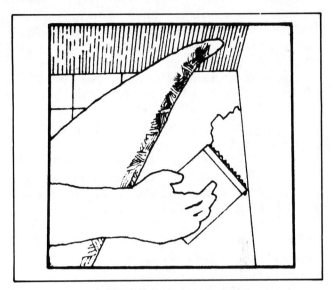

Apply Adhesive at Edges
Figure 8-19

Now use a hand roller to firmly press the flooring onto the fresh adhesive. (See Figure 8-20.) Fold back the flooring on the next wall and spread the 3-inch band of adhesive along the one foot left unspread. Continue spreading around the corner and along the edge to within one foot of the next corner. Again, roll the edge of the flooring onto the adhesive. Continue adhering and rolling like this until the entire perimeter of the room has been fastened with adhesive.

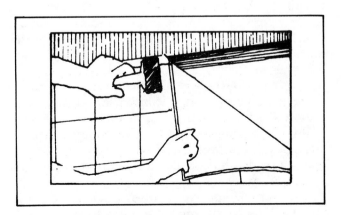

Press Covering Firmly in Place
Figure 8-20

Most sheet vinyl can be applied directly over virtually all types of structurally sound subfloors such as plywood, NPA-approved particleboard, and concrete, whether above, on, or below grade.

Some sheet vinyl can be applied over all kinds of existing floors such as asphalt, vinyl-asbestos tile, rubber tile, vinyl tile, linoleum, sheet vinyl, unglazed ceramic tile, terrazzo, marble, and clay tile. The perimeter on all nonporous surfaces must be roughened. Removal of an entire existing resilient floor, whether it is tile or sheet vinyl, should be done only if absolutely necessary.

Usually you have to replace only those tile that are damaged before installing sheet goods over a worn tile floor. Remove old damaged tile with a putty knife or a scraper. Scrape up the old adhesive and apply new adhesive. The new tile doesn't have to match the old except in thickness.

It's often easier (and more economical) to install new underlayment over the existing floor covering than to remove the existing floor.

Resilient Vinyl Sheet Removal

Should it be necessary to remove an existing sheet vinyl flooring, the following procedures are suggested:

Do not sand existing resilient flooring, backing, or lining felt. These products may contain asbestos fibers that are not readily identifiable. Inhaling asbestos dust may cause Asbestosis or other serious illness.

The wear layer should be cut into narrow strips, taking care not to score the subfloor if it is wood-type subfloor. Peel up the strips from the backing by pulling or by rolling around a core, which will control the stripping angle and create a uniform and more constant tension.

After removing the wear layer, scrape off the remaining felt with a scraper. *Do not sand.* Any unevenness in the subfloor due to scraping should be leveled and smoothed with a latex underlayment mastic.

Resilient Tile Removal

There are several good tile-removing machines and tools. They use vibration or blades to dislodge the tile. Such tools as hand scrapers, long-handled scrapers, ice chippers, and weighted scraping tools will also do the job. Some require more muscle than others.

The dry-ice method involves placing dry-ice inside a 2- by 2-foot wooden frame without a bottom or top. Cover the frame with a piece of carpet, burlap, or some other insulating material to prevent the dry-ice from evaporating too quickly and to direct the freezing action downward. A rope or cord is attached to the box frame to permit easy maneuvering of the frame or "ice box" over the tile floor.

Put the box on the tile you want to remove first. Leave it there for five minutes. Then remove one of the tiles with a scraper or brick chisel. If the tile does not snap loose readily, freeze it some more. Remove the frozen area while the adjacent area is being frozen.

Small areas of tile may be removed with heat. Be careful not to start a fire. Have water or a blanket handy. Use a blow torch or one of the small propane gas torches. An infrared heat lamp also works.

Never use solvents for the removal of asphalt adhesive residue. The solvents will carry the residue deeper into the pores of the subfloor. Later they will rise or bleed to the surface after it is covered with the new flooring.

A grinder, such as a concrete or a terrazzo grinder, is often used to remove asphalt adhesive residue. Wet sand (regular mason's sand) is used with the grinder to prevent contaminating the grinding stones and to aid in removing the residue.

Sub Floors—Underlayment

Some resilient floor coverings cannot be installed over just any particleboard underlayment—the adhesive won't "take." Particleboard underlayment for resilient floor coverings should be approved for that purpose by the National Particleboard Association (NPA).

Single-layer board floors that are not tongued and grooved and that are not over 3 inches wide should be covered with 3/8-inch or heavier plywood in any of the following grades:

1. Underlayment INT-APA (Interior type)
2. Underlayment INT-APA (Exterior glue)
3. C-C Plugged EXT-APA (Exterior type)

Where excessive water spillage is expected such as in a kitchen, bathroom, mud-room, entryway or laundry room, plywood panels of exterior plywood or plywood bonded with exterior glues are recommended.

Single-layer tongue and groove floor boards up to 3 inches wide should be covered with tempered hardboard underlayment or 1/4-inch recommended grade plywood.

Double-layer wood floors with boards 3 inches wide or more are treated the same as single-layer floors.

If the double-layer boards are less than 3 inches wide, renail all loose boards and replace defective or badly worn boards with new boards. Fill the cracks and holes with plastic wood or properly fitted wood pieces. Then install the floor covering. A better way is to install tempered hardboard over this floor.

The important thing to remember about resilient floor covering—whether it is sheet vinyl, vinyl tile, or linoleum—is that a bulge, crack, protruding nail head or hole will show through. The idea is to have a smooth, firm base for the new covering.

If the old wood floor has boards that are warped, cupped, or irregular, then cover it with the right underlayment. It's that simple. If something will show through the new covering—and what won't?—cover it up permanently.

Neither tempered hardboard nor resilient flooring should be installed over "sleeper" constructed subfloors on or below grade without a proper vapor barrier. Crawl spaces over which tempered hardboard is to be installed should be at least 18 inches high and properly ventilated.

Existing Resilient Floors as Underlayment

New resilient floor covering can be installed over an existing resilient surface if it is firmly bonded to the floor, is not sharply or deeply embossed, is free of dirt, grease and oil, and the finish is not stripped. But an old cushion vinyl floor is not a good underlayment. If two resilient floors are already down, don't install a third.

Concrete Subfloors

A concrete subfloor surface should be free of expansion joints, depressions, scale, and foreign deposits of any kind. It should be dry, dense and smooth. Paints, varnish, oil and wax should be removed. A good paint remover for concrete subfloors is trisodium phosphate and hot water. Oakite and Climalene will also do the job.

Paints with a chlorinated rubber or resin base which cannot be removed with trisodium phosphate may be removed by grinding or sanding.

Cracks, expansion joints and score marks in the concrete should be cleaned and filled with latex underlayment mastic (suspended, on, or below grade). Suspended concrete subfloors with a chalky or dusty surface should be swept clean and sized with one coat of primer. A dusty concrete floor on or below grade tells you that moisture is coming through the floor.

A rough concrete floor can often be smoothed with clean, sharp white sand and a terrazzo grinder. To prevent dust, keep the sand and concrete wet while grinding. Don't soak the floor, though. It'll have to be thoroughly dry before installing the flooring. If the floor is too rough for grinding, use a latex underlayment mastic to smooth it out.

If the concrete floor froze before setting up, it will be scaly and cracked. About all you can do is add a top coat of concrete at least 1½ inches thick or install a "sleeper" joist system and put down a new subfloor.

Concrete floors that have been treated with alkali must be neutralized before installing floor tile. A mixture of 1 part muriatic acid to about 8 parts of water will neutralize the floor. Spread the solution over the floor and leave it there for about an hour before rinsing off with clear water. The floor must be thoroughly dry before installation of the new floor covering.

Vinyl-Asbestos Tile

Vinyl-asbestos tile is a blended composition of asbestos fibers, vinyls, plasticizers, color pigments and fillers. The tiles are 9 by 9 inches square and 1/16, 3/32, or 1/8 inch thick. They come in various patterns and textures, are semiflexible, and offer good resistance to grease, alkaline substances, oils, and some acids. The tiles must be placed on a rigid subfloor and can be installed above, on, and below grade. They don't require waxing and can be purchased with a peel-and-stick backing.

Subfloor preparations and requirements are essentially the same as for vinyl sheet flooring.

Border tiles on opposing sides of the room should be the same width. If the border tile on one side of the room is 5 inches, the border tile on the opposite side of the room should be 5 inches. Likewise, the border tiles at one end of the room should be the same as the border tile at the other end.

To lay the starter runs, locate the center of the room by finding the center of each wall. Snap a chalk line on the subfloor across the center of the room and then down the center of the room. (See Figure 8-21.) The room is now divided into four equal parts.

Now lay a row of loose tiles along the chalk line from the center point to one side wall and one end wall. (See Figure 8-22.)

Measure the distance between the wall and the last full tile at B. If this space is less than a half of a tile wide, snap a new line half a tile width closer to the opposite wall. (See Figure 8-23.) Follow the same procedure for row B. This will ensure that the border tiles are not small pieces.

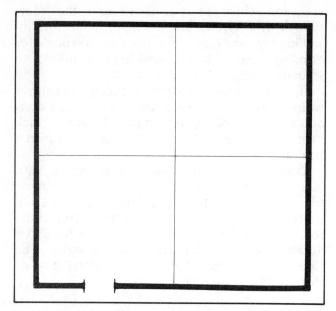

**Lay Room Out Into 4 Equal Parts. Snap
Chalkline on Subfloor
Figure 8-21**

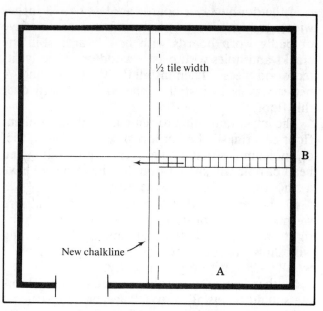

**Shift Tiles to New Chalkline Which is Half of a Tile
Width From Old Chalkline
Figure 8-23**

Lay the A-B rows first, then the balance of this quarter section. Lay the next quarter, starting at the established chalk line and moving toward the walls.

Self-sticking tiles are applied directly to the underlayment without adhesives. For non-self-

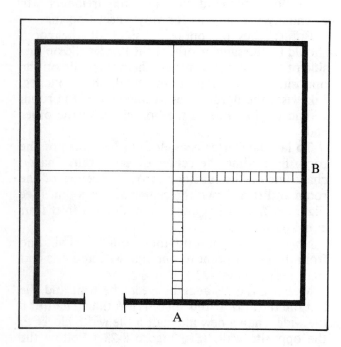

**Lay a Row of Loose Tiles
Figure 8-22**

sticking tile, the adhesive should be applied in a thin coat. It's easy to get too much adhesive on the floor, so watch it. A little is often too much, and the stuff will keep working up between the tiles until the mortgage is paid off. Don't slide the tile into place. This causes the adhesive to pile up at the joint. Set it down flat and firm.

To cut and fit the tile next to the wall, place a loose tile (A, in Figure 8-24) squarely on top of the last full tile closest to the wall. Now place a third tile (B) on top of A and against the wall. Using the edge of the top tile as a guide, mark the tile under it, A, with a pencil. Cut tile A along the pencil line.

Follow the same procedure for marking and cutting tile for corners and irregular shapes. (See Figure 8-25.)

Asphalt Tile

Asphalt tile is a brittle, inexpensive floor covering, ideal for concrete slabs where moisture is a minor problem. But it will stain and deteriorate when contacted with mineral oils and animal fats, and won't recover from indentation. It's a combination of ground limestone, asbestos fibers and mineral pigments with an asphalt binder. The dark colors are the least expensive, with the cost rising as the color lightens. The color pattern extends through the tile for most patterns. Surface patterns wear out rather quickly under heavy use.

The underlayment requirements and installation procedures are basically the same as for vinyl tile.

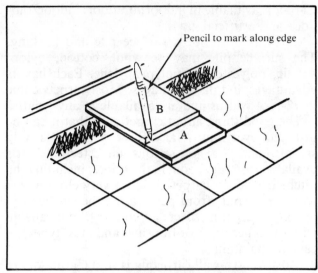

Measuring Wall Border Tile Fit
Figure 8-24

Cork Tile

Cork tile consists of granulated cork bark combined with a synthetic resin as a binder. It's a natural material. The tile has good sound absorption qualities. When coated with vinyl film, it is easily maintained. It comes in the popular 12-inch squares and is available in thicknesses from 1/8 to 1/2 inch. However, the material is affected by grease and alkalies, and maintenance can be a problem. Underlayment requirements and installation procedures are basically the same as for the other tiles.

Rubber Tile

Natural or synthetic rubber, mineral fillers and organic pigments are used to produce a somewhat limited range of patterns and colors. They come in 9- and 12-inch squares.

Indentation recovery is excellent. Use epoxy adhesive for on-grade slab installation when there is no vapor barrier. Underlayment requirements and installation procedures are basically the same as for the other tiles.

Hard Flooring

Hard flooring or non-resilient flooring includes stone, brick, slate, ceramic tile and terrazzo. Installation of these floors requires proper attention to framing and underlayment requirements.

Terra cotta and glazed ceramic tile are a good bet against staining. As glazed tile, they are fairly easy to scratch. The formation of tiny cracks is evidence of age. Small ceramic tile you see so much in bathrooms are called "mosaic" and usually come stuck to a backing sheet, which makes it possible to lay a number of the tile at a time. You grout the joints between each tile after they are set in place. Large 16- by 18-inch size ceramic tile is also available for floor covering.

A special stain-resistant sealant is required for the porous unglazed ceramic tile, flagstone, and

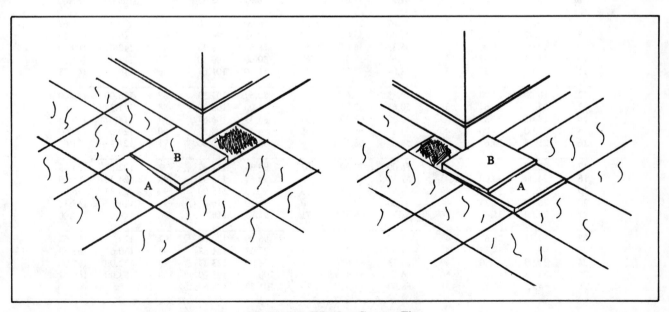

Measuring Tile for Corner Fits
Figure 8-25

slate. Clay or quarry tile, usually unglazed, has a long wear life. The tile is popular in food preparation areas such as in schools and restaurants because of its service life and its resistance to grease, oils and chemicals. It is easy to clean and maintain. The tile comes in reds, buffs, blacks, browns, greys, golds, and variants of these. A semi-glazed type is available in greens, greys and browns. The tile ranges in thickness from 1/4, 1/2 and up to 1½ inches, depending on widths and lengths. Square, rectangular, and geometric shapes are available.

Terrazzo is in a category all its own. It is made of marble chips mixed with portland cement mortar and is ground and polished to a smooth finish. It resists moisture and is fairly easy to maintain.

Hard floors are usually laid by mason tradesmen. The floors are expensive, but their long life may justify the cost on better jobs.

Carpet

Wall-to-wall carpeting can do wonders for a room or house. Cold floors are insulated and sound-proofed; small rooms are made to look larger.

Carpet is available in indoor, outdoor, indoor/outdoor and artificial grass.

Carpets consist of a surface pile and backing. The surface pile may be wool, cotton, nylon, acrylic, polyester, or polypropylene. Each has its advantages and disadvantages. Most carpets come in 12-foot widths and are available in any length.

The good thing about carpeting is that it can be installed over almost any floor that is relatively smooth and free of serious imperfections. It's available for every room in the house, including the kitchen. Kitchen carpeting has a very close weave to prevent spills from penetrating the surface too quickly. Carpeting offers a "comfort" unavailable with the other types of flooring, and most types are easy to maintain.

Most wall-to-wall carpeting is sold by the square yard. To find the square yards needed for a room, multiply the length by the width in feet and divide by 9. Figure 8-26 is a carpet yardage calculator.

The padding or underlay extends carpet life, increases soundproofing and adds underfoot comfort. The "one-piece" and cushion-backed carpeting requires no padding or underlay. Com-

Length Feet	12 Ft. S.Y.	15 Ft. S.Y.	Length Feet	12 Ft. S.Y.	15 Ft. S.Y.	Length Feet	12 Ft. S.Y.	15 Ft. S.Y.	Length Feet	12 Ft. S.Y.	15 Ft. S.Y.
2	2.67	3.33	27	36.00	45.00	52	69.00	86.67	77	102.67	128.33
3	4.00	5.00	28	37.33	46.67	53	70.67	88.33	78	104.00	130.00
4	5.33	6.67	29	38.67	48.33	54	72.00	90.00	79	105.33	131.67
5	6.67	8.33	30	40.00	50.00	55	73.33	91.67	80	106.67	133.33
6	8.00	10.00	31	41.33	51.67	56	74.67	93.33	81	108.00	135.00
7	9.33	11.67	32	42.67	53.33	57	76.00	95.00	82	109.33	136.67
8	10.67	13.33	33	44.00	55.00	58	77.33	96.67	83	110.67	138.33
9	12.00	15.00	34	45.33	56.67	59	78.67	98.33	84	112.00	140.00
10	13.33	16.67	35	46.67	58.33	60	80.00	100.00	85	113.33	141.67
11	14.67	18.33	36	48.00	60.00	61	81.00	101.67	86	114.67	143.33
12	16.00	20.00	37	49.33	61.67	62	82.67	103.33	87	116.00	145.00
13	17.33	21.67	38	50.67	63.33	63	84.00	105.00	88	117.33	146.67
14	18.67	23.33	39	52.00	65.00	64	85.33	106.67	89	118.67	148.33
15	20.00	25.00	40	53.33	66.67	65	86.67	108.33	90	120.00	150.00
16	21.33	26.67	41	54.67	68.33	66	88.00	110.00	91	121.33	151.67
17	22.67	28.33	42	56.00	70.00	67	89.33	111.67	92	122.67	153.33
18	24.00	30.00	43	57.33	71.67	68	90.67	113.33	93	124.00	155.00
19	25.33	31.67	44	58.67	73.33	69	92.00	115.00	94	125.33	156.67
20	26.67	33.33	45	60.00	75.00	70	93.00	116.67	95	126.67	158.33
21	28.00	35.00	46	61.33	76.67	71	94.67	118.33	96	128.00	160.00
22	29.33	36.67	47	62.67	78.33	72	96.00	120.00	97	129.33	161.67
23	30.67	38.33	48	64.00	80.00	73	97.33	121.67	98	130.67	163.33
24	32.00	40.00	49	65.33	81.67	74	98.67	123.33	99	132.00	165.00
25	33.33	41.67	50	66.67	83.33	75	100.00	125.00	100	133.00	166.77
26	34.67	43.33	51	68.00	85.00	76	101.33	126.67			

The Carpet Yardage Calculator
Figure 8-26

mon types of padding are soft and hard-backed vinyl foams, sponge-rubber foams, latex (rubber), and felted cushions made either of animal hair or a combination of hair and jute. The latex and vinyl foams are probably the most practical. Their waffled surface helps to hold the carpet in position. Standard padding is 4½ feet wide.

Foam rubber-backed carpeting is popular. It can be laid in basements and below grade. It is mildew-proof and unaffected by water. It can be laid directly on unfinished concrete floors. The backing is non-skid, and since it is a heavy material, tacks or adhesive are not necessary to hold the carpet in place.

The indoor/outdoor carpeting is recommended for installation over concrete and floor tile surfaces. The carpet backing is made of a closed-pore type of vinyl or latex foam which keeps out moisture. Vinyl and asbestos floor coverings accumulate moisture when covered with carpet. The moisture soaks through into the carpet and will eventually produce a musty odor and cause mildew stains to appear.

Summary
Whether you install the flooring yourself or subcontract the installation, be sure your jobs get good professional quality work. Most floor covering supply houses maintain crews who do nothing but install resilient floor covering (vinyl sheet flooring) and carpeting. Since this is about all the crew does, they usually become experts at it and few problems later arise from such installation.

As a helper once told me, "How many problems can you have with an old floor? If there are more problems than square feet, just tear the whole thing out and start from scratch!"

Old Carpenter's Rule

Lower the ceiling or remove the wall. Brace it well or prepare for a fall.

9

Walls and Ceilings

The interior walls of a house are generally referred to as partitions. They usually carry no weight and serve only to partition off the various rooms of the house. Exterior walls are structural—they hold up the roof. But interior walls can also be structural, so you have to be careful when tearing out interior walls. To remove a load bearing wall is not simple. It can be removed, but beams must be installed to carry the roof load.

Interior partitions can be built with little attention to structural requirements. Purpose, design and function dictate their placement. Partitions can be relocated to separate the various areas of activity into a well-organized plan for orderly living.

In any partition removal, the cavity left in the floor, walls, and ceiling where the partition joined the surface will sometimes present a problem. With plaster or sheetrock walls and ceilings the problem is minimized. Plaster can be spread in the cavity and finished smooth so there is no evidence of a "break" in the wall or ceiling. In sheetrock walls and ceilings, sheetrock of identical thickness is cut to fit the cavity and finished smooth with tape and joint compound.

For wood flooring with strips that run parallel to the removed partition, strips of the same thickness and type can be inserted and finished to match the existing floor. If the run of the flooring is perpendicular to the removed wall, there is a problem. Regardless of what you do, the patch will be noticeable. A new floor covering in this case is the best solution.

Removing a Partition

First, remove the finished material. On the Brown job all the walls are plastered, so removal is fairly easy. In some places just tapping with a hammer causes the material to fall off in little chunks. The more firmly attached plaster may be forced off with a crow bar, narrow spade, or similar tool.

Removing plaster can create a lot of dust. Open a window for fresh air and close off the rest of the house to keep out the dust. The removal of sheetrock and plaster is strictly demolition. The material is not salvageable. Where paneling is involved, you might want to use care in removing it, especially where cavity finish must be done with the panels.

Remove the plaster lath. You are now down to the framing. This is the time to remove any wiring or pipe in the partition. The wiring might end-run into a wall outlet or it might be a circuit run through an outlet. If it's an end-run up through the sole plate from underneath the floor or down through the top plate from overhead, the solution is simple. The outlet and box can be removed after the electricity is cut off at the panel. The wire is pushed back underneath the floor or back into the attic, and the surplus wire cut off. Run the end into a metal junction box, individually cap each wire with a wire nut, tape them, and install a metal cover over the box. Follow the code. If the wire is a circuit run, remove it from the partition and rejoin it so that the circuit continues.

Pipes are removed and the system rejoined as is appropriate.

If this is a non-bearing partition, the framing components—studs, sole plate and top plate—are removed without further ado. The ceiling won't sag and the house won't fall in. The framing members are salvageable and may be reused. Personally, I prefer not to use such materials except for bracing. The wood is hard, the ends often split, and the members are usually a different size than those available at the yards.

To remove a load bearing partition you must first support the ceiling joists. Where attic space above the partition allows working room, a supporting beam can be installed above the ceiling joists in the attic and the joists supported from the beam. The ends of the beam must be supported on an exterior wall, a bearing partition, or a post that will transfer the load to the foundation. The joists are secured to the beam at the point of intersection in the manner shown in Figure 9-1. Keep in mind that getting the beam material up into the attic can be a problem if there is no suitable access hole or attic vent that can be removed. You can install the beam and lock it all together before removing the wall. No temporary support is needed.

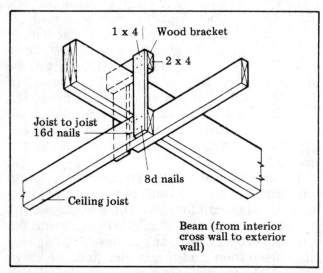

Framing For Flush Ceiling With Wood Brackets
Figure 9-1

I like the exposed beam support approach better. It solves the wall and ceiling cavity problem and ensures support. The ceiling joists are supported on each side of the partition to be removed by temporary "jacks" and adequate blocking. The partition is removed, the beam installed, and the "jacks" removed. The bottom of the beam should be at least 6 feet 8 inches above the floor.

But the exposed beam approach is not always feasible, so other methods are required. One way to handle the problem when no attic space is available, as in the case of a ground floor of a two-story house, is the installation of a beam in the ceiling with the joists butting into the beam. (See Figure 9-2.) Temporary "jacks" and blocking are required on each side of the partition to be removed. In this procedure the joists are cut to allow room for the beam. Use joist hangers to secure the joists to the beam.

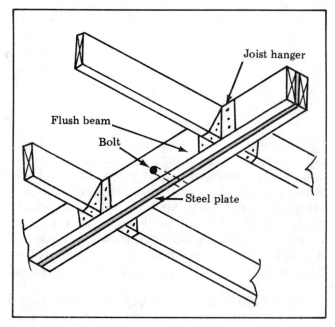

Flush Beam Steel Reinforced
Figure 9-2

The beam span, the span of the joists framing into it, and the material used for the beam will determine the beam size. Since the width of the joists often determines the width of the material used in the beam, use a steel-reinforced beam to ensure adequate structural support. Make certain the steel is thick enough to do the job. Bolt the beam together. The steel sold for this purpose has pre-drilled bolt holes.

Adding a Partition

Not all remodeling is removing partitions. A lot of jobs will require adding a partition. The good thing about adding a partition is that you seldom have to worry about the partition supporting anything except itself. It still must, however, be properly installed. The framing material is usually 2 x 4-inch size, though there are 2 x 3-inch partitions and, occasionally, 2 x 2-inch partitions for closets.

The first step is to mark off the partition on the floor. Snap a chalk line on the floor where the sole plate will go. With a level, draw a plumb line on each wall where the partition will intersect that wall. Mark the plumb line all the way to the ceiling. Next, snap a chalk line on the ceiling from the plumb line of each wall.

Now, nail down the sole plate. If the sole plate is running parallel to the floor joists, you might need to install a double joist directly under the sole plate. If the sole plate is already over a joist, add a joist against it, making it a double joist. If the partition is in the second story over a finished room, skip this procedure. Usually the second story partition can be located a few inches either way to fall over an existing joist. You want to make things as easy as possible. Don't tear up the downstairs ceiling just to install one partition.

The top plate is next. Here you need to secure the plate to something solid in the ceiling. That can be a problem. If it's on the first floor of a two-story house and the finished ceiling is already in place, how are you going to get in there to tie the plate in? You can't. Not without tearing out a section of the ceiling. Let's do it this way. Nail the end studs: the two that go against the walls where you marked the plumb line. There may or may not be an existing stud to nail your stud to. Usually there isn't. So these two end studs will be cut to fit on top of the sole plate and against the finished ceiling. (You did remove the ceiling molding here, didn't you?)

Toenail the studs into the sole plate with 8d box nails. Next, pre-drill for 16d (or longer) nails to go through the end stud you just toenailed to catch the top plate of the intersecting wall. Three nails are sufficient. Do the same thing with the end stud on the other wall.

Now cut the top plate (one continuous 2 x 4) to fit snugly between the two end studs. Toenail it in place with 8d box nails. Cut two jack studs to fit against the end studs under the top plate just installed. Nail these to the top plate and bottom plate with 8d box nails. Use 16d or 12d nails to secure it to the end studs. Cut another 2 x 4, 3 inches shorter than the top plate (one continuous 2 x 4), and nail it in place. Use 12d nails to face nail into the top plate and 8d nails to toenail into the jack studs. (See Figure 9-3.) The double top plate will strengthen the wall span and eliminate any tendency for it to bow later.

If ceiling joists are parallel to the partition and

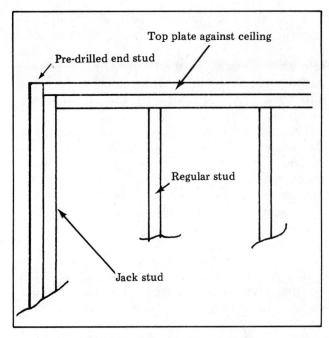

Non-Load Bearing Partition Framing
Figure 9-3

access is available in the attic, install solid blocking over the top plate area at no more than 2-foot spacing (Figure 9-4) and nail the top plate to the blocking. If the ceiling and floor joists are perpendicular to the partition, nail the plates to each joist.

Complete the framing by adding studs 16 or 24 inches o.c. Fit the studs firmly between the plates. Check the required stud lengths at several points. There could be some variation.

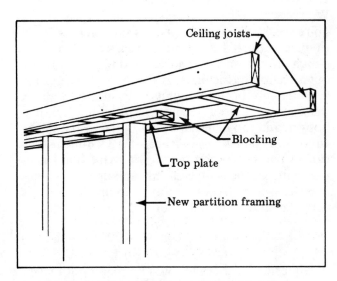

Blocking Between Joists to Which the Top Plate of a New Partition is Nailed
Figure 9-4

A partially assembled partition wall can sometimes be raised when there is something solid to nail the components to. In such a procedure the sole plate is nailed to the floor. The top plate is nailed to the studs and the frame tilted and worked into place. Then the studs are toenailed to the bottom plate.

This method is more likely to cause damage to the finished walls and the ceiling. A fully assembled frame cannot be raised in a room unless the frame height is at least 1/2 inch shorter than the finished room height. Also, wall damage is likely to occur when the framed partition is worked into place.

Special Framing

Some special framing is required to accommodate heating, plumbing and electrical service. Heating ducts, water piping, electrical conduit, and plumbing stacks and drains need a place to run. There is always some modification of standard framing necessary to allow these runs. Plumbing and electrical subs are famous for cutting away entire sections of critical structural members to perform their installations. Most of them are professionals; if they have to do any cutting, they'll reinforce the component. But some are only interested in the quickest and easiest way to get through and get paid. It doesn't seem to make much difference to them that the house might cave in before they can get away. Saws seem to fascinate them. They start cutting and just keep right on going.

Walls With Utility Runs

Walls containing plumbing stacks or vents often require special framing. Four-inch soil stacks will not fit in a standard 2 x 4-inch stud wall. Even the 3-inch plastic is a tight squeeze and requires cutting out a section of the plate. A thicker wall makes for a neater job, both for utility runs and the wall. A thicker wall is usually constructed with 2 x 6-inch bottom and top plates and 2 x 4's placed flat at the edge of the plates. (See Figure 9-5.) 2 x 6-inch studs can also be used in this manner. This method leaves the center of the wall open for running supply lines and drain lines. Another way is to use 2 x 6 studs with the studs notched or drilled for pipe runs.

Interior Wall Finish

The interior finish wall should have a smooth, even surface. That is, plaster walls should be without such flaws as waves, dips or humps. Dry wall, brick, stone, or other masonry-type walls should

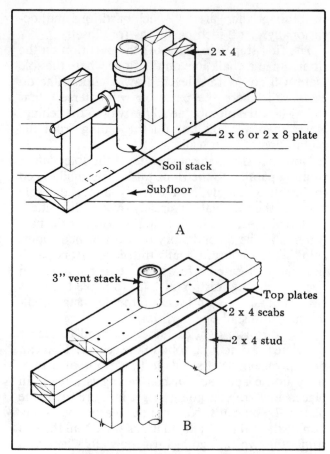

**Framing For Vent Stack: A, 4-Inch Soil Pipe;
B, 3-Inch Stack Vent
Figure 9-5**

also be smooth. In short, they should be professionally installed.

Most minor cracks in plaster or drywall can be patched by filling with a plastic patching mix and sanding. But in cases like Brown's house where the plaster is full of cracks and falling out in places, it should be removed and drywall installed. Drywall installation is usually the best bet where a new wall finish is required.

When we think of drywall, we usually think of gypsum board (usually referred to as sheetrock), hardboard, wood paneling, plywood and fiber board. Drywall is applied to framing or to furring strips over framing or masonry walls, and to existing wall finishes. Drywall can often be glued to a smooth existing wall. If such is the case, the existing finish wall offers a solid backing for the new material, and the thickness of the new material is not as critical as when applied directly to framing or furring strips. Figure 9-6 is a guide to thickness of drywall material applied to framing or furring strips on 16- and 24-inch centers.

Minimum Material Thickness (Inches) With Framing Spaced		
Material	**16" Centers**	**24" Centers**
Gypsum Board	⅜	½
Hardboard	¼	--
Wood Paneling	¼	½
Plywood	¼	⅜
Fiberboard	½	¾

Drywall Thickness and Spacing
Figure 9-6

To eliminate waviness in 1/4-inch plywood or hardboard, install it over 3/8-inch gypsum board.

Installation of new wall material over old is a common practice and a good one provided the wall is smooth. Sometimes it isn't. If the surface is very irregular, the wall should be furred out and shimmed. (See Figure 9-7.) A straightedge placed against the wall will tell you how much shimming is required.

Locate the studs. They'll be on 16- or 24-inch centers. Some houses are built with the studs on non-bearing walls placed from the outside wall toward the center and on load bearing walls from right to left. Others are laid out in the opposite fashion. And others are random, with one room done one way and another room or wall done another way. Different builders and carpenters work in different ways. About the only thing that you can be sure of is that the studs will be on that 16- or 24-inch mark. Sometimes you can't even be too sure about that. But that's the way it's supposed to be. So, when you locate a stud, the next one should be either 16 or 24 inches away.

Locate studs by nail signs or tapping with a hammer for the solid sound of a stud. The stud finder that operates via a magnet pointer might help some people locate studs, but I have tried several and either they were faulty or I was. I found that I could locate 10 studs by sounding with a hammer while someone else was moving the stud finder over the wall looking for a nail head that might indicate the center of the stud.

Furring strips (1 x 2's or 1 x 4's) are applied horizontally on 16- or 24-inch centers, depending on the thickness of the new wall material to be applied. (See Figure 9-8.) All base moldings, ceiling moldings, and window and door casings are removed. Fur out around all openings. The furring strips are nailed to the studs.

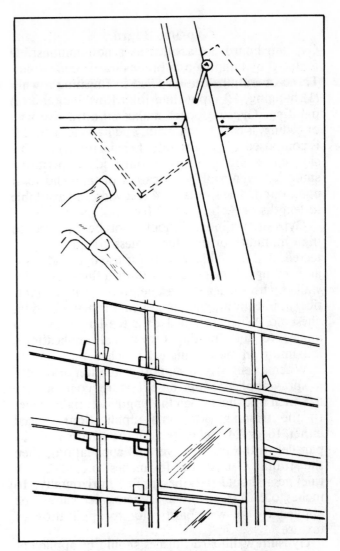

Shingle Shims Behind Furring to Produce a Smooth Vertical Surface
Figure 9-7

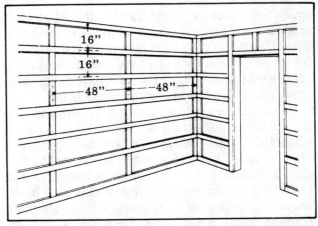

Application of Horizontal Furring to Interior Wall
Figure 9-8

Gypsum Board

Gypsum board is classified as a non-combustible material by building and fire insurance companies. Three operations are involved in finishing a wall: (1) hanging, (2) taping and joint finishing, and (3) painting. Gypsum board makes an attractive wall or ceiling, is easy to maintain, and has a long life. It is composed of gypsum filler faced with paper. The side edges are recessed or tapered to permit a smooth "even-wall" application of tape and joint compound. The sheets are 4 feet wide and available in lengths of 8, 10, and 12 feet.

Gypsum wallboard should not be exposed to high humidity or moisture unless adequately protected. 3/8-inch thickness is for use over old walls and ceilings. When 1/2- or 5/8-inch thickness is installed with the length parallel to the framing (vertical application), maximum frame spacing is 16 inches o.c. When installed with the length at right angles to the framing (horizontal application), maximum frame spacing is 24 inches o.c.

When a ceiling is to support insulation or receive a spray-applied water-base texture coating, use 5/8-inch thickness when the length is at right angles to the framing (horizontal application) spaced either 16 or 24 inches o.c. If the length is to be parallel to the framing (vertical application), then the framing must be 16 inches o.c. 1/2-inch thickness should only be applied horizontally, 16 inches o.c. A pigmented primer-sealer should be applied to the wallboard prior to application of texture.

Gypsum wallboard finishes should be applied at temperatures above 55 degrees, with continuous controlled heating provided at least 24 hours before starting. During periods of high humidity, be sure to provide adequate ventilation.

To ensure proper performance of gypsum wallboard fasteners, wood framing should be good grade 2 x 4's; straight, dry, and uniform in dimension. Figure 9-9 shows the recommended fasteners for single- and double-layer wallboard application using wood furring. (See Figure 9-10 for additional application instructions on the wood frame.)

"Floating angle" application is designed to reduce stress and strain on the wallboard panels when the framing settles. Certain nails or screws are eliminated in interior angles where ceiling and sidewalls meet and where sidewalls intersect. (See Figure 9-11.) Conventional fastening is used in the remaining ceiling or wall areas. Follow standard framing practices for corner fastening. Always install ceilings first. Fit wallboard snugly into all corners.

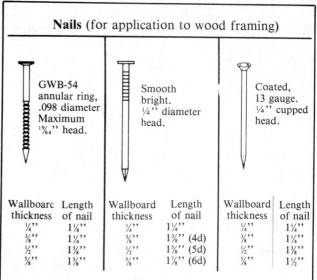

Nails (for application to wood framing)

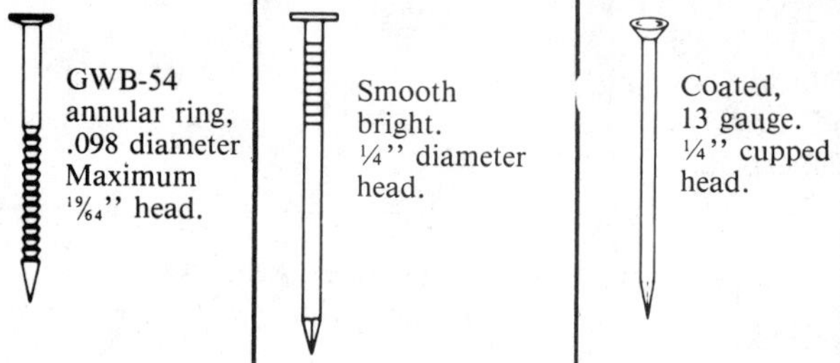
GWB-54
annular ring,
.098 diameter
Maximum
19/64" head.
Smooth
bright.
1/4" diameter
head.
Coated,
13 gauge.
1/4" cupped
head.

Wallboarc thickness	Length of nail	Wallboard thickness	Length of nail	Wallboard thickness	Length of nail
1/4"	1 1/8"	1/4"	1 1/4"	1/4"	1 1/4"
3/8"	1 1/4"	3/8"	1 3/8" (4d)	3/8"	1 1/4"
1/2"	1 3/8"	1/2"	1 5/8" (5d)	1/2"	1 3/8"
5/8"	1 3/8"	5/8"	1 7/8" (6d)	5/8"	1 1/2"

Note: When Parkerhead-type nails are used, follow manufacturer's recommendations.

Recommended Fasteners for Single and Double Layer Wallboard
Figure 9-9

When gypsum board is applied across ceiling joists (at right angles), conventional nailing should be used where the ends of the boards abut the wall intersection. At the long edges of the board running parallel to the intersection, the first nail should be about 7 inches from the wall joint.

When the wallboard is applied in the direction of joists, the long edges of the board abutting the wall should be nailed conventionally. At wall intersections, where the ends of the board meet, the first nail should be approximately 7 inches from the joint. (See Figure 9-12.)

On walls, all wallboard should be in firm contact at the ceiling line to support the ceiling boards previously applied. Along the ceiling intersections, omit nails directly below the ceiling angle. The first nail should be about 8 inches from the ceiling intersection. At all vertical interior angles, omit only the corner nailing of the board first applied; it will be overlapped in the angle. The overlapping board is to be nailed in the conventional manner, 8 inches o.c. (Figure 9-12).

Dry gypsum wallboard may be bent to the radii indicated in Figure 9-13. For shorter radii, thoroughly dampen both sides of the wallboard and allow moisture to penetrate the core before applying. The board will resume its original hardness when dry.

Nails for single nailing should be spaced and driven a maximum of 7 in. o.c. on ceilings, and a maximum of 8 in. o.c. on walls.

**Not less than ⅜"
from edges or ends**

Screws

For wood studs — Phillips head for use with power driven tool. Length: 1¼".

Note: In Multi-ply construction base layer may be fastened with flat staples of 16 gauge galvanized wire, ½" wide and 1" long. **Do not use staples in single layer construction.**

Double Nailing
This method of attachment minimizes nail pops. Nail the field of the board 12" o.c., starting from center and working toward perimeter. Drive second nail 2" from each of the first nails. Conventional nailing is used around the perimeter of panels.

Not less than ⅜"
Approx. 12"
2" maximum
2½" maximum
Maximum 8" side walls
Maximum 7" ceilings

Adhesive-Nail-On Application
Adhesive is applied to studs and joists and wallboard panels placed in position. In sidewall applications space nails no more than 24" o.c. In ceiling applications space nails no more than 16" o.c.

Fastener Spacing With Adhesive or Mastic Application and Supplemental Fastening

Framing Member Spacing	Ceilings		Partitions Load Bearing		Partitions Nonload-Bearing	
	Nail	Screw	Nail	Screw	Nail	Screw
16" o.c.	16"	16"	16"	24"	24"	24"
24" o.c.	12"	16"	12"	16"	16"	24"

Adhesive Application Patterns

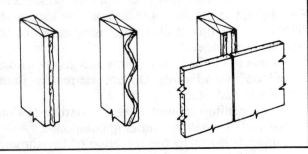

**Wood Frame Application Methods
Figure 9-10**

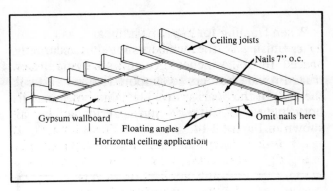

**Ceiling Application
Figure 9-11**

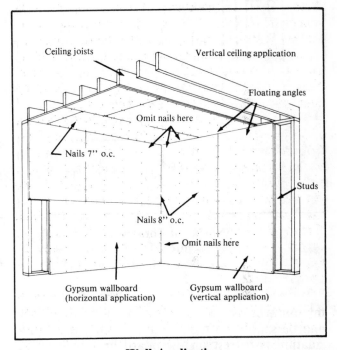

**Wall Application
Figure 9-12**

Wallboard Thickness	Bending Radii (Dry)	
	Width	Lengthwise
1/4"	15'	5'
3/8"	25'	7½'
1/2"	----	20'

**Bending Radii for Dry Wallboard
Figure 9-13**

When framing for gypsum wallboard, as for any other finish wall, the members should be accurately spaced 16 or 24 inches o.c. and set straight and true. Provide headers to support fixtures and cut the ends of the boards. Provide nailing members in both planes of interior and exterior angles as shown in Figure 9-14.

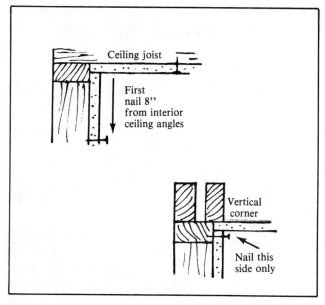

Nailing at Corners
Figure 9-14

Wood furring on ceilings should have a minimum cross section of nominal 2-inch by 2-inch and be spaced no more than 24 inches o.c. If insulation is to rest on the board in a ceiling, the spacing of framing should not be over 16 inches o.c. and the board must be applied at right angles to the framing. You may want to put a vapor barrier between wallboard and the insulation. Unheated spaces above wallboard ceilings must be adequately ventilated.

Install ceilings first. Fit panels snugly against the ceiling and square the first board with the adjacent wall.

Board with square or beveled edges should be applied vertically; the tapered edge board either vertically or horizontally. Bring panels to moderate contact, but do not force them together. Stagger the joints so that the corners of four boards do not meet at one point. On opposite sides of partitions, perpendicular joints should not occur on the same studs.

Fasten wallboard panels by nailing from the center toward the ends and edges. On intermediate and end supports, space the nails 6 to 8 inches apart on walls and 5 to 7 inches apart on ceilings. Keep nails 3/8 inch in from the ends and edges of panels.

Water Resistant Gypsum Board

A specially-formulated back board (B/B) is used as a base for adhesive application of ceramic, metal and plastic wall tile in areas of high moisture such as shower stalls and tub enclosures.

Backer board consists of an asphalt-treated core covered with a heavy water-repellent back paper and ivory face paper which does not require surface sealant before tile is applied. Joints do not require taping. It should not be used as ceiling or soffit material.

Use furring around the tub enclosure and shower stall so the inside face of the fixture lip is flush with the face of the backer board. The top of the furring should be even with the upper edge of the tub or shower pan. (See Figure 9-15.)

On studs spaced over 16 inches o.c., as when walls are to be surfaced with ceramic tile over 5/16 inch thick, locate one row of blocking approximately 1 inch above the top of the tub or shower receptor and another midway between the fixture and the ceiling.

Provide appropriate headers or supports for the tub, plumbing fixtures, soap dishes, grab bars and towel racks. Install the tub, shower receptor or pan before installing the wallboard. Tubs, shower receptors or pans should have an upstanding lip or flange 1 inch higher than the water drain or threshold of the shower. Interior angles should be reinforced with supports to provide rigid corners.

Apply the backer board horizontally to eliminate butt joints. Allow a 1/4-inch space between the lip of the fixture and the paper-bound edge of the board. Space nails 8 inches o.c. and screws 12 inches o.c. If ceramic tile thicker than 5/16 inch is to be used, space the nails 4 inches o.c. and screws 8 inches o.c. Dimple nail heads, using care not to tear the face paper.

In areas to be tiled, treat joints and angles with a waterproof tile adhesive. Do not use regular joint compound and tape.

Caulk openings around pipes and fixtures with a waterproof non-setting caulking compound.

Apply tile down to the top edge of the shower floor surfacing material, to the return of the shower pan, or over the tub lip, as shown in Figure

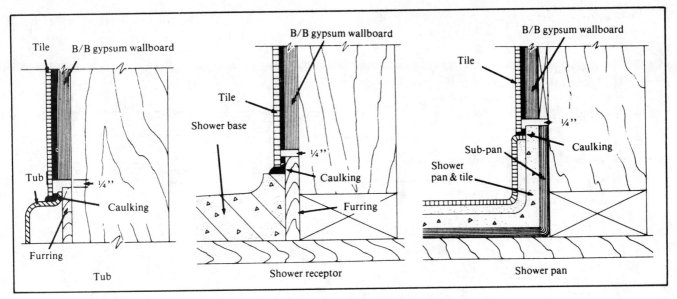

Backer Board Installation Procedures
Figure 9-15

9-15. All joints should be completely and continuously grouted.

Fill the space between fixture and tile with non-setting caulking or tile grout compound.

Sound Control Techniques—Remodeling

In remodeling and rehabilitating multifamily units, one of the major problems is sound control between the units. You can reduce sound transmission by adding an auxiliary wall.

Another method of reducing noise is to apply 5/8-inch gypsum wallboard over the existing partition finish on both sides of the partition. Figure 9-16 illustrates a few tips for quiet conditioning. A wallboard footage table is shown in Figure 9-17.

All nails driven into gypsum wallboard are set into the surface (without breaking the paper) to leave a "dimple." This allows for a smooth covering of the nails with the joint compound.

Ready-mix joint compound comes in a smooth working paste form, ready to use for embedding and taping application.

Sheetrock taping requires a little skill. It takes practice to be able to do a satisfactory job. Sheetrock hanging and finishing is a field all its own. Most builders (including the small ones) sub the job out. A skilled crew can come in and hang and tape the stuff while we're still trying to figure out what goes horizontally and what doesn't. Those fellows scurry around on stilts, slapping it up in no time. They don't need T braces to hold the sheet against the ceiling joists until it's nailed.

However, in case you insist on doing it yourself, here's the procedure:

1. For the first coat use a broad knife (4 inches) to butter the compound evenly over joints of abutting wallboard panels. Butter one complete joint run at a time. Apply tape to the full length of the joint, centering it over the joint. Holding the broad knife at about a 45-degree angle, use moderate pressure on the knife to embed the tape into the compound. Remove excess compound. Treat all butt joints and interior angles (corners) in a similar manner. Fill depressions around nail heads with the first coat of joint compound.

2. The second coat is applied after the first coat is dry. Feather out at least 2 inches on each side of the tape. Apply a second coat over the nail heads.

3. When the second coat is dry, apply a finish coat of compound to joints, nail heads and nicks. When dry, sand smooth.

Butt joints are not tapered, so use care. Don't build up a mound. Feather the compound well out on each side of the joint. The final application should be 14 to 18 inches wide.

A metal corner bead is used on outside corners. Nail it on and finish it with compound in three applications the same as other applications. Feather each coat a couple of inches further out.

After the joint treatment is thoroughly dry, seal or prime all surfaces with a vinyl or oil base primer. This will equalize the absorption differences bet-

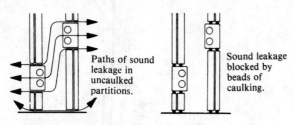

Caulking in the right places—at perimeter wall openings, electrical boxes, other wall openings or outlets—is essential to maintenance of designed STC value.

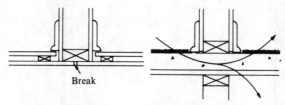

Avoid continuous structural elements which provide a flanking path for noise transmission. By utilizing a "break" in the subfloor, impact noise between adjacent rooms can be reduced.

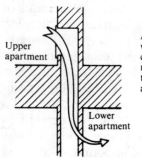

Western framing can effectively block sound transmission between floors.

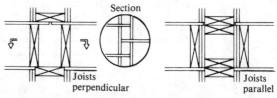

Double blocking with joists perpendicular allows electrical and plumbing facilities to penetrate plates without the necessary caulking required to prevent flanking noises. Parallel joist construction incorporating double sills and plates reduces sound transmission while providing an areaway for utilities.

A flanked path for noise occurs when ducts penetrate a sound conditioned floor. Thin walled metal ducts passing vertically through party walls should be avoided.

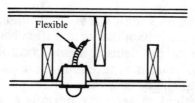

To prevent "short circuiting" a staggered joist floor-ceiling assembly, do not connect lighting fixtures across adjoining joists. While possible, use non-metallic or BX wiring.

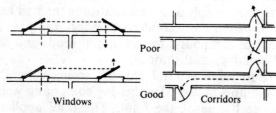

Doors opening on a common corridor and exterior casement windows should be positioned carefully in order to avoid creation of sound "reflectors."

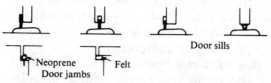

Door openings can destroy the value of sound conditioned partition systems unless treated properly. Use solid core doors rather than the hollow core type. Provide soft weather stripping around the head and jamb areas, plus automatic threshold closers, for an effective seal against sound "leaks."

Tips For Quiet Conditioning
Figure 9-16

Pieces	4' x 6'	4' x 7'	4' x 8'	4' x 9'	4' x 10'	4' x 12'	4' x 14'	4' x 16'	2' x 8'	2' x 10'	2' x 12'	Bdls.	3/8"	1/2"
	BOARD PRODUCTS — SQUARE FEET											16" x 48" Lath		
2	48	56	64	72	80	96	112	128	32	40	48	1	32	21.33
4	96	112	128	144	160	192	224	256	64	80	96	2	64	42.67
6	144	168	192	216	240	288	336	384	96	120	144	3	96	64.00
8	192	224	256	288	320	384	448	512	128	160	192	4	128	85.33
10	240	280	320	360	400	480	560	640	160	200	240	5	160	106.67
12	288	336	384	432	480	576	672	768	192	240	288	6	192	128.00
14	336	392	448	504	560	672	784	896	224	280	336	7	224	149.33
16	384	448	512	576	640	768	896	1,024	256	320	384	8	256	170.67
18	432	504	576	648	720	864	1,008	1,152	288	360	432	9	288	192.00
20	480	560	640	720	800	960	1,120	1,280	320	400	480	10	320	213.33
22	528	616	704	792	880	1,056	1,232	1,408	352	440	528	11	352	234.67
24	576	672	768	864	960	1,152	1,344	1,536	384	480	576	12	384	256.00
26	624	728	832	936	1,040	1,248	1,456	1,664	416	520	624	13	416	277.33
28	672	784	896	1,008	1,120	1,344	1,568	1,792	448	560	672	14	448	298.67
30	720	840	960	1,080	1,200	1,440	1,680	1,920	480	600	720	15	480	320.00
32	768	896	1,024	1,152	1,280	1,536	1,792	2,048	512	640	768	16	512	341.33
34	816	952	1,088	1,224	1,360	1,632	1,904	2,176	544	680	816	17	544	362.67
36	864	1,008	1,152	1,296	1,440	1,728	2,016	2,304	576	720	864	18	576	384.00
38	912	1,064	1,216	1,368	1,520	1,824	2,128	2,432	608	760	912	19	608	405.33
40	960	1,120	1,280	1,440	1,600	1,920	2,240	2,560	640	800	960	20	640	426.67
42	1,008	1,176	1,344	1,512	1,680	2,016	2,352	2,688	672	840	1,008	21	672	448.00
44	1,056	1,232	1,408	1,584	1,760	2,112	2,464	2,816	704	880	1,056	22	704	469.33
46	1,104	1,288	1,472	1,656	1,840	2,208	2,576	2,944	736	920	1,104	23	736	490.67
48	1,152	1,344	1,536	1,728	1,920	2,304	2,688	3,072	768	960	1,152	24	768	512.00
50	1,200	1,400	1,600	1,800	2,000	2,400	2,800	3,200	800	1,000	1,200	25	800	533.33
52	1,248	1,456	1,664	1,872	2,080	2,496	2,912	3,328	832	1,040	1,248	26	832	554.67
54	1,296	1,512	1,728	1,944	2,160	2,592	3,024	3,456	864	1,080	1,296	27	864	576.00
56	1,344	1,568	1,792	2,016	2,240	2,688	3,136	3,584	896	1,120	1,344	28	896	597.33
58	1,392	1,624	1,856	2,088	2,320	2,784	3,248	3,712	928	1,160	1,392	29	928	618.67
60	1,440	1,680	1,920	2,160	2,400	2,880	3,360	3,840	960	1,200	1,440	30	960	640.00
62	1,488	1,736	1,984	2,232	2,480	2,976	3,472	3,968	992	1,240	1,488	31	992	661.33
64	1,536	1,792	2,048	2,304	2,560	3,072	3,584	4,096	1,024	1,280	1,536	32	1,024	682.67
66	1,584	1,848	2,112	2,376	2,640	3,168	3,696	4,224	1,056	1,320	1,584	33	1,056	704.00
68	1,632	1,904	2,176	2,448	2,720	3,264	3,808	4,352	1,088	1,360	1,632	34	1,088	725.33
70	1,680	1,960	2,240	2,520	2,800	3,360	3,920	4,480	1,120	1,400	1,680	35	1,120	746.67
72	1,728	2,016	2,304	2,592	2,880	3,456	4,032	4,608	1,152	1,440	1,728	36	1,152	767.00
74	1,776	2,072	2,368	2,664	2,960	3,552	4,144	4,736	1,184	1,480	1,776	37	1,184	789.33
76	1,824	2,128	2,432	2,736	3,040	3,648	4,256	4,864	1,216	1,520	1,824	38	1,216	810.67
78	1,872	2,184	2,496	2,808	3,120	3,744	4,368	4,992	1,248	1,560	1,872	39	1,248	832.00
80	1,920	2,240	2,560	2,880	3,200	3,840	4,480	5,120	1,280	1,600	1,920	40	1,280	853.33
82	1,968	2,296	2,624	2,952	3,280	3,936	4,592	5,248	1,312	1,640	1,968	41	1,312	874.67
84	2,016	2,352	2,688	3,024	3,360	4,032	4,704	5,376	1,344	1,680	2,016	42	1,344	896.00
86	2,064	2,408	2,752	3,096	3,440	4,128	4,816	5,504	1,376	1,720	2,064	43	1,376	917.33
88	2,112	2,464	2,816	3,168	3,520	4,224	4,928	5,632	1,408	1,760	2,112	44	1,408	938.67
90	2,160	2,520	2,880	3,240	3,600	4,320	5,040	5,760	1,440	1,800	2,160	45	1,440	960.00
92	2,208	2,576	2,944	3,312	3,680	4,416	5,152	5,888	1,472	1,840	2,208	46	1,472	981.33
94	2,256	2,632	3,008	3,384	3,760	4,512	5,264	6,016	1,504	1,880	2,256	47	1,504	1,002.67
96	2,304	2,688	3,072	3,456	3,840	4,608	5,376	6,144	1,536	1,920	2,304	48	1,536	1,024.00
98	2,352	2,744	3,136	3,528	3,920	4,704	5,488	6,272	1,568	1,960	2,352	49	1,568	1,045.33
100	2,400	2,800	3,200	3,600	4,000	4,800	5,600	6,400	1,600	2,000	2,400	50	1,600	1,066.67
200	4,800	5,600	6,400	7,200	8,000	9,600	11,200	12,800	3,200	4,000	4,800	60	1,920	1,280.00
300	7,200	8,400	9,600	10,800	12,000	14,400	16,800	19,200	4,800	6,000	7,200	70	2,240	1,493.33
400	9,600	11,200	12,800	14,400	16,000	19,200	22,400	25,600	6,400	8,000	9,600	80	2,560	1,706.67
500	12,000	14,000	16,000	18,000	20,000	24,000	28,000	32,000	8,000	10,000	12,000	90	2,880	1,920.00
600	14,400	16,800	19,200	21,600	24,000	28,800	33,600	38,400	9,600	12,000	14,400	100	3,200	2,133.33
700	16,800	19,600	22,400	25,200	28,000	33,600	39,200	44,800	11,200	14,000	16,800	200	6,400	4,266.67
800	19,200	22,400	25,600	28,800	32,000	38,400	44,800	51,200	12,800	16,000	19,200	300	9,600	6,400.00
900	21,600	25,200	28,800	32,400	36,000	43,200	50,400	57,600	14,400	18,000	21,600	400	12,800	8,533.33
1,000	24,000	28,000	32,000	36,000	40,000	48,000	56,000	64,000	16,000	20,000	24,000	500	16,000	10,666.67

Wallboard Footage Table
Figure 9-17

ween the sheetrock surface and the joint compound surface, which then permits a uniform texture over the entire wall or ceiling surface.

Plywood and Hardboard

Plywood and hardboard panels are 4 x 8 feet. The panel is finished to have the appearance of boards or planks, of equal or random widths. The plywood is made in a number of species and finishes. There are low-price to expensive panels, depending on the species, finish and quality. Hardboard imprinted with a wood grain pattern is the least expensive. The better hardboard paneling uses a photograph of wood to provide the wood grain effect. All the panels, therefore, usually have the same wood grain pattern. Real wood panels all have different grains—something that may or may not be important to the customer. Plywood and hardboard are also available with a durable plastic finish. Hardboard is also made with vinyl coatings in different patterns and colors.

Generally, the surface preparation necessary for good sheetrock application will do nicely for plywood and hardboard panels. The wall should be fairly even, without dips, humps or waves. If it isn't, then furring is recommended. Fur out spacing the same as for sheetrock. Shim when necessary. Loose plaster should be removed and the wall prepared for neat panel application by furring, if required. Always leave existing sheetrock on the wall unless it, or the framing, is in such a state that it prevents a good paneling job. Remove all molding and trim from the walls and from around doors and windows. Use care if the molding and trim are to be reused. In most cases, the doors and windows will require extending out to the same thickness as the panel. Use a spacer to finish the windows and doors neatly.

Plywood panels often vary in color and texture. You should stand the panels up in the room and arrange them in the order that provides the best color or texture pattern for the room. Mark or restack the panels according to the order in which the sheets will be installed.

The panels can be nailed or glued to the stud framing. Where required, a vapor barrier, such as polyethylene or building paper, can be applied against the studs.

In installing the paneling, determine the stud run and centers. Some prefer to nail up all the whole sheets first and then drop back and catch the corner spaces requiring less than a full sheet. Others prefer to cut for the corners first and go from there, completing each wall as they work around the room. If stud spacing from a corner allows the use of a full sheet, that's fine. Plumb the thing and go. The practice of installing all the full sheets first sometimes can result in saving a couple of sheets of paneling.

Where the corners are true, the installation can be accomplished without inside corner molding. Some prefer to use cove molding for all corners.

After the first plywood sheet is plumbed and nailed or glued in place, the next sheet is butted to it. Hardboard paneling edges should touch very lightly. Sometimes the edge of a panel may not exactly fit the edge of the adjoining panel, leaving the stud showing. Use a black felt marker and mark the stud where the panels butt. Do this before nailing the panel up.

Match-color nails are available for most paneling. Space the nails 8 to 10 inches apart on edges and at intermediate supports.

Use a hand or power saw to cut the panels. With a hand saw, place the finish side of the panel up. With a power saw, place the back side up on the workbench. The panel face is always cut into. Use a combination hollow ground blade on power saws. If you must cut with the face up, score the saw line with a knife. This prevents splintering. Keep the blade on the waste side of the line.

When measuring for window and door openings, you can often hold the panel in position while someone marks the opening on the back of the panel with a pencil. Return the panel to the workbench and adjust the saw line, allowing for jamb thickness, and cut.

For wall outlets and switch boxes, mark along the edges of the box with a pencil or chalk. Place the panel in position and tap the panel against the box with your hand. This will mark the correct position of the box on the back of the panel. Return the panel to the bench, back side up, and drill starter holes at each corner of the box. Saw out the piece with a keyhole or sabre saw.

Paneling a Basement

Let's do this step-by-step, as shown in Figure 9-18. Basement conversion to a den or game room is a remodeling job that most small builders do quite frequently. Below ground masonry walls often pose a moisture problem. If moisture is a problem, the walls and floor should be treated with a sealant to resist the moisture. Follow the manufacturer's instructions.

Use a vapor barrier where paneling is to be installed over a masonry wall. The material can be either foil or plastic film such as polyethylene.

For furring strip installation, apply the strips at the top and bottom of the wall, horizontally and vertically, 16 inches o.c. Use concrete nails or a stud driver.

A free standing wall is assembled on the floor and then tilted into place. If this wall is preferred over furring strips, start by placing the top and bottom plates on the floor.

Where the free standing wall fits into a corner, two studs are used. One stud creates a surface for attaching the paneling. The second stud provides a place to tie-in at the corner for the next free standing unit.

Measure the actual width of one of the 2 x 4's, and mark this distance from the end of the plates. Then square a line across the plates. Measure in from opposite edges of the two plates, the thickness of a 2 x 4, and mark as shown by the heavy black "X". At right angles from the X's, mark each plate with an X. This establishes the marking points for nailing the corner studs.

Starting from the first line you squared across the plates, mark at 16-inch intervals and square a line. These are your guides for nailing the studs. Lay out the marked plates (top and bottom pieces) before nailing. Cut studs ½ inch less than the actual measurement to allow for tilting the wall into place.

Start nailing at the marked ends of the plates, using the X's as guides. To set the studs in place, match the ends to the X's. Nail the ends or corner studs securely. Then move to the 16-inch interval marks and nail the remaining studs.

Tilt the free standing wall into place. If you've allowed the one-half inch in height, the wall should fit easily.

Check that the wall is plumb by pushing the stud wall flush. Move the wall out at the top or bottom, as necessary. Temporarily tack the wall into place at the top while you check for true plumbness. When the wall is plumb, mark a guideline on the floor. You're now ready to secure the wall to the floor. Use concrete nails or a stud driver.

Check the wall again for plumbness. At the top, drive narrow wood wedges, from the opposite side of the joist, to make a holding wedge. The wall can now be nailed to the joists.

For window openings place needed extra studs on each side of the opening. Find the proper location for sill studs and install short studs below the opening, picking up the spacing 16 inches o.c. Frame these in.

Electrical wiring and junction boxes should be installed after all free standing walls are firmly fix-ed in place. Install insulation on the exterior walls.

An open stud wall is used for dividing areas (partitions). This wall can be assembled and tilted into place. Allow one-half inch of height for tilting in.

Door openings are framed in. Double-studding at the top and sides of the opening provides rigidity.

When the paneling is delivered, stand them on their long edge in the area to be paneled at least 48 hours before installation. This lets the panels adjust to room temperature and humidity.

Match the grain and color before installation by standing the panels around the walls. This way you can select the best pattern for the room.

When cutting panels with a circular power saw, cut from the back side of the panel. Mark and saw from the surface side of panel when using a hand saw.

Use color-matching nails. They require no setting or filling.

Deduct one-half inch for the cutting height of the panels. The installed panel should be ¼ inch above the floor and below the ceiling. Start at the corner; one nail will hold the panel when testing for plumb. Place a level on the edge of the panel to find the true plumb. This is important, since the first panel establishes the vertical alignment for the panels to follow. A wood shim or wedge placed at the bottom of the panel holds the panel plumb and makes it easier to move.

Tack the paneling at the top to hold it in place. (Drawing a black line with a felt pen down the stud where panels will meet keeps the stud surface from showing.) Start nailing at the corner and move down the stud. Edge nails should be 8 to 10 inches apart, or closer if required. Intermediate stud nails should be about 8 to 10 inches apart.

Don't force panels into place; their edges should touch very lightly. For electrical outlets, measure from the floor up and from the edge of the previously installed panel. Mark the panel to be cut. Be sure to allow for the ¼ inch that panel will be off the floor.

Use as a pattern an actual junction box that matches the one installed in the wall. This helps avoid incorrect measurements. Drill holes in the corners of the pattern. Keep drilled holes within the pattern area to avoid over cutting. Use a saber saw or keyhold saw to cut out the opening. The hole may be cut ⅛ inch larger than the pattern line since the faceplate will cover the area.

For window openings measure from the last installed panel to the edge of the opening and from the floor to the opening. Allow ¼ inch between the floor and panel bottom.

Layout and cut the panel. Remember to cut on the back side when using a circular saw, and face side when using a hand saw. Check the panel for fit before nailing.

When using adhesives to install panels on a stud wall, wire brush all stud nailing surfaces. A clean stud surface provides a better bonding.

Use a regular caulking gun. Trim the applying end of the adhesive cartridge to apply a ⅛-inch continuous strip of adhesive at panel joints, and at the top and bottom plate surfaces. Place 3-inch long beads of adhesive, 6 inches apart, on intermediate studs. Do not skimp on adhesive. Beads must be at least 3 inches long.

Put the panel in place, setting it ¼ inch from top. Tack it at the top, using a color-matching nail. Double check to be sure the panel is properly placed. Place the panel on the adhesive, using firm, uniform pressure. This spreads the adhesive bead evenly between studs and panel. Grasp the bottom of the panel at the edges, and slowly pull the panel out and away from the studs. After the required time (follow the adhesive manufacturer's directions) press the panel again at all stud points. After the required time, recheck the panel. Go over all intermediate stud areas and edges, applying pressure to ensure firm adhesion and an even panel surface.

When paneling around a door opening, follow the same procedure as for window openings. Lay out the door opening pattern on the panel and cut it out. (See Figure 9-19.) After the panel is cut, check for fit. Moldings provide the finishing touch.

Vapor Barrier Goes Up

Begin Wall Assembly On Floor

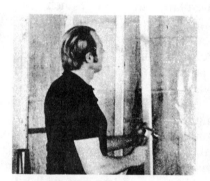
Apply Furring Strips

Two Studs For Corner

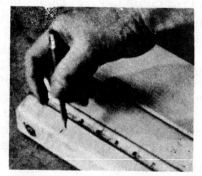
Measure 2 x 4 Actual Width

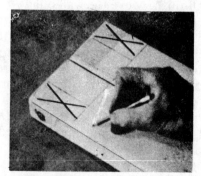
Mark Positions

Paneling a Basement
Figure 9-18

Mark For Corner Studs

Mark Off Centers

Layout Marked Plates

Start Nailing Corners

Nail Corners Securely

Lift Wall Into Place

Check Plumbness

Tack Temporarily

**Plumb and Mark Guideline
on Floor**

**Paneling a Basement
Figure 9-18 (continued)**

Recheck Plumbness

Frame Windows

Install Electrical Boxes

Add Insulation

Install Partition

Frame Doorways

Adjust Paneling to Room Temperature

Match Panels

Power Saw Back.

Paneling a Basement
Figure 9-18 (continued)

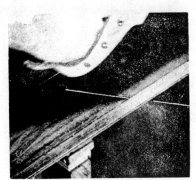

Handsaw Front

Start at Corner

Find True Plumb

Use a Wedge

Tack in Place

Measure Outlet Cut-Out

Use Box as Pattern

Drill Holes

Cut-Out Opening

Paneling a Basement
Figure 9-18 (continued)

Measure Window Cut-Out

Check For Fit

Wire Brush Studs

Apply Adhesive

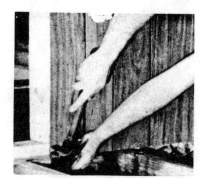

Tack Panel Into Place

Press Panel To Contact Adhesive

Paneling a Basement
Figure 9-18 (continued)

Measure For Door Cut-Out
Figure 9-19

Wood and Fiberboard Paneling

Wood and fiberboard paneling are tongued and grooved and available in various widths and patterns. Wood is limited to 8 inches in width. Fiberboard is available in 12- and 16-inch widths. This "plank" paneling is usually applied vertically but can be applied horizontally for special effects. When vertically applied, 1 x 4-inch nailing strips or 2 x 4-inch blocking at 24-inch intervals is recommended. The wood paneling can be sanded and finished after installation.

Paneling over masonry requires furring or nailing strips as previously discussed for sheetrock and plywood panels. The strips may be applied with concrete nails or a stud driver. The strips or furring material should be treated with a wood preservative to prevent deterioration by moisture.

Horizontally installed paneling may be nailed directly to the studs. The tongue and groove panels are blind nailed through the tongue. The wider "planks" may require face nailing in addition to blind nailing. The appropriate size staple may be used in the tongue of fiberboard in lieu of nails. Where adhesive is used, the only nailing required is the blind nailing.

Ceiling Tile

Ceiling tiles and panels for suspended ceiling

systems come in several sizes. The 12 x 12-inch size seems to be the most popular and practical for the average size room. Ceiling tile can be installed directly to an existing ceiling with cement, over an existing ceiling, or directly to the ceiling joists with furring strips. Ceiling panels are suspended on bars or strips. We'll look at each method.

Cement Method

Determine the number of 12 x 12-inch tiles you'll need for the room. Multiply the length of the room by the width to determine the number of tiles needed and add one extra tile or 12 inches to each dimension. This allows for cuts and any required fitting. For example:

Room length:	12' + 1' = 13'
Room width:	10' + 1' = 11'
	(13' x 11' = 143)
Total tiles needed	143

If the room doesn't measure an exact number of feet, use the next highest number, as follows:

	Actual Dimensions	Multiply
Room length	12'-8"	13' + 1' = 14'
Room width	10'-5"	11' + 1' = 12'
		(14' x 12' = 168')
Total number of tiles needed:		168

For a balanced ceiling appearance, the border tiles on opposite sides of the room should be the same width, preferably more than one-half the width of the 12-inch tile. To determine border tile widths, measure the short wall of the room first. If the wall does not measure an exact number of feet, add 12 inches to the odd inches left over and complete the simple calculations as shown in Figure 9-20. To install the tile, follow the procedures in Figure 9-21.

Furring Strip Method

Whether you use the cement or furring strip method, you'll want to know how to "square" the room. Some rooms are badly out of square and "squaring" the room before you start will ensure an acceptable ceiling.

Once you've been through a job using the cement procedure, you're ready to tackle something a little more difficult. The procedure to follow when using the furring strip method is shown beginning with Figure 9-22.

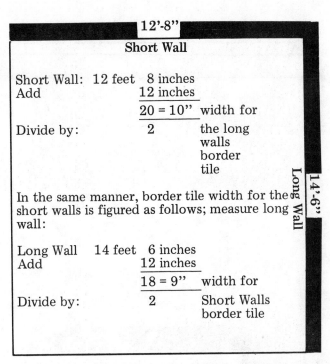

Short Wall

Short Wall:	12 feet	8 inches	
Add		12 inches	
		20 = 10"	width for
Divide by:		2	the long walls border tile

In the same manner, border tile width for the short walls is figured as follows; measure long wall:

Long Wall	14 feet	6 inches	
Add		12 inches	
		18 = 9"	width for
Divide by:		2	Short Walls border tile

Determine the Size of the Border Tile
Figure 9-20

Ceiling tiles should be cemented only to a sound, level ceiling. Remove any loose wallpaper or flaking.

Installing Ceiling Tiles
Figure 9-21

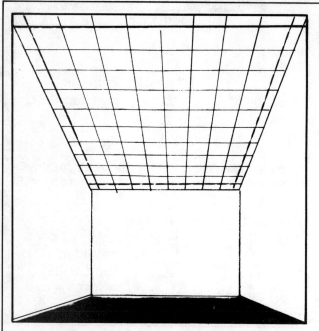

Measure the room to ensure that border tiles will be the same width on opposite sides.

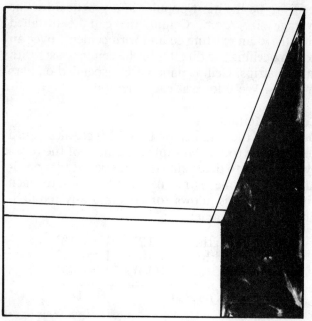

Snap a chalk line to align the first row of border tiles along the long and short walls. (At this time, we'll assume that the room is square. We'll deal with out-of-square rooms shortly.)

Cut your first border tile to size. Cut face up. Use a sharp utility knife.

Place 5 thin daubs of cement on the first border tile. Follow the cement manufacturer's instructions.

Installing Ceiling Tiles
Figure 9-21 (continued)

Fit the border tile into position in the corner of the room. Note the position and run of tongue and flange. If installation is a checkerboard pattern, separate the tile into two stacks according to the grain/tongue and flange run. To avoid contraction and expansion problems, store the tile in the room 24 hours before installation.

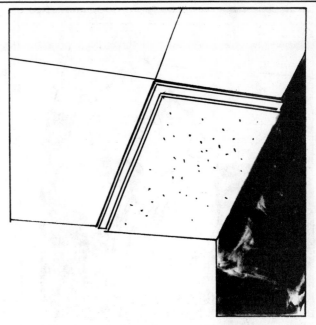

Make sure the stapling flange is lined up on the chalk line.

Staple each flange to hold the tile in position while the cement dries.

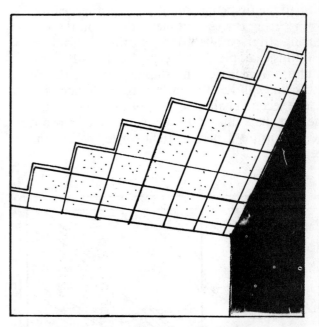

Work across the ceiling, installing two or three border tiles at a time. Fill in between the border tiles with full-size tiles.

Installing Ceiling Tiles
Figure 9-21 (continued)

When you reach the opposite wall, measure and cut each tile to fit.

Once all the tiles are in place, nail up the molding to finish the ceiling.

Installing Ceiling Tiles
Figure 9-21 (continued)

Snap the first chalk line on the existing ceiling (if you're using the cement method), or on the wood furring strips parallel to the starting sidewall.

In this example, the sidewall border is 10 inches. Adding ½ inch for the stapling flange, the first chalk line will be 10½ inches from the sidewall.

The second chalk line must be at a right angle to the first chalk line or the tiles won't line up properly. To ensure a perfect right angle, follow these steps:

Measure the width of the end wall border tile plus ½ inch, and mark Point A on your first chalk line. (See Figure 9-23.) Next, measure exactly 3 feet from Point A, and mark Point B.

Now measure exactly four feet from Point A and strike an arc. (See Figure 9-24.) From Point B, measure exactly five feet toward the first arc, and strike a second arc. The point where the two arcs intersect is Point C. Snap the second chalk line through Points A and C. Check the squareness of

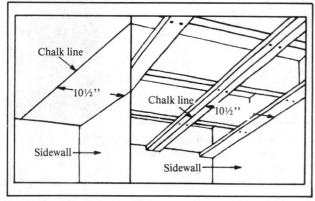

Measure for Furring Strips
Figure 9-22

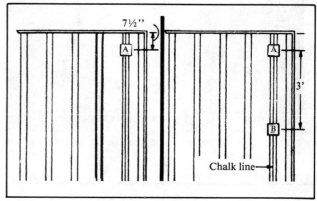

Second Chalkline Right Angle to First
Figure 9-23

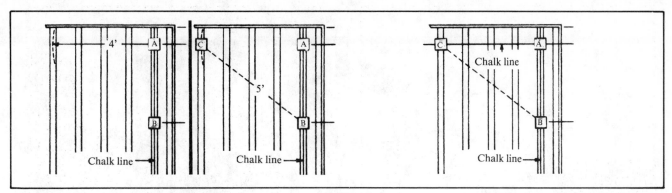

Strike the Arc
Figure 9-24

the two chalk lines with a carpenter's square.

Nail furring strips to the ceiling, perpendicular to the ceiling joists, 12 inches o.c. from the starter course furring strip. Install the last strip against the sidewall. Now measure the first border tile. (See Figure 9-25.) Cut the tile using a carpenter's square and a utility knife, as shown in Figure 9-26.

Fit the border tile in position in the corner of the room. Make sure the stapling flange is lined up on the chalk line, and staple the tile into place. Be sure the run of your pattern is even. Work across the ceiling, as shown in Figure 9-27. Where necessary, measure and cut the tile to fit electrical ceiling boxes (Figure 9-28).

Once the tiles are in place, nail up the wall molding to finish the ceiling. (See Figure 9-29.)

Suspended Ceilings

Remove ceiling panels from the package 24 hours before installing. This allows them to adjust to the normal conditions of the area where the installation will take place.

Measure the room and draw a diagram on a layout sheet (Figure 9-30). With a dotted line, indicate locations of ceiling joists (Figure 9-31). Decide which way you want the 2 x 4-foot panels to run. Usually, the length of the panel is installed parallel to the short wall, as shown in Figure 9-32. Convert the room's short-wall dimensions to inches. Divide this measurement by 48 inches if the panel length will run parallel to the short wall. Divide by 24 inches if the panel length will run parallel to the long wall. (For 2 x 2-foot panels, divide by 24 inches.)

First Border Tile Measured for Cut
Figure 9-25

Cutting the Tile
Figure 9-26

Work Across the Ceiling
Figure 9-27

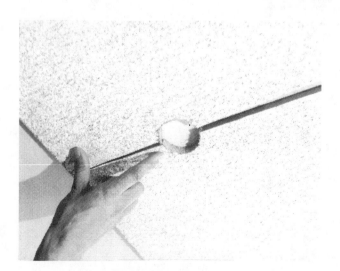

Cut-Out for Ceiling Box
Figure 9-28

Ceiling Molding Finishes the Job
Figure 9-29

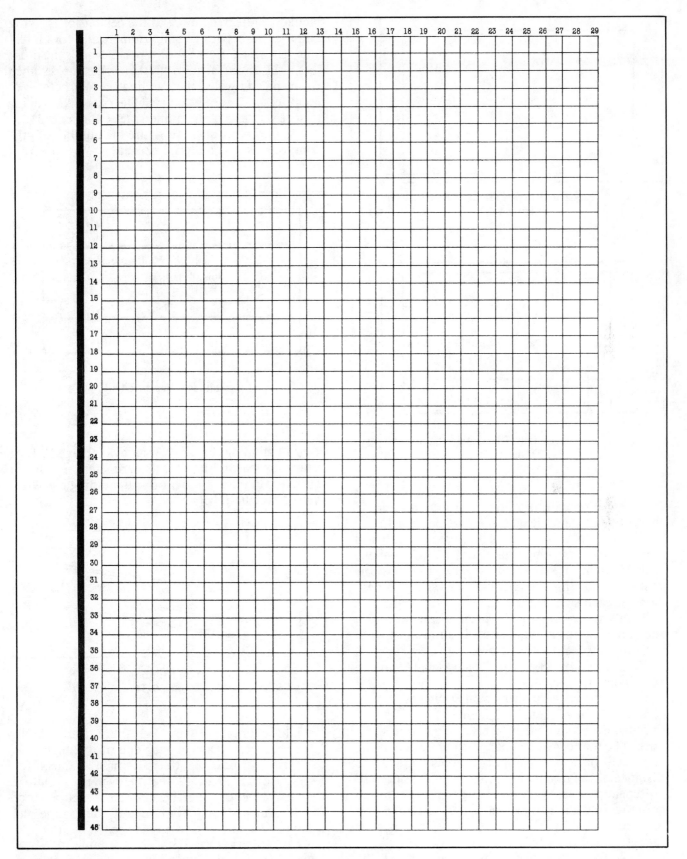

Room Layout Sheet
Figure 9-30

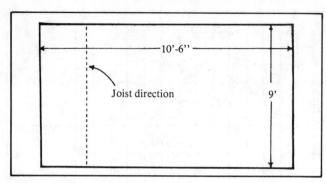

Lay Out Border Tile and Run
Figure 9-31

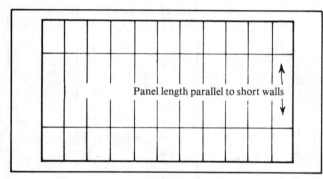

Lay Out Tile Runs
Figure 9-32

Take the remainder of this division and add 48 inches if the panel length will run parallel to the short wall. Add 24 inches if the panel length will run parallel to the long wall. (For 2 x 2-foot panels, add 24 inches.) Round off to the closest whole number. Half this figure equals the border dimensions at each side of the room.

Here's an example. For a room 9 feet wide with the panel length running parallel to the short wall:

Convert 9 feet to 108 inches.

108 divided by 48 equals 2 panel lengths with a remainder of 12 inches.

12 inches plus 48 inches equal 60 inches.

60 divided by 2 equals 30 inches.

Thus, the border panels at each side will be 30 inches. The number of full panels running across the room will be one. This dimension also equals the distance of the first main runner from the sidewall.

Repeat these calculations using the length of the room to find the end border panel size. Now you are ready to draw the gridwork pattern on your layout. (See Figure 9-33.)

Draw the first and last main runners at the border tile distance from the sidewalls and perpen-

dicular (90-degree angle) to the ceiling joists. Now add the "in-between" main runners at 4-foot intervals. Using a different-color pencil, mark the cross tees—this time starting with the border tile distance from the end walls, then adding the "in-between" cross tees at 2-foot intervals. The cross tees will intersect with the main runners. (See Figure 9-33.)

Use the layout sheet to figure the number of grid components and material needed.

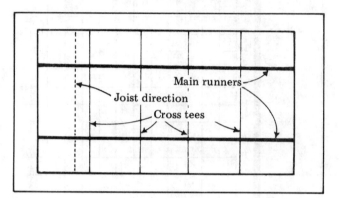

Main Runner Lay Out
Figure 9-33

Wall molding (Figure 9-34) is available in 10-foot lengths. Measure the room's perimeter, and divide by 10 to find the number of wall molding pieces required for the job.

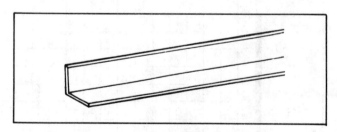

Wall Molding
Figure 9-34

Main runners (Figure 9-35) are available in 12-foot lengths. Tabs at each end of the main runner make it possible to join main runners for lengths longer than 12 feet. However, no more than two sections can be cut from each 12-foot main runner. Using your layout, determine the number of main runner pieces needed for the job.

Cross tees (Figure 9-36) are available in 2-foot and 4-foot lengths with connecting tabs at each end. Only two border cross tees can be cut from a

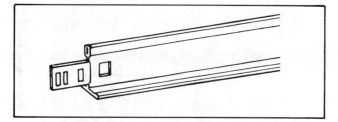

Main Runner
Figure 9-35

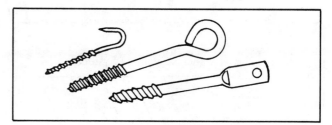

Wire Fasteners
Figure 9-38

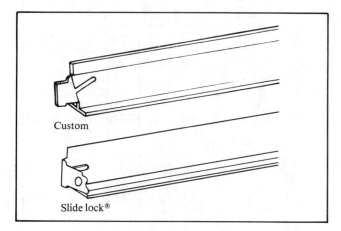

Custom

Slide lock®

Cross Tees
Figure 9-36

standard-length cross tee. Count the number of cross tees indicated on your layout.

You will need enough hanger wire (Figure 9-37) to suspend main runners every four feet. The wire at each suspension point should be 6 inches longer than the distance between the support joint and new ceiling. For most residential jobs, 16-gauge wire will do the job. For light commercial jobs, use the heavier 12-gauge wire.

Wire fasteners (Figure 9-38) are needed at each support point to attach wire to the existing ceiling or joist. Make sure fasteners are long enough to

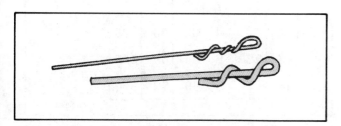

Hanger Wire
Figure 9-37

provide strong support. Place them every four feet and on each side of every main runner joint.

Count the number of ceiling panels indicated on your layout. Allow for cutting border panels. The panels are not designed to support insulation materials.

Figures 9-39A through 9-39D give charts for material requirements for various size rooms.

To begin the installation, measure the new ceiling height in each corner of the room. Allow a minimum of 3-1/2 inches below the level of the existing ceiling or exposed joists. If light fixtures are to be used above the new ceiling, allow for this. Snap a chalk line through the measurement marks. Be sure to check the levelness of the chalk line.

Next, nail wall molding to the wall along the chalk line mark. Treat corners as shown in Figure 9-40.

Following the layout, locate the position of the first main runner. At this location, snap a chalk line on the old ceiling or joists. Continue across the room, snapping main-runner chalk lines every 4 feet, parallel to the first main-runner line. Fasten wire fasteners (screw eye) into the joist every four feet where the ceiling joists intersect with the main-runner chalk line. Attach hanger wire to each screw eye. Securely wrap the wire around itself at least three times.

Pick the corner where you'll start your installation. Now, at the border-panel distance from each wall, stretch two reference strings, each at a perfect 90-degree angle to the other. Be sure to attach the strings right below the wall molding. These reference strings will serve as a guide for cutting main runners and cross tees. (See Figure 9-41.)

Using reference string AB as a guide, measure each main runner individually (do not use the first main runner as a pattern for cutting) and cut the main runner so that the first reference string falls on a cross tee slot. (See Figure 9-42.)

Now locate the wire-support hole (a small, round hole at the top of the main runner) farthest from

2' x 4' Panel Suspended System

Room Length (Feet)	10' Wall Moldings	12' Main Runners	4' Cross Tees	Number of Panels	Square Feet
6	4	1	6	9	72
8	4	2	9	12	96
10	4	2	12	15	120
12	5	2	15	18	144
14	5	3	18	21	168
16	6	3	21	24	192
18	6	3	24	27	216
20	6	4	27	30	240
22	7	4	30	33	264
24	7	4	33	36	288
26	8	5	36	39	312
28	8	5	39	42	336
30	8	5	42	45	360
32	9	6	45	48	384
34	9	6	48	51	408
36	10	6	51	54	432
38	10	7	54	57	456
40	10	7	57	60	480
42	11	7	60	63	504
44	11	8	63	66	528
46	12	8	66	69	552
48	12	8	69	72	576

For a 10-Foot Wide Room
Figure 9-39A

2' x 4' Panel Suspended System

Room Length (Feet)	10' Wall Moldings	12' Main Runners	4' Cross Tees	Number of Panels	Square Feet
6	4	1	6	9	72
8	4	2	9	12	96
10	5	2	12	15	120
12	5	2	15	18	144
14	6	3	18	21	168
16	6	3	21	24	192
18	6	3	24	27	216
20	7	4	27	30	240
22	7	4	30	33	264
24	8	4	33	36	288
26	8	5	36	39	312
28	8	5	39	42	336
30	9	5	42	45	360
32	9	6	45	48	384
34	10	6	48	51	408
36	10	6	51	54	432
38	10	7	54	57	456
40	11	7	57	60	480
42	11	7	60	63	504
44	12	8	63	66	528
46	12	8	66	69	552
48	12	8	69	72	576

For a 12-Foot Wide Room
Figure 9-39B

2' x 4' Panel Suspended System

Room Length (Feet)	10' Wall Moldings	12' Main Runners	4' Cross Tees	Number of Panels	Square Feet
6	4	2	8	12	96
8	5	2	12	16	128
10	5	3	16	20	160
12	6	3	20	24	192
14	6	4	24	28	224
16	6	4	28	32	256
18	7	5	32	36	288
20	7	5	36	40	320
22	8	6	40	44	352
24	8	6	44	48	384
26	8	7	48	52	416
28	9	7	52	56	448
30	9	8	56	60	480
32	10	8	60	64	512
34	10	9	64	68	544
36	10	9	68	72	576
38	11	10	72	76	608
40	11	10	76	80	640
42	12	11	80	84	672
44	12	11	84	88	704
46	12	12	88	92	736
48	13	12	92	96	768

For a 14-Foot Wide Room
Figure 9-39C

2' x 4' Panel Suspended System

Room Length (Feet)	10' Wall Moldings	12' Main Runners	4' Cross Tees	Number of Panels	Square Feet
6	5	2	8	12	96
8	5	2	12	16	128
10	6	3	16	20	160
12	6	3	20	24	192
14	6	4	24	28	224
16	7	4	28	32	256
18	7	5	32	36	288
20	8	5	36	40	320
22	8	6	40	44	352
24	8	6	44	48	384
26	9	7	48	52	416
28	9	7	52	56	448
30	10	8	56	60	480
32	10	8	60	64	512
34	10	9	64	68	544
36	11	9	68	72	576
38	11	10	72	76	608
40	12	10	76	80	640
42	12	11	80	84	672
44	12	11	84	88	704
46	13	12	88	92	736
48	13	12	92	96	768

For a 16-Foot Wide Room
Figure 9-39D

the starting end wall and closest to the last hanger wire. Mark the hole on the first main runner. (See Figure 9-43.)

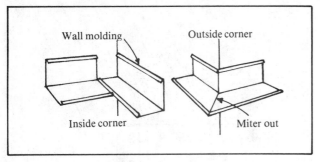

Corner Treatment
Figure 9-40

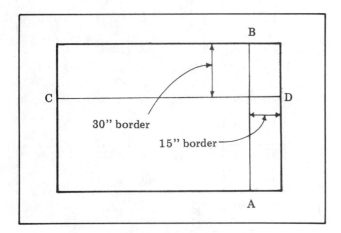

Border Tile Layout
Figure 9-41

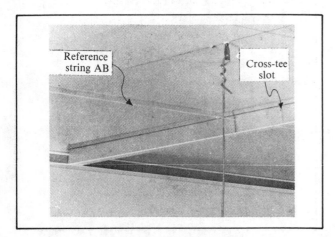

Use Reference String as Guide
Figure 9-42

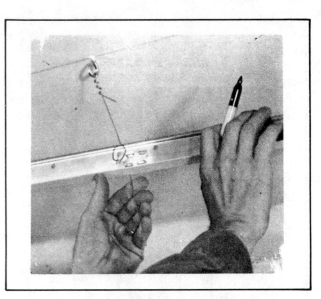

Locate Wire Support Hole
Figure 9-43

Carry the main runner to a sidewall and lay it on the wall molding with the cut end butting against the end wall. Mark the sidewall through the top of the marked support hole. Remove the main runner and drive a nail into the point just marked.

Repeat on the opposite side of the room. Stretch a string from nail to nail, and align each hanger wire so it intersects with the string. (See Figure 9-44.) Make a 90-degree bend in hanger wires where they meet the string.

Now rest the cut end of the first main runner on

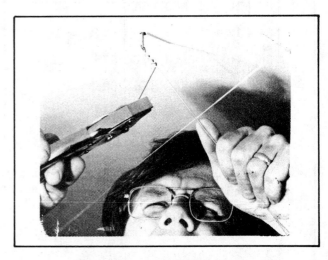

Make 90° Turn in Wire Where it Touches String
Figure 9-44

the end-wall molding. Take hold of the prebent hanger wire and run it through the support hole. Secure it by bending it up sharply and twisting the wire around itself at least three times. (See Figure 9-45.) Align the rest of the hanger wires to the appropriate support holes, make a 90-degree bend where they intersect the top of the support holes, and attach as before.

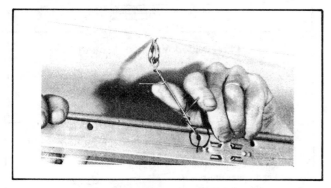

Secure the Main Runner
Figure 9-45

Connect as many main-runner sections as needed to the other end of the room. Cut the excess main runner with snips, and use this piece to start the next row. Repeat this same procedure for the remaining rows, bending and attaching hanger wires as before. Be sure to cut each main runner so that a cross tee slot aligns with reference string AB. For better support, make sure a hanger wire is located close to the point where two main runners are connected. Position an additional hanger wire at the connecting point if necessary. Check to ensure that main-runners are level. (See Figure 9-46.)

When installing cross tees, begin by placing the full 4-foot cross tees in rows away from the borders. When you have them in place, lay in a few full ceiling panels. This will help to stabilize and square the entire grid system as you continue to work.

Cut the first row of cross tees for the border tile, measuring each individually and using reference string CD as a guide. To measure each cross tee, align the edge of the first main runner with the reference string under it. Measure from the sidewall to the near edge of the main runner. Cut the cross tee to this dimension and install. Continue in this manner, cutting each cross tee individually along the first wall. Follow the same procedure to cut tees for the last row along the opposite border.

Make Certain Main Runners are Level
Figure 9-46

Attach the 4-foot cross tees to main runners across the room at 2-foot intervals. Lock them in. (See Figure 9-47.) For 2 x 2-foot panels, attach cross tees at the midpoints of 4-foot cross tees. (See Figure 9-48.)

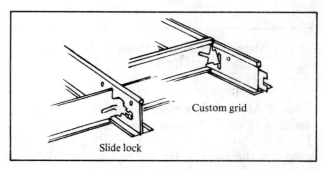

Custom grid

Slide lock

Connect Cross Tees to Main Runner
With Either Type of Tee
Figure 9-47

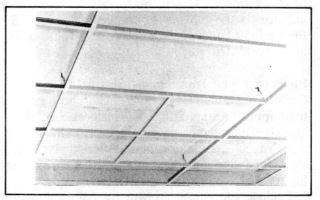

For 2 x 2 Foot Tile Connect Cross Tees
Figure 9-48

Measure and cut border panels individually. Using a leftover cross tee or main runner section as a straightedge, cut panels face up with a very sharp utility knife. To drop ceiling panels into position, tilt slightly, lift above the framework, and gently rest the edge on the cross tee and main runner.

If translucent plastic lighting panels are to be installed, the panels can be cut by scoring repeatedly with a sharp knife until the panel is completely cut into. The panel is installed with the glossy or smooth side up. Where fluorescent lighting fixtures are used above the grid, the tubes should be centered over the panel. Surface-mounted fixtures should be installed as shown in Figure 9-49.

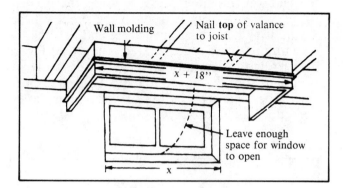

**Three-Sided Valance Around
Basement Window
Figure 9-50**

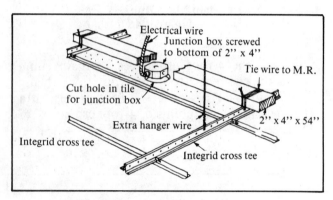

**Method for Surface Mounted Fixtures
Figure 9-49**

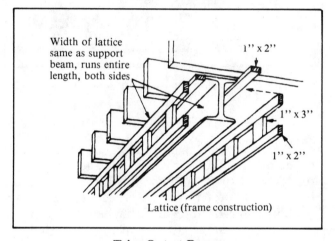

**Trim Out at Beam
Figure 9-51**

Boxing Around Basement Windows
Build a three-sided valance around each window. 1/4-inch plywood for the top, and 1 x 6-inch white pine for the three sides are recommended. Make sure that the valance is wide enough to allow the window to open and long enough to provide for open drapes. A length of 18 inches is usually sufficient for the drapery—about 9 inches for each side of the window. Attach the top of the completed valance to the bottom of the ceiling joists. (See Figure 9-50.) Install the appropriate wall molding at the level desired.

Boxing Around Iron Support Beams
Construct wooden lattices to attach to both sides of the support beam. Use 1 x 2-inch wooden strips and 1 x 3-inch center supports spaced 16 inches o.c. to construct each lattice. Nail lattices to 1 x 2-inch cleats installed on each side of the beam. (See Figure 9-51.)

Enclose the support beam by nailing a finish material that matches the room's walls to each lat-

tice. Attach the same material to the exposed face of the beam by driving nails into the base of the lattice frames. To finish the box, attach the corner moldings as illustrated in Figure 9-52.

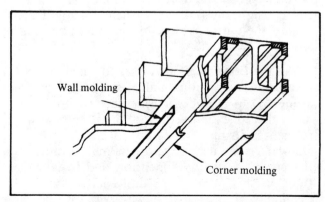

**Finish Treatment of Beam
Figure 9-52**

Snap a chalk line on the finished box at a point of the new ceiling height, and nail appropriate wall molding along the line.

In houses with heating and cooling air ducts, the ducts often run parallel to the support beams. Lattice work may be constructed to enclose both support beam and ducts. (See Figure 9-53.)

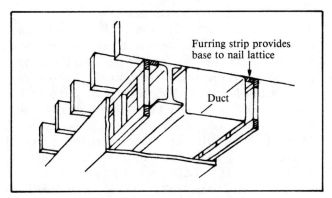

Enclosing Beam and Duct
Figure 9-53

Boxing Around Basement Stairways

Nail 1 x 3-inch cleats into the ceiling joists. The distance from the old ceiling to the new ceiling will determine the width of the valance material. 1 x 4-inch or 1 x 6-inch white pine is most commonly used.

Nail the valance into the cleats. Cover the seam with standard molding. Install the wall molding at the height desired, as shown in Figure 9-54.

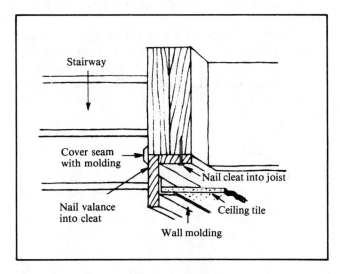

Finishing at Stairwell
Figure 9-54

Interior Trim

Interior trim is the "finish" of a room. It "molds" the breaks and intersections into one unit. Figure 9-55 shows the basic moldings and where they are used.

Many of the older houses have molding long ago discontinued. Matching it will be impossible in most cases. So, if you're planning to remove any of it for reuse, proceed carefully.

When selecting molding for the job, choose smooth, close-grained materials. Northern white pine, spruce, mahogany, and ponderosa pine are good. Gum, yellow poplar and birch are hard, resist rough treatment, and have a smooth texture. Birch, ash, oak, maple, cherry and walnut are excellent woods for a natural finish. They are uniform in color, density and hardness. Prefinished PVC molding is shaped and finished to resemble wood. It is vinyl and will not crack or chip. PVC molding has good flexibility and impact resistance. Various colors are available to match prefinished paneling.

The miter box and coping saw are used to cut molding where one molding meets the profile of another molding, usually creating a 90-degree angle (inside and outside corners at ceiling or floor line; at door and window casings.) To form a tight right angle, set the miter box at 45 degrees and trim each of two mitering members at opposite angles. The resulting profile of each molding member should fit together perfectly.

Ceiling or crown molding and base molding are best mitered and coped for a good fit in inside corners. The first piece to go into the corner is cut straight (90 degrees) and the end is butted to the wall. The second piece is mitered and coped and fitted snugly against the first piece. Outside corners are mitered at 45 degrees and fitted. In-between joints are mitered 45 degrees for butting together. When mitering ceiling molding, place it in the miter box upside down and reverse to the way it goes up.

Door and window casing is cut at 45 degrees. The top piece of the casing is cut 45 degrees on both ends. The side pieces are cut 45 degrees on one end, with the other end cut 90-degrees to fit flush against the floor or against the window stool.

The shoe and base molding fit against the base of the wall, flush with the floor line. The shoe, because of its flexibility, is usually nailed to the base to conceal uneven floors. Where carpeting is installed on the floor, shoe molding is not used.

Corner guard moldings, which cover and protect raw panel edges at inside and outside corners, are trimmed 90 degrees and applied with nails or

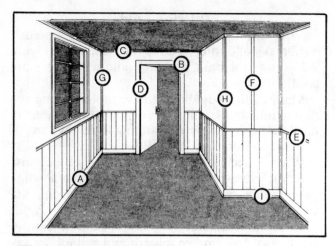

Where to Install Moldings

Note: The letters below are keyed in the illustration to
to indicate where each molding type can be used.

A	Base	To trim along floor line; protects paneling from bumps.
B	Casing	For trimming doors, windows, and other openings.
C	Ceiling	To trim along ceiling line.
D	Stop	To prevent doors from swinging through frame and to hold windows in their tracks.
E	Plycap	Also called "chair rail"; used for trimming the top edge of panels in wainscoting.
F	Batten	Used with ungrooved paneling to conceal joints.
G	Inside Corner Guard	Covers seam where two walls come together.
H	Outside Corner Guard	Caps raw edges of panels, protects corners from impact damage.
I	Base Shoe	Decorative finishing molding applied at intersection of base molding and floor.

Where to Install Moldings
Figure 9-55

adhesive to fit flush against the ceiling and base moldings.

Moldings can be used as a decorative trim for walls, doors and cabinets. A variety of types are available for many possible decorative treatments.

Summary

There are interior finish materials I didn't touch on. Stone, brick and plaster are three. If you're skilled in one of these areas, you'll do these types of walls when such work is required. If you're not, then I can't help you. I've never plastered a wall or put stone on one. And I'm not about to attempt giving instructions in these areas. Unless the work fascinates you, sub it out. A man can work himself to death trying to do everything. Besides, if you do try doing it all, you'll have to carry around so many different tools that your truck will look like a traveling hardware store.

Applying wallpaper is another subject I didn't touch on. It's no problem. Any builder can learn to hang wallpaper with a little practice. Wallpaper comes and goes in popularity. Right now it is used more than it was a few years ago. Tomorrow, who knows? It may one day return to every room in the house, plus the ceiling. Wallpaper makes a pretty wall and offers some beautiful decorative themes. When used in a bathroom properly, it can really spruce things up. The dining room is another room where it can be used.

Now that you've come this far, do you think you can put up sheetrock or paneling? How about that fancy suspended ceiling? Think you can handle that?

If not, go back and study this chapter again. There's not a ceiling or a wall job around that you can't handle. The most important thing is laying the job out, knowing in advance what you're going to do and then doing it right. That's the key to it all.

Old Carpenter's Rule

Old carpenters never quit. They just forget to finish.

10

Doors and Windows

Ever notice how an ordinary looking woman can put on a pair of those big designer glasses and look like a model in a glamor magazine—a little shift of emphasis and a transformation is achieved.

You can do the same thing with a house—transform it by just shifting the emphasis. Change the exterior doors, rearrange the windows, install larger or smaller windows—even just taking out the old, worn-out windows and doors vastly improves the looks of the place.

People notice things like doors and windows. They set the tone of the house and help others characterize it as attractive or unattractive. That's why windows and doors are a good place to start when remodeling any home.

Doors

Doors can be a problem. They can stick, fail to latch, or drag over the floor. Who knows why a door properly installed and operating nicely one day will bind or stick the next? But this can be remedied. If the door binds, locate the point of bind and plane it down. The top and the side can be planed without removing the door. Plane off enough to do the job. You don't want to have to redo it after you've re-finished it.

If the bind is near the latch, you might have to remove the door and plane off the hinge side. There's not much planing you can do in the latch area because the latch cannot be set back any appreciable distance without extending the hole for the lock-set. And you can't do much of that because the coverplate will only cover so much hole.

Go ahead and plane off the hinge side. You might have to rout a deeper set for the hinges, but that's less trouble than trying to readjust the lockset. Now, if you set the hinges too deeply, the door will bind on that side. If the hinge-rout is too deep, take the hinge off and add a cardboard filler or shim under it to bring it up. If the door latch does not catch when closed, check to see if it's working properly. It might be binding and not extending out fully. If it's working properly, the problem is in the strike plate. Take it off and shim it out. Not too far, or it might bind against the door. Use a cardboard shim.

Frequently, the hinge screws have been in and out so many times they have lost their grip. If so, use a daub of epoxy glue on the screws before you reset them. (If the screw holes are wallowed out excessively, mix a little fine sawdust with the epoxy glue.)

You can spend a lot of time taking the kinks out of old doors and windows. Use caution in setting a price for this type of work or you'll end up working for less than the minimum wage.

Exterior doors take a lot of punishment when they're exposed to the weather. The exterior door should be the solid-core or panel type. Hollow-core does not stand up to the elements and should be used for the interior only.

Standard height for doors is 6 feet 8 inches. The

main entrance door should be 3 feet wide. The kitchen or back door should be at least 2 feet 8 inches wide, though the 2-foot 6-inch width is often used.

You're going to replace many exterior doors. Many people like to change their exterior doors now and then because it offers a "new look" to the old homestead. A door is like a photograph—it needs a nice frame. A crooked frame detracts from the photo. The same thing happens when an attractive door is set in a warped frame. Besides, trying to fit a new door in an off-square frame can be a frustrating task. You'll wear out yourself and the door by hanging and re-hanging the thing.

Stick in a new frame. You can get the door and frame prehung and ready to install. (See Figures 10-1A and 10-1B.) Notice the construction details.

If you're framing a new door opening in a wall, the rough opening should be the height of the door plus 2¼ inches above the finished floor. The width should be the width of the door plus 2½ inches. You can frame the door as illustrated in the figures.

After sheathing and panel siding are placed over the framing, hang the frame. Use a non-hardening caulk on each side and over the opening. When the frame is installed, the exterior casing will fit over the ribbon of caulk.

Where panel siding or a similar exterior material is used, caulk under the sill and cover with appropriate molding, such as quarter-round or 1 x 4-inch board.

Prehung units are just that. The door is hung to the frame on hinges. The holes for the lock-set are already cut. The lock-set is not included in the cost of the unit, but is easily installed by following the instructions included in the carton.

Prehung door units are the best thing to happen to carpentry since the invention of the power saw. It's more economical to run down to your supply house and get a prehung unit than to attempt to build the frame and hang the door yourself. And the job will probably be a whole lot neater, too.

Those little cardboard pieces that are stapled on the inside of the door jambs between the door edge and the jamb on prehung units are there for a purpose. They ensure that you don't fit the door too tightly within the frame when you hang it. They're spacers. After you hang the door and nail it in place you can remove them. Bend the staples down or put the spacers in a trash container; the staples are sharp as needles and will stick through the sole of a shoe easily.

The prehung door and frame is installed as a unit. The door is not removed. The whole unit is installed plumb and square. Secure it in place by nailing through the side and head casing. Nail the hinge side first.

Most prehung exterior door units are made so that the unit may be installed resting on the subfloor. The underlayment and finished floor are fitted to the sill. Brick row-lock is laid to the underside of the sill on the exterior side in brick veneer construction.

There are low-cost plain doors and expensive, hand-carved doors. In fact, there is a door of almost any type and design that one might desire. The most common exterior doors are the panel and flush types. Both are available with lights. And there are insulated doors with insulated tempered glass lights. Figure 10-2 illustrates a few styles that are available.

Figure 10-3 illustrates the proper door clearance and hardware placement. There should be a 3/4-inch clearance at the bottom of the door if there is a forced air heating and cooling system. The framing for interior doors should have an opening which is the height of the door plus 2 inches. The opening width should be the door width plus 2½ inches. Figures 10-4 through 10-9 should be helpful for door framing, installation, and finishing.

Windows

What was said about off-square doors also applies to windows. Old windows that have deteriorated should be replaced. Odd-size window openings exist in most older houses. It's generally better to have windows custom made to fit such odd openings than to stud out for the new window. Decreasing the size of the opening to fit a new window is a lot of extra work, including interior and exterior wall finishing.

There are three basic window designs to choose from: sliding, fixed, and swinging. The window should be selected for view, amount of light and ventilation. The day of cheap energy is over. Cooling on hot days is fine but many are turning off the cooling units and opening the doors and windows for fresh air. Screens are back in style.

Storm doors and windows are essential in cold winter areas to keep the heating cost down. Weatherstripping and caulking is essential to eliminate heat loss and drafts.

Wood and metal windows are available in many types and styles. Select the best quality window and you can't go wrong. If you install a cheap-grade window, you might have to start repairing it before the job is finished. Figure 10-10 will help to select the type of window best suited for the job.

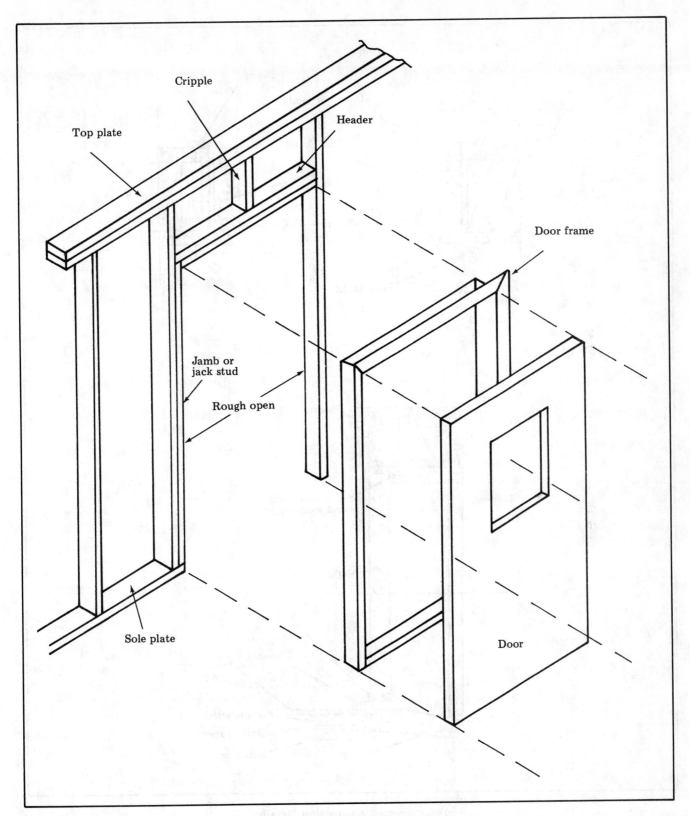

Prehung Units Eliminate Errors
Figure 10-1A

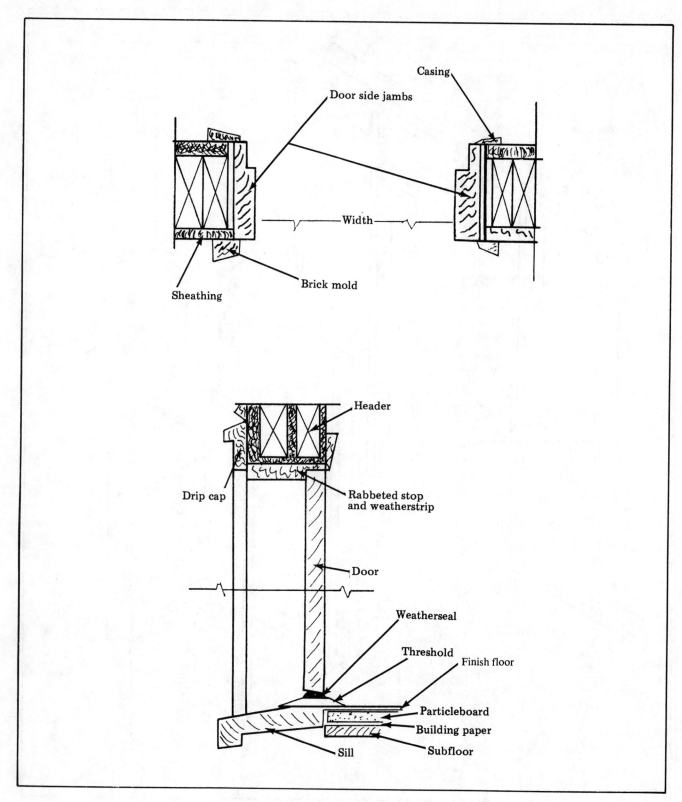

Door Frame Construction Details
Figure 10-1B

Exterior Doors: Panel Type and Flush Type
Figure 10-2

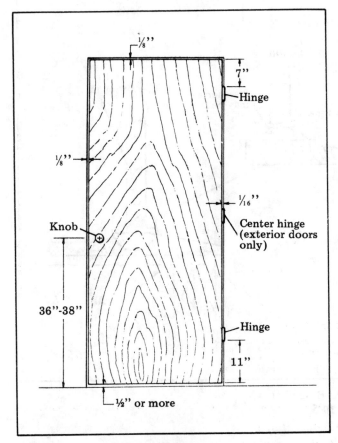

Door Clearances
Figure 10-3

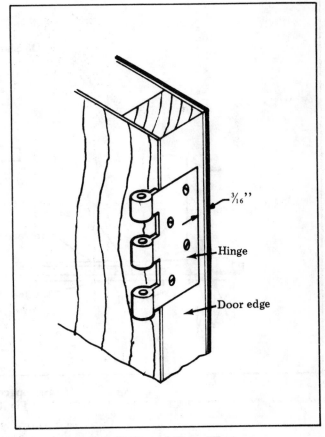

Installation of Door Hinges
Figure 10-4

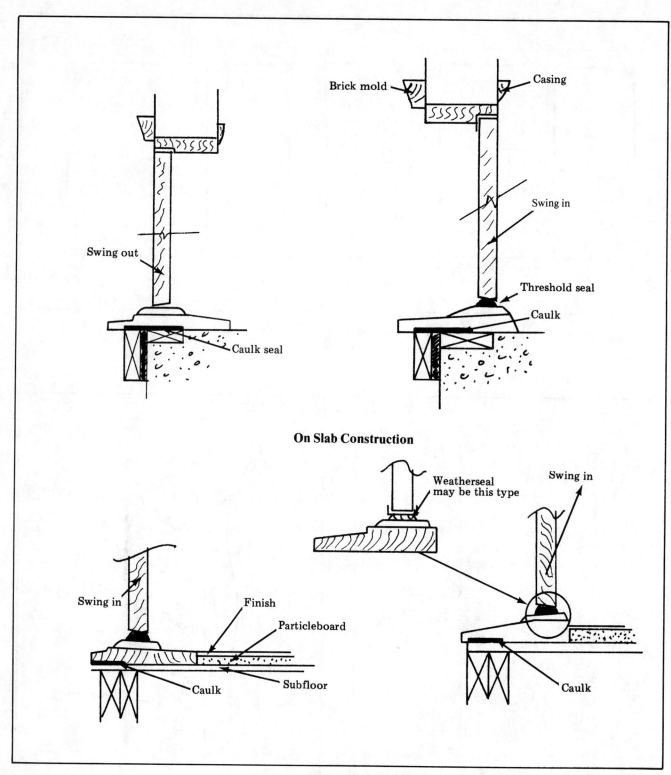

On Slab Construction

Crawl Space or Basement
Figure 10-5

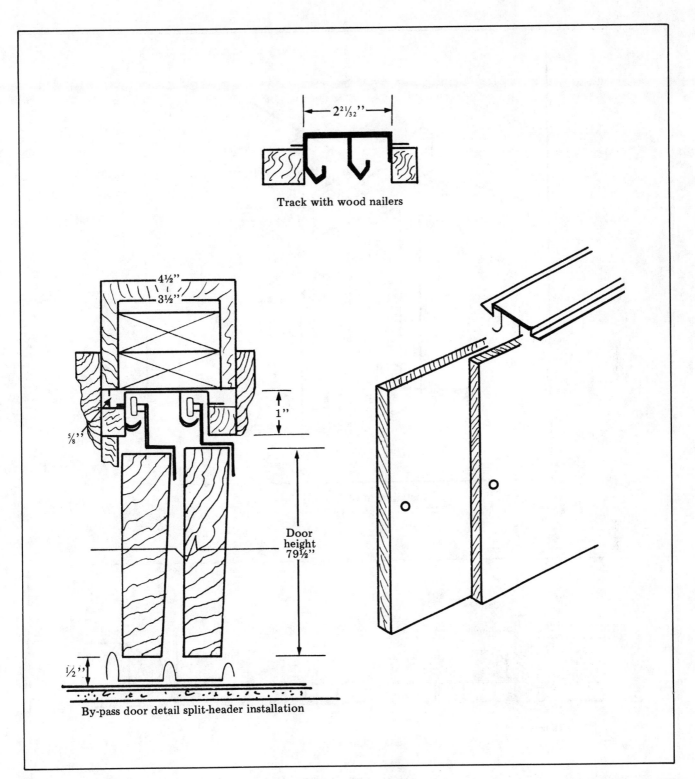

Track with wood nailers

By-pass door detail split-header installation

Double Sliding Interior Doors
Figure 10-6

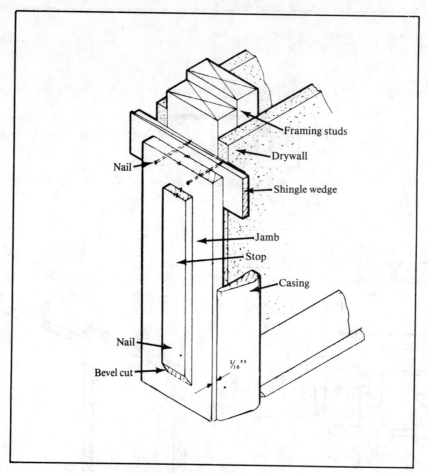

Installation of Door Trim
Figure 10-7

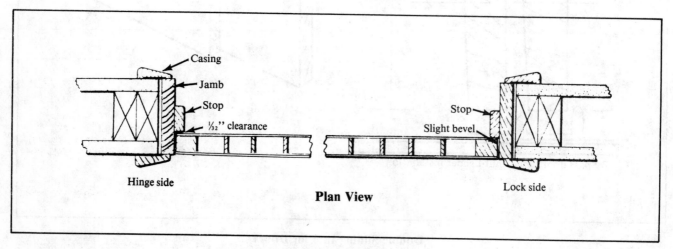

Door Stop Clearances (Plan View)
Figure 10-8

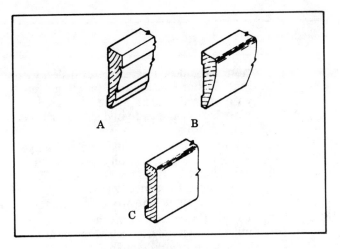

Styles of Door Casings
A, Colonial; B, Tear Drop; C, Plain
Figure 10-9

ing details for a window, and details of a double-hung window frame.

New windows are usually purchased as a complete unit, including sash, frame and exterior trim. Screens are usually available at extra cost. As a general rule, rough opening sizes are as follows:

- Double hung window (single unit)
Rough opening width—glass width plus 6 inches.
Rough opening height—total glass height plus 10 inches.
- Casement window (two sashes)
Rough opening width—total glass width plus 11¼ inches.
Rough opening height—total glass height plus 6⅜ inches.

Another method to determine rough opening is based on the sash opening size. Figure 10-14 illustrates this procedure.

If in doubt, always have the window on the job site *before* framing the rough opening. Allow one inch of play to square the window and you can't go wrong.

Figures 10-11 through 10-13 describe the materials used in window construction and the advantages and disadvantages of each, typical framing details for a window, and details of a double-

Window Type	How does it operate?	Is it easy to clean?	How is it for viewing?	How is it for ventilation?
Fixed	Does not open, so requires no screens or hardware	Outside job for exterior of window	No obstruction to view or light	No ventilation Minimum air leakage
Sliding Double-Hung	Sash pushes up and down. Easy to operate except over sink or counter	Inside job if sash is removable	Horizontal divisions can cut view	Only half can be open
Horizontal-Sliding	Sash pushes sideways in metal or plastic tracks	Inside job if sash is removable	Vertical divisions cut view less than horizontal divisions	Only half can be open
Casement (Side Hinged)	Swings out with push-bar or crank Latch locks sash tightly	Inside job if there is arm space on hinged side	Vertical divisions cut view less than horizontal	Opens fully. Can scoop air into house
Awning (Top Hinged)	Usually swings out with push-bar or crank. May swing inward when used high in wall	Usually an inside job unless hinges prevent access to outside of glass	Single units offer clear view. Stacked units have horizontal divisions which cut view	Open fully. Upward airflow if open outward; downward flow when open inward
(Bottom Hinged)	Swings inward, operated by a lock handle at top of sash	Easily cleaned from inside	Not a viewing window, usually set low in wall	Airflow is directed upward
Jalousie	A series of horizontal glass slats open outward with crank	Inside job, but many small sections to clean	Multiple glass divisions cutting horizontally across view	Airflow can be adjusted in amount and direction

Guide to window selection
Figure 10-10

Glass Area		Glass Area	
Single-strength glass	Suitable for small glass panes. Longest dimension—about 40 inches.		tion. Should be treated to resist decay and moisture absorption. Painting needed on outside unless frame is covered with factory-applied vinyl shield or other good coating.
Double-strength glass	Thicker, stronger glass suitable for larger panes. Longest dimension—about 60 inches.	Aluminum	Painting not needed unless color change is desired. Condensation a problem in cold climates unless frame is specially constructed to reduce heat transfer. Often less tight than wood frames.
Plate glass	Thicker and stronger for still larger panes. Also more free of distortion. Longest dimension—about 10 feet.		
Insulating glass	Two layers of glass separated by a dead air space and sealed at edges. Desirable for all windows in cold climates to reduce heating costs. Noise transmission is also reduced.	Steel	Painting necessary to prevent rusting unless it is stainless steel. Condensation a problem in cold climates.
Safety glass	Acrylic or plexiglass panels eliminate the hazard of accidental breakage. Panels scratch more easily than glass. Laminated glass as used in automobiles also reduces breakage hazard.	Plastic	Lightweight and corrosion-free. Painting not needed except to change color.
		Hardware	Best handles, hinges, latches, locks, etc. usually are steel or brass. Aluminum satisfactory for some items but often less durable. Some plastics and pot metal are often disappointing.
Wood	Preferable in cold climates as there is less problem with moisture condensa-		

Guide for window materials
Figure 10-11

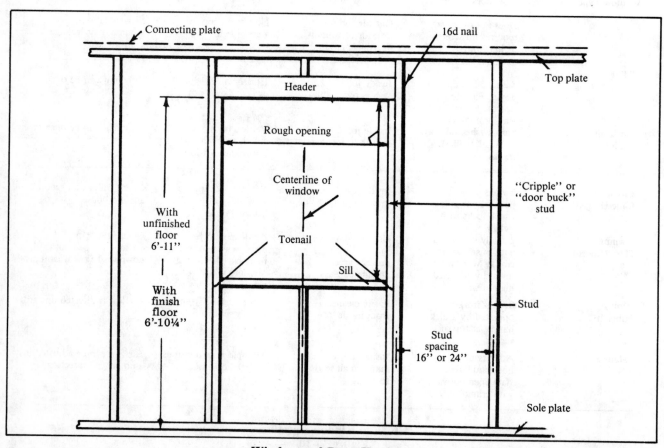

Window and Door Headers
Figure 10-12

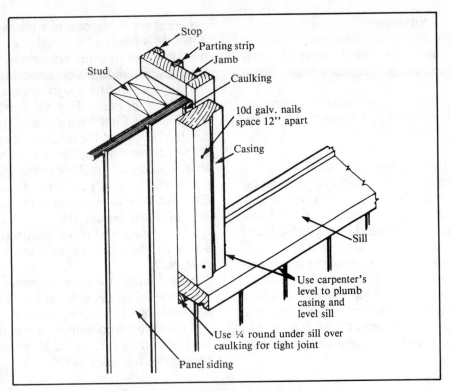

Installation of Double-Hung Window Frame
Figure 10-13

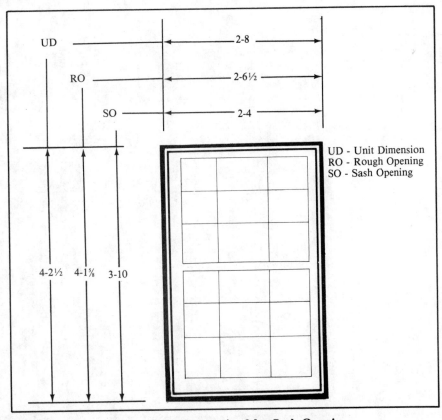

UD - Unit Dimension
RO - Rough Opening
SO - Sash Opening

Rough Open Determined by Sash Openings
Figure 10-14

Summary

The window height in conventional construction is the same as the door height. That is, your door and window headers should be the same height from the floor.

Repairing old, worn-out windows is about as profitable for the remodeler as hauling a fare one block is for the cabbie. Sometimes repairing windows isn't worth the putty you use.

Window and door work is one of those items that can be profitably handled during a remodeling job. As a single item it can result in a loss if you aren't careful about figuring the job. What basis do you use for figuring the manhours? I wish there were a sure method. There isn't, so base it on your own appraisal of the job and your own experience.

This doesn't mean that you should shy away from a job of replacing the windows or doors in a house. But replacing a single window or door is a one-block haul. Unless you need the exercise or experience, you'd be better off taking a nap as far as profit is concerned.

One more thing before we leave doors and windows. It is never wise to take someone else's word on the rough opening requirements for a window. Once I was building a house and the framing was at the stage where the rough opening size of the windows was needed. The windows were a special order and had not been shipped from the manufacturer. The man at the building supply house searched somewhere in his files and came up with a size that he said was the RO size for that particular window style. So that's the size I framed. Guess what? Exactly! When the windows arrived on the job not one of them would fit. The RO was too small! Each opening had to be changed. Have you ever wanted to brush someone's teeth with a 16-ounce hammer?

Now I either measure a sample window or have sample window sizes on the job. The same type and size window sometimes requires a different RO, depending on who manufactures it. The difference is in how they extend the side jamb below the sill and above the head jamb.

11

Exterior Siding

Hardboard is made from logs, chips and saw-mill by-products. It's a popular exterior siding because of its availability, reasonable cost, and because it's fairly easy to install without special tools. It has some advantages over wood: it doesn't split under nail pressure, is free of knots and is uniform in thickness, density and appearance. Since it is a wood product, hardboard requires maintenance just as wood siding does and is subject to the same destructive forces, such as termites and weather. However, there is no grain to raise or check.

Hardboard needs to be protected from the weather prior to installation and priming. It should be evenly stacked, protected from grease and dirt, and handled carefully for a neat job. A properly installed vapor barrier (1 perm or less rating) such as polyethylene film or foil-backed gypsum board, is required on the warm side of the building. This will stop damaging condensation from forming within the walls.

Hardboard may be applied over foam plastic sheathing. The following special application and construction techniques are recommended:

• Adequate bracing of the wall is required.

• Nail lengths must be increased to compensate for the greater thickness of this sheathing. (For 3/4-inch foam-lap siding use 10d nails. For panel siding use 8d nails. For 1-inch foam-lap siding use 12d nails, and for panel siding use 10d nails.) Use care to avoid crushing the sheathing during nailing.

• It is important to use a continuous unbroken vapor barrier, such as 6-mil polyethylene film, on the interior face of the walls to reduce the possibility of moisture accumulation inside the wall cavities. In some cases it may be necessary to vent these cavities to the outside.

Cutting

Whenever possible, do the cutting and marking on the back side. If the finished surface must be cut, use clean tools and place heavy paper or cardboard beneath the saw. Use a fine-toothed hand saw or a power saw with a combination blade.

Exposed Nails

For exposed nailing, finish nail heads with matching touch-up paint or use color-matched nails. Special color-matched nails, caulk and touch-up paint are available. When using the colored nails to install prefinished panels, use the plastic hammer head cap furnished with the nails.

Wall Construction

Siding may be applied over most existing siding, sheathed or unsheathed walls with studs spaced not more than 16 inches o.c. Adequate bracing of the wall with corner bracing or sheathing is necessary. To re-side walls that have old shingle siding, it's best to remove the existing siding before installing the hardboard. Allow at least 6 inches between the bottom of the lap siding and the ground.

When applying siding directly to studs or over wood sheathing, use building paper or felt directly under the siding. For siding and outside metal corners, use 8d galvanized box nails. For the heavy gauge vinyl strip and inside metal corners, use 6d or longer galvanized siding or box nails.

Masonry Construction

Whenever siding is applied over masonry, the wall must be furred out, with the framing spaced no more than 16 inches o.c. and thick enough to accept the full length of the recommended nail. A continuous vapor barrier of 6-mil polyethylene must be placed between the siding and the masonry. The recommended method is to place the vapor barrier between the framing and the masonry wall. Insulation may then be installed between the framing members. Install a vent strip at the top of the siding under the eaves to vent the back of the siding.

Joints

On sheathed and unsheathed walls, all butt joints should fall opposite a stud. Use metal joint molding at all butt joints, inserting it from the top into the gap at the siding butt joint.

Corners

Use prefinished metal outside and inside corners and trim. Wood trim around doors and windows should be at least 1⅛ inches thick. (See Figures 11-1A and 11-1B.)

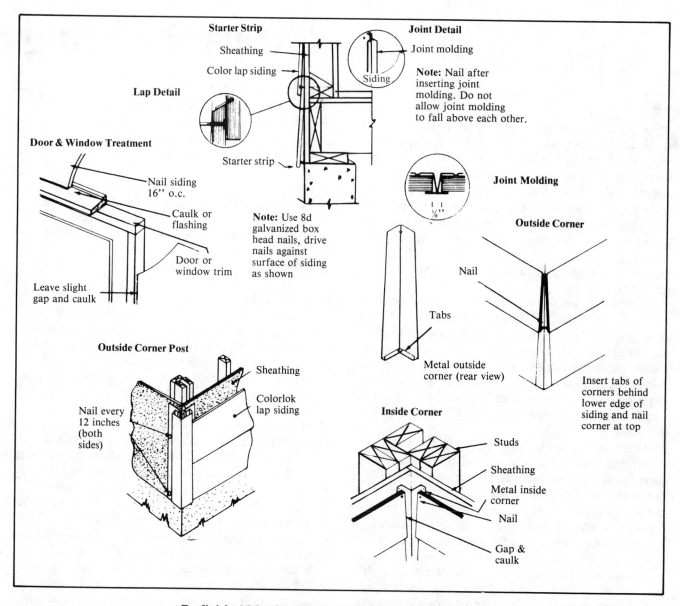

Prefinished Metal Inside and Outside Corners and Trim
Figure 11-1A

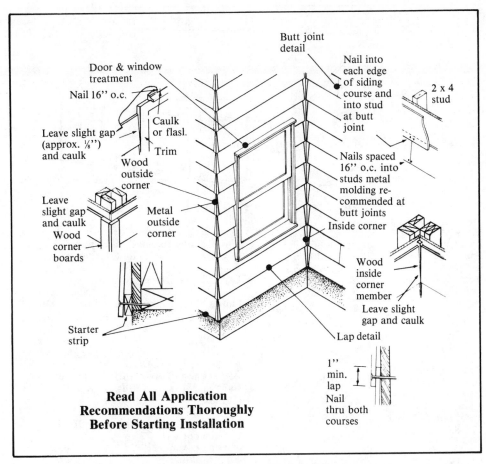

Door and Window Treatment Corner
Treatment - Starter Strip
Figure 11-1B

Finishing

Prime-coated panels must be painted within 120 days after installation. If they have been exposed for a longer period, reprime the siding with a good quality exterior-grade oil-base primer. Unprimed panels must be primed before finish painting. Quality exterior alkyd, oil or latex finish may be used. For all paints, follow the manufacturer's recommendations for using special primers or undercoats, the rate of spread, and application procedures.

All exposed areas must be painted. Two coats of paint over the primed surface are recommended. Only siding with wood-like surface textures is suitable for staining. Use quality exterior opaque acrylic latex stains. Do not use oil base stains. Multiple coats of stain will ensure the longest service life. Prestained panel siding does not require further staining.

Step-By-Step Installation

Let's put up some panels. The gable ends will be lap siding. It makes a nice looking job by combining panels and horizontal siding in this fashion. The first thing you have to do is figure the material requirements. Figure 11-2 shows how to estimate siding requirements. The procedure to use for installing panels is shown in Figure 11-3.

Now let's try hardboard lap siding. The lap siding requirements are estimated in almost the same way as panel siding. (See Figure 11-4.)

Figure 11-5 gives detailed procedures taking you to the finished job. (See Figure 11-6.)

Use Figure 11-7 as a guide when you are estimating a siding job.

Hardboard Shakes

Apply shakes over sheathing or directly to the studs. No building paper is needed over sheathing in a vertical wall application unless it's required by your building code. When it is required, apply building paper over the studs before direct application of the siding to the framing. Shakes may also be applied to sloped framing less than 15 degrees

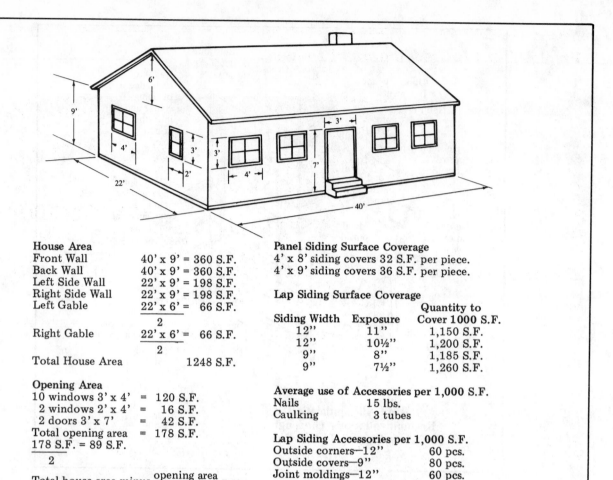

House Area		
Front Wall	40' x 9'	= 360 S.F.
Back Wall	40' x 9'	= 360 S.F.
Left Side Wall	22' x 9'	= 198 S.F.
Right Side Wall	22' x 9'	= 198 S.F.
Left Gable	$\frac{22' \times 6'}{2}$	= 66 S.F.
Right Gable	$\frac{22' \times 6'}{2}$	= 66 S.F.
Total House Area		1248 S.F.

Opening Area

10 windows 3' x 4'	=	120 S.F.
2 windows 2' x 4'	=	16 S.F.
2 doors 3' x 7'	=	42 S.F.
Total opening area	=	178 S.F.

$\frac{178 \text{ S.F.}}{2} = 89 \text{ S.F.}$

Total house area minus $\frac{\text{opening area}}{2}$ = required siding.

1248 S.F. minus 89 S.F. = 1159 S.F.

Panel Siding Surface Coverage
4' x 8' siding covers 32 S.F. per piece.
4' x 9' siding covers 36 S.F. per piece.

Lap Siding Surface Coverage

Siding Width	Exposure	Quantity to Cover 1000 S.F.
12"	11"	1,150 S.F.
12"	10½"	1,200 S.F.
9"	8"	1,185 S.F.
9"	7½"	1,260 S.F.

Average use of Accessories per 1,000 S.F.

Nails	15 lbs.
Caulking	3 tubes

Lap Siding Accessories per 1,000 S.F.

Outside corners—12"	60 pcs.
Outside covers—9"	80 pcs.
Joint moldings—12"	60 pcs.
Joint moldings—9-3/8"	80 pcs.

To estimate how much siding is needed to cover a house you must first calculate its total surface area. Multiply the length by the height of each side. To calculate the area of a gable, multiply the length by the height and then divide the result in half. Add these square foot figures together plus the areas of any other elements that will be sided such as dormers, bays and porches.

You now have the total surface area of the house. But since the windows and doors will not be covered, their surface areas must be allowed for. To do this multiply the width by the height of each opening. Add the results together. Since some waste is involved in cutting the siding to fit around these openings, the total area of the openings is divided in half. This figure is then subtracted from the total area of the house to arrive at the total square footage of siding required to cover the house.

Estimating Siding Requirements
Figure 11-2

from vertical (45/12 pitch). In this case, apply the sheathing on the framing and cover with building paper before applying the shakes.

Corner bracing is required when the siding is applied directly to the studs. Use a good quality, non-hardening sealant where the siding meets the windows, doors, vertical and horizontal trim, etc. Shiplap edges are not sealed. When siding rests on a wood or concrete sill, there must be a good slope for drainage and the juncture must be sealed with a permanent sealant.

Always leave a slight gap at joints, junctions, corners and openings. Never spring panels into place. Leave a 1/32-inch gap between shiplap edges and a 1/16-inch gap at junctures. Lower edges should never be less than 8 inches above a grade having adequate drainage.

Vapor barriers such as polyethylene must be applied on the room side of exterior walls, regardless of the type of sheathing and even if there is no sheathing. Omitting the barrier may cause buckling.

Preparation for installing panel siding

Carefully stack siding on 2'' x 4'' wood stringers spaced no more than 3' apart to prevent contract with the ground. Keep the siding protected from dirt and unnecessary damage by covering with heavy plastic film or other suitable material.

Ordinary carpenter tools are all that is needed for most installations. Check all power tools, ladders and extension cords for safe operation. A hard hat and safety glasses are also recommended.

Remove downspouts. If a new fascia will be installed, remove the complete gutter system.

Remove old caulking from around window and door trim. If window sills and drip caps extend beyond the outside edge of trim, cut them back flush with the trim to form square corners.

Old corner boards, roof edging and other trim such as cove molding and half-round must be removed at this time.

Secure all loose existing siding by re-nailing. Where necessary, replace broken or rotted boards.

Installation Procedures for Panel Siding
Figure 11-3

Installing panel siding over a flat surface

Where old corner boards have been removed, it is necessary to build-out the surface flush with the face of existing siding. Plumb the new corners.

After removing corner boards, find the *lowest point of the old siding* by using a line level on all sides of the house. Or, measure down from a reasonably uniform point such as the soffit. Do this at all corners and note the longest measurement. Mark this measurement on the bottom of the wall at the corner from which you'll start. Remember: *Always work from* left to right.

Making sure the corner is plumb, measure from the corner to the right, the width of the new panel siding. Tack a nail at the mark in order to hang a chalkline. Use a level to make sure the line is plumb, then snap the line. Slide the first panel into position making sure the underlap edge contacts the chalking along the entire edge, and meets the line level or slightly covers the mark previously made when measuring down from the soffit. Tack the panel in place.

Check the panel with a level. Be sure it's plumb and nail in place. It's important to nail along the left edge first. Space nails along edges every 6" o.c. *Always nail from top to bottom, left to right! Do not nail in panel grooves. Do not nail the leading right-hand edge at this time.* Nailing the shiplap joint of the second panel will secure this edge.

Put two rows of vertical nails 16" o.c. in the body of the panel. Space nails 12" o.c. vertically. Remember: Always nail from top to bottom, left to right.

Installation Procedures for Panel Siding
Figure 11-3 (continued)

Installing panel siding over a flat surface (continued)

Metal "J" trim installed at all window and door openings receives the cut edge of the siding, providing a neat, finished appearance.

Determine the window area to cut out of a siding panel by using a long level, or line level, as a reference point for measuring. Measure from this point, which will be the bottom of the new siding, to the bottom of the window sill. Then measure to the top of the window.

Take the horizontal measurement from the previously installed panel to the window, measuring from the inside edge of the shiplap joint to the inside of the "J" trim.

If you're using a circular power saw, transfer the measurements to the backside of the panel siding, taking care to measure from the *bottom* of the panel and from the *left-hand* edge to the window cut-out area.

Installation Procedures for Panel Siding
Figure 11-3 (continued)

Installing panel siding over a flat surface (continued)

After cutting, install the panel by sliding it around the window and against the "J" trim with moderate contact.

Once the panel is in place, check the shiplap joint of the two panels for proper engagement. Since the preceding panel is already plumb, engaging the shiplap edge automatically plumbs the second panel.

Shiplap and butt joint detail

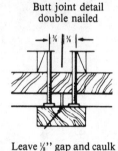

| Butt joint detail double nailed | ⅜" shiplap detail double nailed | ¾" shiplap joint detail single nailed |

Leave ⅛" gap and caulk   ⅜"   ⅝"   Nail ⅜" from panel edges

Installing panel siding over beveled lap siding using furring strips

When installing new panel siding over existing beveled lap siding it may be necessary to use furring strips, particularly if the old walls are wavy. Use 1" x 3" or 1" x 4" furring. First, nail horizontal strips along the entire length of the wall, top and bottom, assuring a nailing base for the top and bottom edges of the siding. Then nail vertical strips in place, 16" on center. Plumb furring and nail into the high points of the beveled siding for solid anchoring. *Make sure furring is positioned where panel edges will join.*

The use of furring strips requires building out the window trim. Do this with wood brick molding, top and sides. Be sure to use a drip cap over the top of the brick mold. Extend the window sill by nailing new lumber on the sill face before building the window out.

Installation Procedures for Panel Siding
Figure 11-3 (continued)

Installing panel siding over beveled lap siding using furring strips (continued)

The panel siding butts to the brick mold with moderate contact (left). It is then caulked along the edge (right).

Installing panel siding over beveled lap siding without furring strips

Panel siding can be applied directly over old beveled lap siding when the wall is straight and true. As an alternate method of vertically aligning the first panel, a plumb butt-board can be used. Allow the butt board to extend past the corner of the wall. Be sure it is plumb, then temporarily nail in place without driving the nails "home," as this board will be removed after securing the first panel.

Install the panel by butting it to the temporary butt board, and tack in place. Check the right edge of the new siding with a level for plumb. Nail along the left edge from top to bottom, 6" o.c. When nailing across the body of the panel be sure to locate nails on the high points of the old beveled siding. This will be easier if a level is used to mark the nailing line.

If the top edge of the panel siding does not fall on a high point of the old siding, install a shim board behind the panel to keep it in the same finished plane as the rest of the wall.

When applying panel siding above panel siding be sure to leave at least one inch of the shim board revealed above the panel. This will provide a nailing base for the top piece of siding. In this instance, metal "Z" flashing is used between the upper and lower pieces of siding to provide a weather seal.

Installation Procedures for Panel Siding
Figure 11-3 (continued)

Optional joint method

An optional method of joining panel siding above panel siding is to butt the two pieces together using moderate contact. Nail as recommended and caulk the joint.

Finish the joint by installing a 1'' x 4'' trim board stained to match the siding.

Install new lap siding over old beveled lap siding

When new lap siding is applied over old beveled lap siding, shim where necessary with horizontal strips to provide an adequate nailing base and a uniform wall. "J" trim extending the full length of the gable can be installed to receive the new siding and provide a finished appearance. When applied over shim material, metal "J" trim is nailed directly to the shims.

An accurate device for determining roof pitch is made by nailing two 1 x 2's together at one end. Then spread the arms of the tool to line-up with the angle on the bottom of the fascia. With a pencil, mark the angle on one arm so you'll be able to transfer the correct angle to the new siding.

Installation Procedures for Panel Siding
Figure 11-3 (continued)

Installing new lap siding over old beveled lap siding (continued)

Transfer the roof pitch to the new siding by opening the arms of the tool to line-up with the reference mark and draw a cut-line. You now have an accurate angle line so each piece of siding will fit correctly when cut.

Install the first lap after making all necessary cuts. To provide a weather seal, allow this piece to lap the panel siding below by 1''. Nail the bottom of the lap siding ½'' above the bottom edge, making sure the nails pass through the panel siding. Again, nail from left to right every 16'' o.c. Do not nail the top of the siding as your next lap will secure both courses.

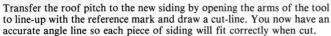

Applying lap siding over a flat surface

Each successive piece laps the preceding lower course by a minimum 1''. If a joint occurs in a lap course, butt together with moderate contact. Nail both sides of the joint at the bottom. On the next course, nail both sides of the joint, as shown. Stagger joints for a neat appearance. A handy "siding crutch" made from a piece of 1'' x 4'' with a cutback angle helps to hold the siding and set the proper lap exposure.

Although steps for applying lap siding over a flat surface are similar to those described in application technique above, furring strips are usually not required. Be sure to use a minimum 1'' lap over the panel siding below. Again, nail from left to right on the bottom edge only.

Installation Procedures for Panel Siding
Figure 11-3 (continued)

Finishing touches

Install 1" x 4" outside corners, stained to match the prefinished siding. On inside corners, one pre-stained board 1" or more can be butted into the corner to cover joints.

Preformed aluminum fascia is available as a neat appearing finish trim.

Use caulking under windows and in other areas where a cut edge of the siding may be exposed.

Nail Chart

Nail Size*	Nail Spacing
Panel Siding 8d (2½" nominal) when applied over existing siding or furring strips.	**Panel Siding** Edges Intermediate Locations — 6" o.c.; 12" o.c. Note: When nailing across the panel from edge to intermediate locations, nail spacing should not exceed 16" o.c.
Lap Siding 8d (2½" nominal) when applied direct to furring strips. 10d (3" nominal) when applied over existing siding.	**Lap Siding** Nails must be spaced no more than 16" o.c.

*Use galvanized nails with a head at least $\frac{3}{16}$" in diameter. Checkered head nails are preferred with textured siding.

Installation Procedures for Panel Siding
Figure 11-3 (continued)

Here's an example:

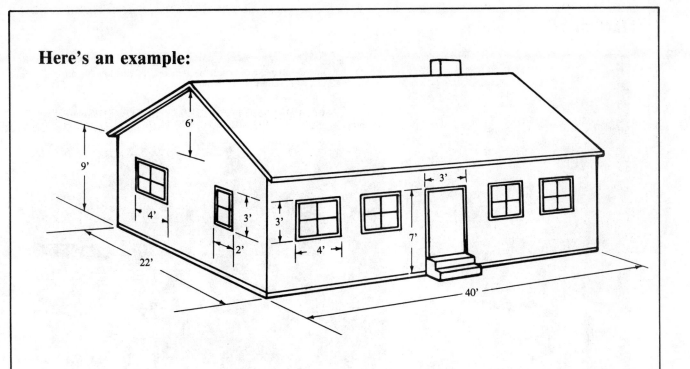

House Area

Front wall	40' x 9' = 360 sq. ft.
Back wall	40' x 9' = 360 sq. ft.
Left side wall	22' x 9' = 198 sq. ft.
Right side wall	22' x 9' = 198 sq. ft.

Left gable $\dfrac{22' \times 6'}{2}$ = 66 sq. ft.

Right gable $\dfrac{22' \times 6'}{2}$ = 66 sq. ft.

Total house area $\qquad$ 1248 sq. ft.

Opening Area

10 windows 3' x 4'	= 120 sq. ft.
2 windows 2' x 4'	= 16 sq. ft.
2 doors 3' x 7'	= 42 sq. ft.
Total opening area	= 178 sq. ft.

$\dfrac{178 \text{ sq. ft.}}{2}$ = 89 sq. ft.

Total house area minus $\dfrac{\text{opening area}}{2}$ = required siding.
1248 sq. ft. minus 89 sq. ft. = 1159 sq. ft.

9-3/8" siding *covers* approximately 66 sq. ft. per carton.

Average* use of Accessories per 1000 SFT of siding

Starter Strip	100 lin. ft.
Outside Corners—12"	60 pcs.
Outside Corners— 9-3/8"	80 pcs.
Inside Corners—10'	2 pcs.
Joint Moldings—12"	60 pcs.
Joint Moldings— 9-3/8"	80 pcs.
J-Trim—12'	200 lin. ft.
Utility Trim—12'	30 lin. ft.

*Allow 10% extra on all accessories to allow for miscalculations and piece miscount, etc.

Estimating Lap Siding Requirements
Figure 11-4

Hardboard Lap Siding

Step 1. Nail all loose existing siding firmly in place and replace all broken or other bad boards. Remove old caulking as necessary to allow for proper fit of accessories around doors, windows, etc.

Step 2. Remove old trim and decorative moldings around doors, windows and other openings.

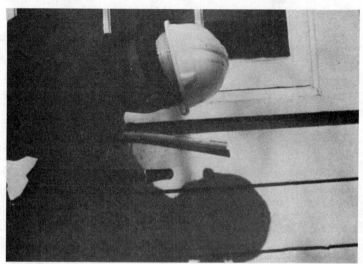

Step 4. Cut "ears" off windows and doors to allow proper fit of accessories.

Step 5. Measure distance from lower edge of existing siding to the top of the wall at all corners. The top point should be a uniform height such as the soffit or trim board. Using the longest measurement obtained, measure downward at each corner and mark the low point at the bottom of the wall. It is very important that the bottom starter strip be installed parallel with the top of the wall to avoid an irregular or wedge shaped top piece of siding, next to the soffit.

Step 3. Remove downspouts. If new fascia will be installed, remove gutters, straps, hangers, etc.

Installation Procedures for Hardboard Lap Siding
Figure 11-5

Step 6. Make a second mark 2'' above the mark already made at each corner. This locates the top of the starter strip. Stretch a chalk line from corner to corner and snap the reference line to mark starter strip location. It is important that the starter strip meets at all corners. Siding extends one inch below the starter strip to conceal minor irregularities. The starter strip should be installed completely around the house before installing any siding.

Step 7. Because few older homes have straight walls, it may be necessary to install furring strips behind the starter strip to insure that the new siding walls will be straight. Install the starter strip nailing 16'' o.c. If continuous inside or outside corners are used, the starter strip should end approximately 2'' from the corner to allow for installation of these accessories.

Step 8. Install continuous inside corners. If continuous outside corner posts are used, they should also be installed at this time.

Step 9. Apply caulking around existing window or door trim and seat J-trim in caulking and nail in place 12'' o.c. Install J-trim wherever it is necessary to cover and protect the ends of the laps. An alternative method used if the opening trim protrudes beyond the siding, is to butt the siding against the trim and then caulk the joint. This method eliminates the need for J-trim.

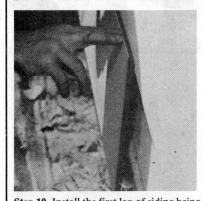

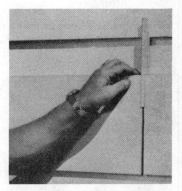

Step 10. Install the first lap of siding being sure to engage the interlocking spline with the starter strip. Nail ½'' below the top edge of the siding using a galvanized hardboard siding nail that will penetrate at least 1¼'' into the existing wood siding. Pull or push downward slightly when nailing to insure the spline is properly engaged. Drive nails firmly into place but do not countersink.

Step 11. When measuring and fitting pieces which fit end to end, allow a ⅛'' gap between pieces to allow insertion of butt joint molding. Also allow ⅛'' gap around all trim to be caulked later.

Step 12. Insert joint molding into ⅛'' gap from the top; after insertion, nail the ends of both pieces of siding in place ½'' from the top edge. For best appearance, stagger these joints at least three feet apart horizontally or three laps vertically.

Installation Procedures for Hardboard Lap Siding
Figure 11-5 (continued)

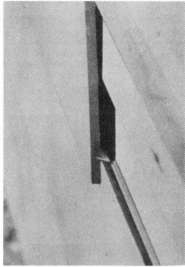

Step 13. When going over old horizontal siding it may be necessary to install horizontal furring strips to provide a straight nailing base and a straight uniform appearing wall. This also brings nailing base out far enough to insure proper locking of the spline with the next siding lap. (Note . . . see steps 20-22 for optional furring methods.)

Step 14. Be sure to always engage the plastic mounting spline on the beveled top edge of the siding, pressing down somewhat firmly on each piece of siding to insure proper seating. (Note horizontal furring strip.)

Step 15. Individual outside corners may be installed as the work progresses, or can be installed after the entire wall is sided. Slide the corners up into place with the top slipping under the course below, seat tabs under the lower edge. When corner is in position drive nail through exposed hole at top of corner. You can then proceed to install the next corner.

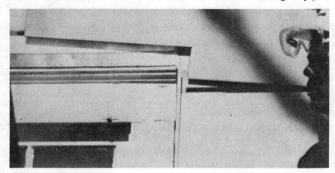

Step 16. Around windows and doors, the siding may have to be cut or notched to fit each opening. When cutting more than 1'' or 2'' into the siding it is usually easier to make a joint as shown. When fitting under windows, cut as needed and caulk later with matching caulk.

Step 17. On gable ends, install J-trim along existing or new trim boards. Make a template to match angle of roof and cut shorter pieces to install along the angle. Then measure the length needed between the two end pieces, allowing for two butt joint moldings, and cut a piece to fit. Again remember to stagger joint moldings for best appearance.

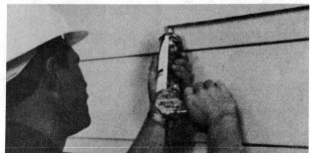

Step 18. Openings for faucets, outlets, etc., can be easily drilled in siding. Turn off the utility, remove the fixture, locate and drill or cut the hole. Then caulk around the opening, replace fixture and turn the utility back on.

Step 19. Caulk wherever the cut edge of the siding may be exposed such as over or under windows. It is not necessary to caulk at joint moldings or individual outside corners, If it's necessary when the job is finished wash dirt off siding using liquid household cleaner. Also touch up as necessary using matching siding touchup paint.

Installation Procedures for Hardboard Lap Siding
Figure 11-5 (continued)

Step 20 - Optional Furring Method
An optional method that is commonly used over masonry walls, and occasionally over wood siding, is to install vertical furring strips over the entire wall. With any type masonry, it is necessary to install a 6 mil polyethylene film between the furring strips and the masonry. It is also necessary for these strips to be well anchored and of sufficient thickness to supply adequate nail holding power.

Step 21 - Optional Furring Method
Installation will proceed as described earlier, except that all joint moldings must be located at a furring strip location.

Step 22 - Optional Furring Method
It is usually necessary to build out the existing wood trim using a new wood brick mold. J-trim can then be used next to the new brick mold or the siding can be cut, butted up to the brick mold and then caulked.

Installation Procedures for Hardboard Lap Siding
Figure 11-5 (continued)

To cut any hardboard product, use a fine-tooth hand saw or a power saw with a combination blade. Be sure that the cutting is into, or toward, the exposed (finish) side. Always prime the cut edge before painting.

When nailing the shakes into place, make certain the nails are long enough to penetrate the studs at least 1 inch. Nailing should be 16 inches o.c. on a line 1 inch from the bottom edge, and set ⅜ inches in from the side edge of the shiplap. At least two nails should be used on any small trimmed pieces. Use 8d or 10d corrosion resistant nails.

Shim to avoid deforming the siding above and below windows or other openings, and to avoid deformation caused by uneven walls. Do not over-drive nails. That is, don't take that last swing with the hammer.

Step-By-Step Procedures
1. Apply siding horizontally over framing spaced up to 16 inches o.c.

2. On horizontal runs of 50 feet or more, use expansion joints or other means of breaking the continuity of the wall.

3. Install a starter strip (wood lath, 3/8 x 1½ inches), using a chalk line to position the strip at the bottom plate parallel to the top plate or soffit. Nail the starter strip to the bottom plate at stud locations with 6d galvanized siding nails.

The Finished Look
Figure 11-6

4. Level and install the first course of siding with the bottom edge at least 1/2 inch below the starter strip as follows:

a. Install inside corner wood members, 1⅛ x 1⅛-inch, and outside corners made up of 1⅛ x 3-inch and 1⅛ x 4-inch wood members. Textured metal outside corners may also be used.

b. Leave a 1/16-inch gap between the siding and the corner boards. Begin at the lower left corner, with the first board trimmed on its left-hand edge if necessary to ensure that the right edge falls over a stud. Then do the balance of the first course, using full pieces until you reach the next corner. There, trim the siding to leave a 1/16-inch gap.

c. Begin the second course with a piece trimmed from the left edge 16 inches shorter than the first course. Overlap the first course a minimum of 1½ inches.

d. Begin the third course with a piece 32 inches shorter from the left edge than the first course. Overlap as in *c*, above.

e. Repeat steps *b, c,* and *d* for the fourth and succeeding courses. (See Figure 11-8.)

Painting and Staining
Painting should be done in favorable weather, between 50 and 80 degrees. Follow the manufacturer's instructions.

All exposed surfaces and bottom edges of the siding must be painted or stained within 30 days after installation. Application by brush or roller is recommended to ensure proper penetration of the coating. Opaque good quality exterior acrylic latex stains, or good quality exterior acrylic latex paints are recommended as finish paint. Use at least two coats.

Airless spray application may be used, but be careful to dry brush the applied coating after application. This works it into the textured crevices. Two coats are necessary, the first to be applied at maximum coverage just short of runs or sags, and then dry brushed immediately after application. The second coat is sprayed and dry brushed after the first coat is dry to the touch.

It's probably easier to use a brush in the first place.

Vinyl Siding
Vinyl siding is available for both vertical and horizontal applications. A vinyl soffit system is also available. All the accessory equipment such as posts for inside and outside corners, starter strips and dividers is available. Horizontal siding usually comes in 12-foot 6-inch long boards with a single-board, 8-inch exposure. Vertical siding is usually 10 feet long with 12-inch-wide exposure. The color is in the vinyl and is long lasting.

Vinyl siding is installed in much the same manner as other sidings. When re-siding a house, secure and nail loose boards. Remove downspouts, lighting fixtures, moldings, and old caulking around windows and doors. Check to see that the sidewalls and base are level and plumb. Fur them out where necessary.

Vinyl siding is loose fitting, and does not provide resistance to air infiltration. Seal air leakage paths prior to installing the siding. The requirements for vapor barrier are the same as for the other sidings discussed above.

Metal snips (tinners and aviation snips) readily cut vinyl panels and accessories. Use a square to mark lines across the panel. Cut the top interlock first, cutting toward the bottom of the panel. The panel should be well-supported on a firm surface. Snip through the back of the butt and then bend the panel and cut at the bottom of the butt. You can also cut panels with a hand saw, hand power saw, or table saw.

The teeth on the blade should be fine and have little or no set. Reversing the blade gives the best results. This is a must in cold weather. (See Figure 11-9.) Using a straightedge and a score knife or

Sketching Area

Miscellaneous

Total house area _____ Sq. Ft.   _____ Cans Touch Up (8 oz.)

Minus ½ opening area _____ Sq. Ft.   _____ Tubes Latex Caulk (matching)

_____   _____ Cartons Colored Nails (Lbs.)

Required siding _____ Sq. Ft.   _____ Rolls Matching Coil Stock (50')

_____ SFT Siding 9-3/8"_____ 12" _____ Color _____

_____ LFT Starter Strip (10')

_____ Ea. Outside Corners 9-3/8" _____ 12" _____ 10' Post _____   SoffIT

_____ Ea. Inside Corner Posts (10')   SoffIT J

_____ LFT J Trim (12')   COIL

_____ LFT Utility Trim (12')   CAVLKING   INSULATION

Job Estimating Sheet
Figure 11-7

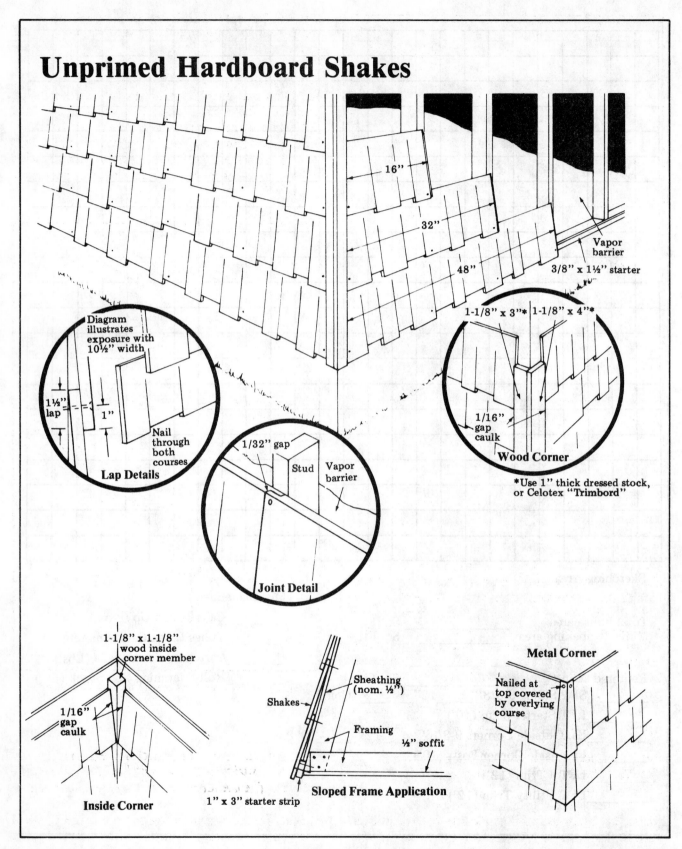

Unprimed Hardboard Shakes

16"

32"

48"

Vapor barrier

3/8" x 1½" starter

Diagram illustrates exposure with 10½" width

1½" lap

1"

Nail through both courses

Lap Details

1/32" gap

Stud

Vapor barrier

Joint Detail

1-1/8" x 3"* 1-1/8" x 4"*

1/16" gap caulk

Wood Corner

*Use 1" thick dressed stock, or Celotex "Trimbord"

1-1/8" x 1-1/8" wood inside corner member

1/16" gap caulk

Inside Corner

Sheathing (nom. ½")

Shakes

Framing

½" soffit

1" x 3" starter strip

Sloped Frame Application

Metal Corner

Nailed at top covered by overlying course

Unprimed Hardboard Shakes
Figure 11-8

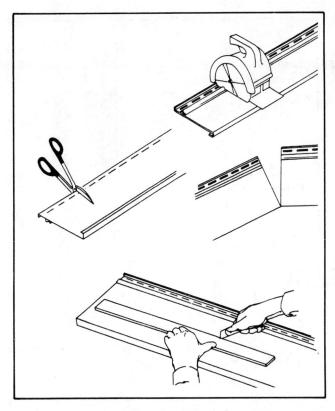

Cutting Procedure
Figure 11-9

through the nailing slots. Leave 3/4 inch between the corner post nailing flange and the end of the starter strip.

Install the first panel in the interlock of the starter strip (Figure 11-11), making sure it is locked. Then nail on 12-inch to 16-inch centers in the center of the nailing slots. Don't drive the nails tight. The panels must be free to move laterally. Apply succeeding courses in the same way. Avoid vertical alignment of joints. Use furring strips in low spots, under windows, and whenever necessary

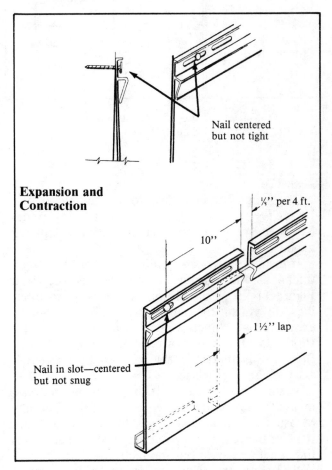

Nailing to Provide for Expansion and Contraction
Figure 11-10

roofer's knife, score the siding panels, then bend. This is an effective and fast way to make lengthwise cuts.

Solid vinyl siding expands and contracts with temperature changes and must be free to move back and forth on the nail shank. Nails shouldn't bind the vinyl and must be near the center of the nailing slot so the siding can move. Panels must be cut short between two solid objects such as two windows or a window and a corner post. Always allow 1/8 inch for expansion for each four-foot length of panel. Also, lap panels 1½ inches to allow 1/2 inch for movement (1/4 inch per 12-foot 6-inch panel).

Nails should be corrosion-resistant metal with a 1/8-inch diameter shank and a 5/16-inch head. Use 1½-inch to 1¾-inch nails for non-insulated jobs. For insulated jobs, use nails long enough to penetrate the framing or sheathing 3/4 inch. (See Figure 11-10.)

After installing corners and window trim, start the siding by marking a chalk line for the starter strip around the house parallel to the top of the foundation wall. Then align the starter strip with this line and nail it in place on 8-inch centers

to maintain the line or pitch of the panels. If you don't maintain the line or pitch of the panels, the job will give the illusion of color difference because of light reflection.

Start installing the siding panels at the back of the house and work toward the front, overlapping the panels toward the entrance ways. Lap your cuts under the factory ends to create the best ap-

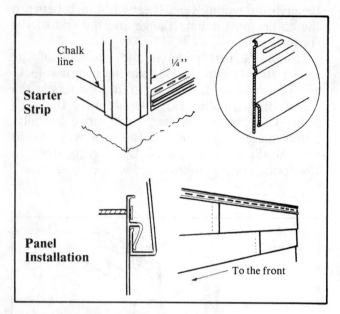

**Starter Strip — Panel Installation
Figure 11-11**

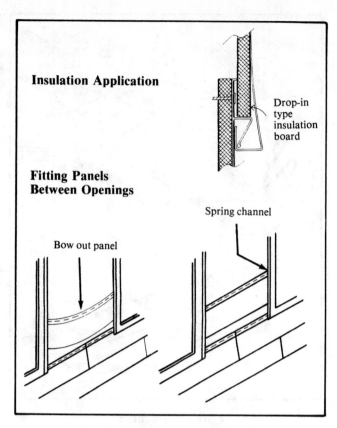

**Insulation Application
Fitting Panels Between Openings
Figure 11-12**

pearance.

To ensure neat end laps, do not nail closer than 10 inches to the end of the overlapping panel. Never stretch the panels vertically or end laps will gape.

If insulation is used, it must not exceed 3/8-inch thickness and should be installed as shown in Figure 11-12. When fitting panels between openings, always cut the panels short to leave room for expansion—1/8 inch for each 4 feet (1/32 inch per foot). Panels can be bowed slightly to get them to fit between trim or the trim can be deflected to help get the panels in place.

A small piece of scrap siding can be used as a measure and guide for cutouts below openings. Measure both sides of the opening to ensure a good fit. Cut the panel to the opening width. Then score the panel and snap out the section.

VFT finishing trim is used under windows to trim the cut edge and provide a lock for the siding panel. After cutting off the siding nailing flange at the desired location, use a Malco Snaplock Punch to provide a lock for the trimmed panel. Install as shown in Figure 11-13, using furring strips as needed to maintain the pitch of the panels.

Window or door head flashing is installed as the siding progresses. Install appropriate flashing if it has not already been done. Then install VJ channel, using mitered or square-edge corners. Cut the panel to fit; use the Malco Snaplock Punch to provide the lock. Attach the required length of VFT

finishing trim. Insert it into a J channel, over the furring when required. (See Figure 11-14.)

Vertical Siding and Combination Applications

Install inside and outside corners by nailing on 12-inch centers. Use a 3/4-inch opening post. When insulation fits in the opening, use a 1⅛"-inch opening.

Install VDDC dual cap or VDC vertical base flashing over the window and door head (Figure 11-15). Window and door trim (J channel) should be installed around all windows and doors, nailing 12 to 16 inches o.c. Use a 3/4-inch opening or a 1⅛-inch opening when insulation fits in the opening.

Before starting to apply the siding, install VJ channel (3/4- or 1⅛-inch) below the window sill and at the top of the wall at the eaves and gables.

If the wall is higher than the panel length, the end of the lower course is capped with a J channel. Then vertical base flashing is installed before the upper course is applied.

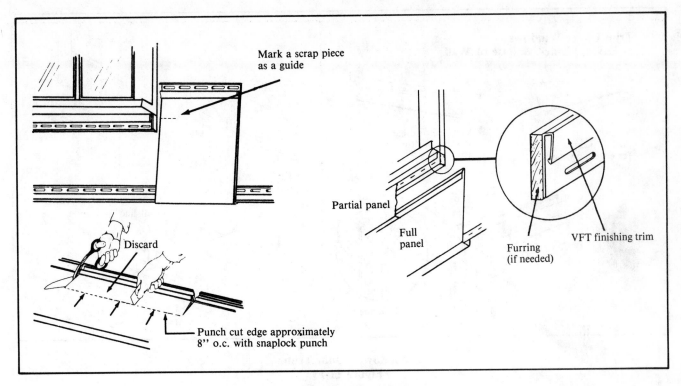

Mark a scrap piece as a guide

Discard

Punch cut edge approximately 8'' o.c. with snaplock punch

Partial panel

Full panel

Furring (if needed)

VFT finishing trim

Fitting Panels Under Openings
Figure 11-13

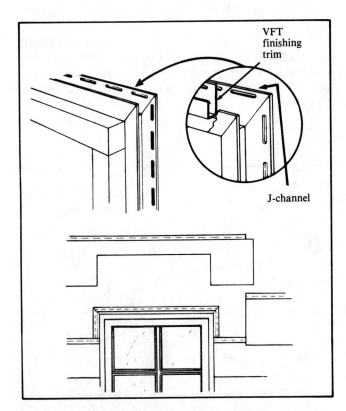

VFT finishing trim

J-channel

Fitting Panels Over Openings
Figure 11-14

Mark a chalk line around the top of the foundation wall where the siding is to be installed. Next, install the vertical base flashing along this reference line. Leave a 1/4-inch space between the corner post nailing flange and the base flashing. A VJ channel may be used at the base of the wall in lieu of vertical base flashing. (See Figure 11-16.)

When installing vertical siding over horizontal siding, place the horizontal siding to the point where the vertical and horizontal panels meet. Cut the top interlock off the horizontal siding and snap it into the finishing trim to hold the siding in place. Do not nail through the siding. Next, install VDC. Install window and door trim, eave or gable trim and cut the panels of vertical siding to length. Follow vertical siding directions for panels as previously discussed. (See Figure 11-17.)

Butting vertical and horizontal siding: Apply two 3/4-inch VJ Channels back-to-back at the point where the vertical and horizontal siding will meet. These channels should be on a vertical line, which can be established by dropping a plumb line from the eave at the desired distance from the corner. Follow previously discussed instructions for installing both types of siding. (See Figure 11-18.)

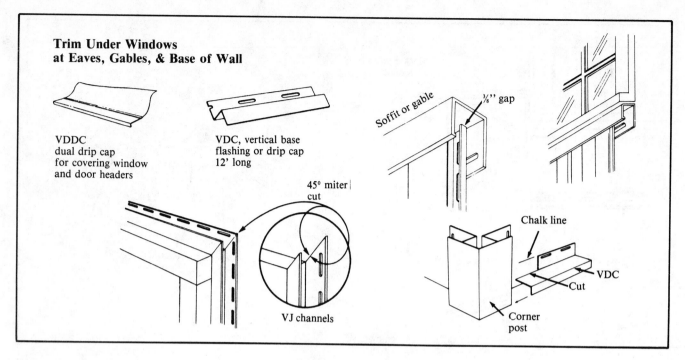

Window & Door Trim
Figure 11-15

Soffit and porch ceiling application: Install VSJ soffit channels at the wall and at the edge of the fascia. Nail soffit channels to the overhang, placing the nails at the corner of the nailing slots on approximately 10-inch centers. (VSF frieze runners may be used at the wall as shown in Figure 11-9. This is needed where outlookers are not a part of the overhang).

Do not drive nails so tight that they bind the vinyl. Nail in the center of the nailing slot so that the soffit can move freely as it expands and contracts.

The 12VN (non-perforated) panels and 12VNP (perforated) panels should be cut to the dimension required for the overhang, and to fit between the two inner channels. Do likewise with 7VN and/or 7VNP panels.

The panels should be approximately 1/8 inch

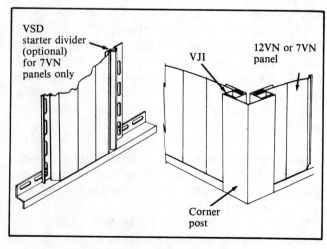

Starter Divider & Panel Installation
Figure 11-16

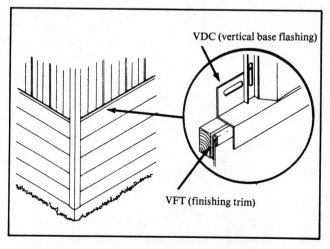

Vertical Siding Over Horizontal
Figure 11-17

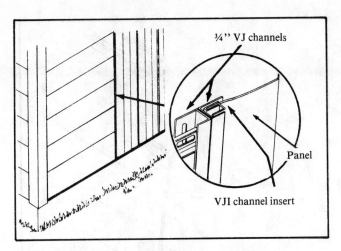

Butting Vertical & Horizontal Siding
Figure 11-18

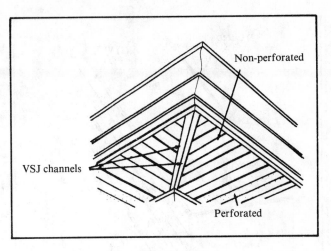

Other Considerations
Figure 11-20

shorter than the inside dimensions from one soffit channel (VSJ) to the opposite one.

Wide overhangs: If the overhang is more than 16 inches from the wall to the edge of the fascia, intermediate nailing is required on the soffit panels so that there is no unsupported span of more than

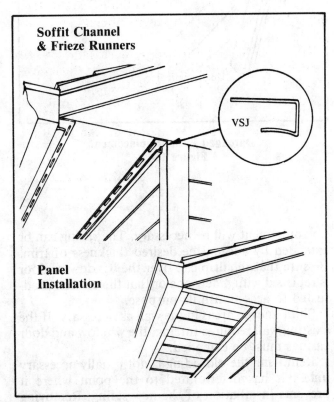

Soffit Channel & Frieze Runners Panel Installation
Figure 11-19

16 inches. A furring strip may be used.

Directional change: When a soffit changes direction, two soffit channels (VSJ) are installed back-to-back to form an "H" member.

Intermixed panels: All panels for the soffit system can be either non-perforated, perforated, or intermixed. The most common soffit system combines both style panels. (See Figure 11-20.)

Fascia treatment: Handle as follows:

(1) Install VFT (finishing trim) to the upper portion of the fascia board.

(2) Cut the fascia panel to the dimensions required to cover the fascia.

(3) Notch the fascia panel end when lapping two panels and where the panel meets the fascia corner.

(4) Punch the fascia with a snap lock tool, then push the fascia up into the VFT trim until it locks.

(5) When fascia corners are required, install them first. Always lap panels and corners away from the front of the house. Allow 3/8-inch overlap for expansion. (See Figure 11-21.)

Damaged panel replacement: To remove a panel that has been damaged, insert an unlocking tool at the end joint or at the end of a panel where it joins a flashing member. (See Figure 11-22.) Hook the tool to the bottom lock and pull downward, deflecting the panel and freeing it from the upper lock. Now remove the damaged panel, install the new one and relock the loosened overlap panel. The unlocking tool is used in a similar manner to reinstall the new panel.

If you're thoroughly confused with all the VJ's and VDC's, Figure 11-23 should clarify the various items we've been discussing.

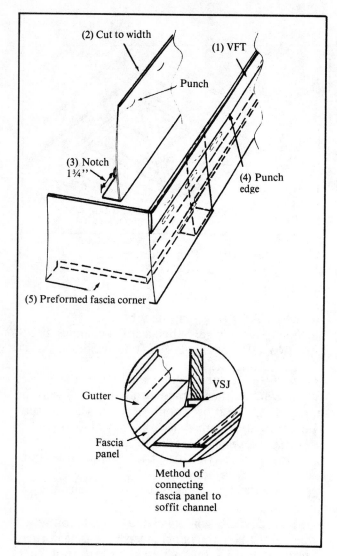

Fascia Detail
Figure 11-21

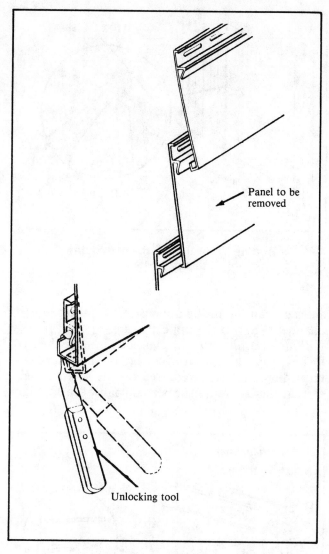

Damaged Panel Replacement
Figure 11-22

Wood Siding

Wood siding has been around a long time. Many of the problems connected with wood siding can be directly attributed to things other than the wood. Paint failure is often caused by moisture from condensation or by water flowing down the wall from the lack of a roof overhang or clogged or faulty gutters.

Vertical and horizontal wood siding is available and is installed using the same procedures as for installing hardboard siding.

The installation of wood siding usually requires the adjustment of window and door trim. The window sills on most houses extend far enough so that no adjustment will be necessary. The casing can be extended by adding the desired thickness of trim. Be sure that the drip cap over the window or door is replaced with a new one or that the old one is adjusted to accomplish its purpose.

Shim uneven areas in the wall as necessary. If the existing wall requires furring, the window and door casings must be extended.

Removing the old siding is not usually necessary unless it has deteriorated to the point where it prevents installation of the new siding. Removing old siding is an expense to the homeowner that should be avoided whenever possible.

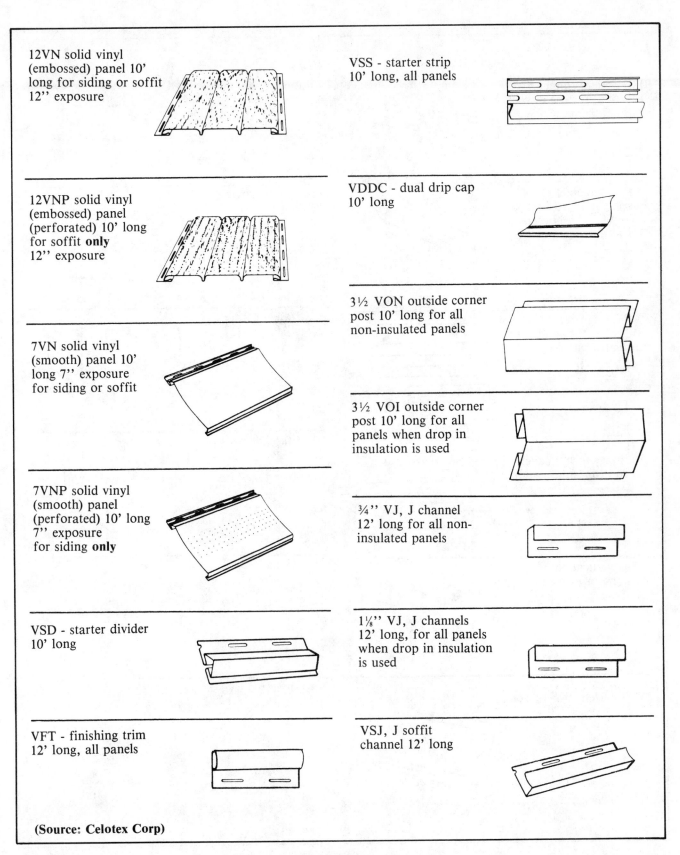

12VN solid vinyl (embossed) panel 10' long for siding or soffit 12'' exposure

12VNP solid vinyl (embossed) panel (perforated) 10' long for soffit **only** 12'' exposure

7VN solid vinyl (smooth) panel 10' long 7'' exposure for siding or soffit

7VNP solid vinyl (smooth) panel (perforated) 10' long 7'' exposure for siding **only**

VSD - starter divider 10' long

VFT - finishing trim 12' long, all panels

VSS - starter strip 10' long, all panels

VDDC - dual drip cap 10' long

3½ VON outside corner post 10' long for all non-insulated panels

3½ VOI outside corner post 10' long for all panels when drop in insulation is used

¾'' VJ, J channel 12' long for all non-insulated panels

1⅛'' VJ, J channels 12' long, for all panels when drop in insulation is used

VSJ, J soffit channel 12' long

(Source: Celotex Corp)

Solid Vinyl Accessories
Figure 11-23

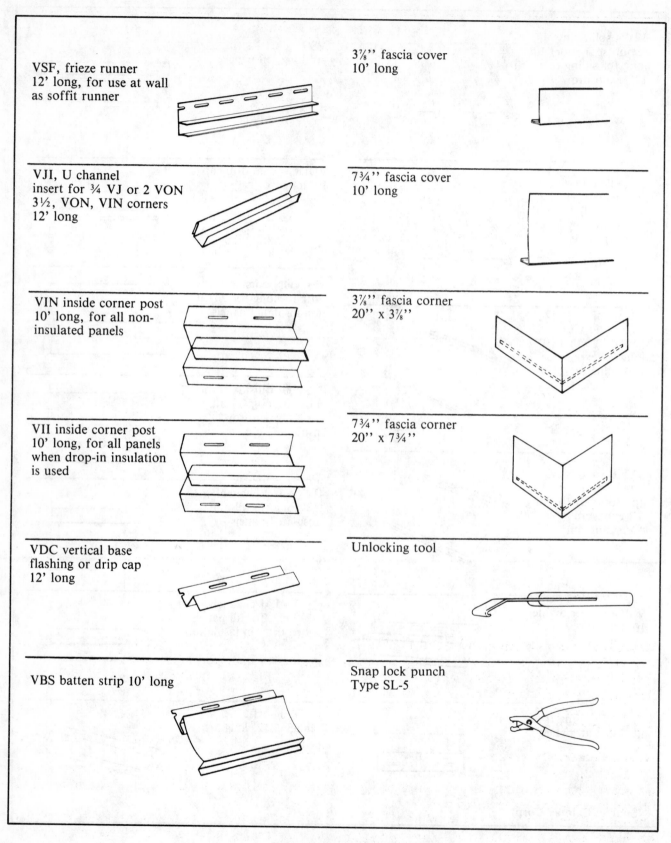

VSF, frieze runner
12' long, for use at wall
as soffit runner

VJI, U channel
insert for ¾ VJ or 2 VON
3½, VON, VIN corners
12' long

VIN inside corner post
10' long, for all non-
insulated panels

VII inside corner post
10' long, for all panels
when drop-in insulation
is used

VDC vertical base
flashing or drip cap
12' long

VBS batten strip 10' long

3⅞'' fascia cover
10' long

7¾'' fascia cover
10' long

3⅞'' fascia corner
20'' x 3⅞''

7¾'' fascia corner
20'' x 7¾''

Unlocking tool

Snap lock punch
Type SL-5

Solid Vinyl Accessories
Figure 11-23 (continued)

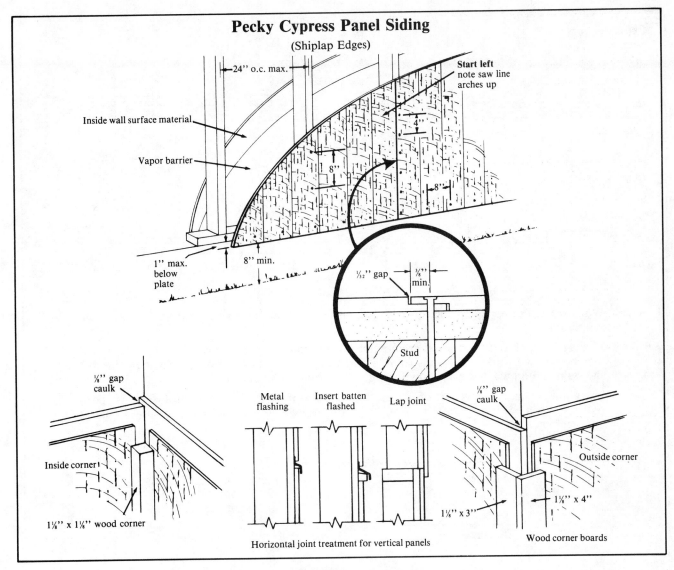

Pecky Cypress Panel Siding
Figure 11-24

Medium Density Overlaid Plywood (M.D.O.)

M.D.O. is manufactured by permanently bonding a resin impregnated overlay sheet to exterior plywood. It's engineered particularly for exterior siding application where a superior painting surface is desired.

General Constructions Notes

• Shiplap-edged panels are intended for use without battens over joints. Panels can be applied directly to the studs without sheathing paper.

• Apply them parallel to the framing, covering the sill and top plate. Panels are 4 feet wide, and most styles are available in 8-, 9- and 10-foot lengths.

• Always leave a slight gap (1/32 inch) at shiplap edges and 1/8 inch at junctions for expansion.

• Vapor barrier requirements are the same as for hardboard.

• Apply sealant to junctions at windows, doors, corners, etc.

• Install wood outside corner boards nailed 12 inches o.c.

• Apply sealant to joints at inside corners and install a 1 x 1-inch wood corner post, nailing 12 inches o.c.

Construction details for shiplap edge panels and nailing requirements and cross section views of plywood panels are shown in Figures 11-24 and 11-25.

Cross Sections:

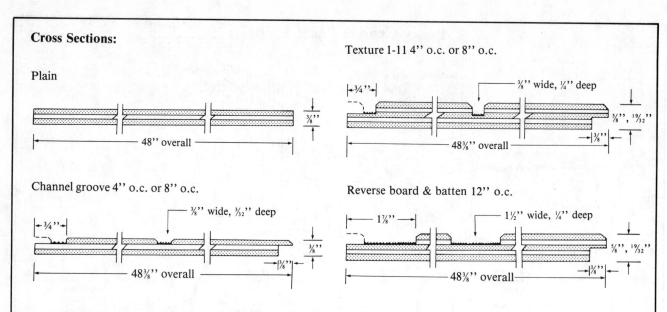

When scarfed or epoxy patched panels are finished, certain factors should be taken into consideration.

If there is no objection to wood grain color show through, epoxy patches and scarfed panels may be finished with a semi-transparent stain. **The patches and scarfed line will show through the stain. Wood patches also show through semi-transparent stains.**

For a more complete coverage of contrasting face features of plywood sidings (such as epoxy patches, wood patches or scarfed joints), the use of heavy bodied stains, opaque stains or paints are recommended.

Thickness	Maximum Spacing of Studs
³⁄₈''	16'' on center
⁵⁄₈''	24'' on center

Thickness	Nail Size and Type*
³⁄₈''	6d casing or siding
⁵⁄₈''	8d casing or siding

Thickness	Nail edges
³⁄₈'' - ⁵⁄₈''	6'' on center

Thickness	Nail Spacing on Intermediate Studs
³⁄₈'' - ⁵⁄₈''	12'' on center

*Use galvanized, aluminum, or other non-corrosive nails; nails may be color coated. Use same schedule for siding over sheathing. Battens if used, can be applied with 8d non-corrosive casing nails spaced 12'' o.c. (staggered).

(Source: Georgia Pacific)

Plywood Siding Cross Sections and Nailing Schedule
Figure 11-25

Aluminum Siding

Aluminum siding came onto the scene shortly after World War II. Wood is still hard to beat, but aluminum siding has the edge where the recurring chore of painting is involved. It is not, however, a siding that will last "forever" as some people seem to believe.

The most common siding pattern currently in use

is the horizontal type with an 8-inch exposure. Double 4- and double 5-inch are also available. Vertical siding is similar in appearance to board and batten or V-groove siding and is used as a design break or for emphasis. Sometimes it's used for gable ends.

Aluminum siding application is much the same as vinyl siding application.

As a small builder, you might prefer to sub out aluminum siding projects to one of the outfits who specialize in this type of work. Keep in mind that you can pick up about five more R's of insulation by installing aluminum siding over breather foil, on furring strips over insulation board, or by using foil-faced insulating backer board.

Summary

Use hardboard, vinyl, wood or aluminum siding to change the looks of any old, run-down house. Each of the sidings discussed in this chapter offers some advantages and disadvantages. The choice is your customer's. Your job is to install it in a professional manner.

No book can cover every detail of all types of sidings available, but we've touched on it enough to give you an idea of what to do and how to do it. If you're trying to learn enough to branch out on your own and think this is too complicated, don't worry. Building *is* complicated. But practically anyone can master construction techniques. It just takes a little study and on-the-job practice.

In panel siding of any style or type, the most important thing is to start right. Get that first sheet or piece up wrong and you throw the whole thing off. Remember to adjust the installation to the house. If the house is off-square, off-plumb or off-level, the best that you can do is "split the difference," that old rule that most builders live by. It's a compromise between perfection and acceptability—between good and bad.

Even new houses are often products of this compromise. Why this is true, I don't know. There's no excuse for new construction to be off-kilter, unless it was built during a wind storm. But it sometimes happens. So if you're called on to throw up siding on brand new construction, don't start grinning until you check it out thoroughly. You just might have to split some differences. If you don't realize that little fact before you nail the first piece up, the only thing you'll split is your profit.

12

Insulation

Every house should be insulated—when it's originally built. That's when doing the job right is easiest and least expensive. But few built before 1974 are insulated as well as we know they should be. The next best time to insulate is during remodeling. That makes insulation part of nearly every significant remodeling and room addition project you will have. Any time you open up an exterior wall, don't close it up again without filling it with plenty of insulation. Your customers will appreciate your professionalism.

Heat escapes from a building by conduction and infiltration. Conduction causes the shady side of a block wall to heat up after the opposite side has been in the sun for a few hours. The rate of loss depends on the size of the surface, the length of time the heat flow occurs, the difference in the temperature between the two surfaces and the conducting qualities of the material.

Infiltration is heat or cold carried by air sneaking in through cracks and openings. Every building constantly exchanges air, and this is good to a point. Outside air leaks in. Inside air leaks out. When an outside door is opened, the exchange occurs. What we try to control is the rate (speed) of the exchange. The more cracks there are around the windows and doors, the faster the exchange.

Insulation is rated by its effectiveness in two ways. One is by the resistance it offers to the flow of heat under known conditions. This resistance is designated by the letter R. The second way is by determining the amount of heat that will pass through it under known conditions. This is designated by the letter C or U.

The R value is the designation you are most interested in. The R factor of a concrete wall 18 inches thick is about the same as 1 inch of mineral wool insulation. The thickness of a material has little to do with its insulating qualities.

When deciding on the insulation to use, consider R factor and job requirements. For example, a 3-inch-thick batt of one type of insulation might have an R10 value. A 3-inch-thick batt of another type insulation might have an R9 rating. Or, a 2-inch-thick rigid foam material might have an R10 rating.

Most insulating materials are stamped with an R value. Where two insulating materials are used together, such as a 2-inch rigid foam material and a 3-inch batt, determine the R value of the combined two materials by addition. If a foam board is R10 and a batt is R10, you have a total value of R20.

Where the material is doubled, the R value is doubled. 2-inch-thick insulation with an R4 value would have an R value of 8 if doubled. Figure 12-1 gives the R value of various materials commonly used in construction.

Insulation Materials

The five most common forms of building insulation are the following:

(1) Batts
(2) Loose fill
(3) Rigid board

Material	Thickness	"R" Value
Air Film and Spaces		
Air space		
Bounded by ordinary mat.	¾" or more	0.91
Bounded by aluminum foil	¾" or more	2.17
Exterior surface resistance	---	0.17
Interior surface resistance	---	0.68
Masonry		
Sand and gravel conc. block	8"	1.11
Sand and gravel conc. block	12"	1.28
Lightweight concrete block	8"	2.00
Lightweight concrete block	12"	2.13
Face brick	4"	0.44
Concrete cast in place	8"	0.64
Building Materials		
Wood sheathing or subfloor	¾"	1.00
Fiberboard insulating sheathing	¾"	2.10
Plywood	⅝"	0.79
Plywood	½"	0.63
Plywood	⅜"	0.47
Bevel lapped siding	½" x 8"	0.81
Bevel lapped siding	¾" x 10"	1.05
Vertical tongue & groove board	¾"	1.00
Drop siding	¾"	0.94
Asbestos board	¼"	0.13
⅜" gypsum lath and ⅜" plaster	¾"	0.42
Gypsum board	⅜"	0.32
Interior plywood panel	¼"	0.31
Building paper	--	0.06
Vapor barrier	--	0.00
Wood shingles	--	0.87
Asphalt shingles	--	0.44
Linoleum	--	0.08
Carpet with fiber pad	--	2.08
Hardwood floor	--	0.71
Windows and Doors		
Single window	--	Approx. 1.00
Double window	--	Approx. 2.00
Exterior door	--	Approx. 2.00

Insulation value of common materials
Figure 12-1

(4) Reflective
(5) Foam

Batts

There are two types of batts or blankets: glass fiber and rock wool. Rock wool is denser and has a slightly higher R value per unit of thickness than glass fiber (3.7 for rock wool as opposed to 3.1 to 3.4 for glass fiber). The two materials can be used interchangeably. Both types of batts and blankets are sold on the basis of their R value, and they cost about the same. Rock wool is relatively heavy, which increases the shipping cost.

Insulation batts and blankets are usually encased in paper, one face of which is a vapor barrier. This may be an asphalt paper or a paper with reflective metal foil backing. Proper installation of the vapor barrier batt and blanket is most important.

The R values for various thicknesses of glass fiber and rock wool batts and blankets are shown in Figure 12-2.

Batts and blankets are available for 16- and

Rated R	Thickness	Kraft paper faced	Foil faced
7	2	x	x
11	3	x	x
13	3⅝	x	-
19	5¼	x	x
22	6	x	x

Rock wool

Rated R	Thickness	Kraft paper faced	Foil faced	Unfaced
3.5	1	-	-	x
4	1⅛	-	-	x
5	1½	-	-	x
7	2¼ - 2¾	x	x	-
11	3½ - 4	x	x	x
13	3⅝	-	-	x
14	5	-	-	x
19	6 - 6½	x	x	x
21	7	x	x	-

Glass fiber

R-values of Rock Wool & Glass Fiber
Figure 12-2

24-inch spacing to fit between studs. Batts are usually 4 feet long. Blankets come in 50-foot lengths packed in a roll. Batt and blanket insulation is normally used during initial construction where it can easily be fitted between the structural members of the floor, walls and ceiling.

Loose Fill

There are four types of loose fill materials:
(1) Glass fibers
(2) Rock wool
(3) Cellulose fiber
(4) Vermiculite

Loose fill is blown or poured in place. Pouring can be done by hand. Blowing is done by blowing machines. All four of these materials can be poured. All except vermiculite can be blown. The R value per inch of thickness of each of these materials is shown below:

	R Per Inch	
	Pouring	Blowing
Glass fiber	2.2	2.2
Rock wool	2.2 - 2.8	2.8
Cellulose fiber	3.7	3.7
Vermiculite	3.0	---

Most of the vermiculite material is sold under the name of Zonolite. It's a mixture of vermiculite and polystyrene beads and has an R value a little better than pure vermiculite.

Pour-in loose fill is designed primarily for application to easily accessible horizontal spaces in floors or ceilings. Loose fill insulation is blown into walls through holes (approximately 2 inches in diameter) drilled at appropriate intervals to ensure application between studs.

Where insulation is applied, moisture problems may develop. This should always be considered when planning insulation of non-insulated houses. Insulation saturated with condensation loses most of its R value.

Rigid Board

Rigid board insulation is available in urethane, styrofoam, glass fiber and polystyrene. As shown below, rigid board has a high R value per inch:

	R Per Inch
Urethane	7.14
Styrofoam	5.41
Glass fiber	4.30
Polystyrene	4.17

Rigid boards resist transmission and absorption of moisture. Polystyrene, styrofoam and urethane boards are flammable and give off noxious smoke. In some areas there may be code restrictions on their use.

Reflective

Rigid insulation boards 1 to 2 feet wide and 1 to 4 inches thick are used as perimeter insulation around the outside edges of houses with concrete slab floors. Since this insulation must be moisture resistant, it is generally made of foamed glass, or foamed plastic such as urethane or styrofoam.

Plastic foam insulation board is also made in

larger tongue-and-groove panels which are glued to the sheathing to cover all exterior wall surfaces. One-inch urethane insulation panels glued to the exterior of an insulated R11 2 x 4-inch stud wall, raises the insulating value to R19. It also cuts down air infiltration through the wall. Plastic foam board is popular because it increases the insulating value of a wall to current standards without altering conventional construction practices.

Ureaformaldehyde and urethane foams using isocyanate compounds should be restricted to those touches surrounding materials, it's ineffective as an insulator and actually becomes a good conductor.

Foam
Foams are used mostly for filling cavity space in existing houses. Special equipment is required for application. Only experienced contractors should handle this stuff.

Which is the Right Material?
The thickness of the material to be installed under

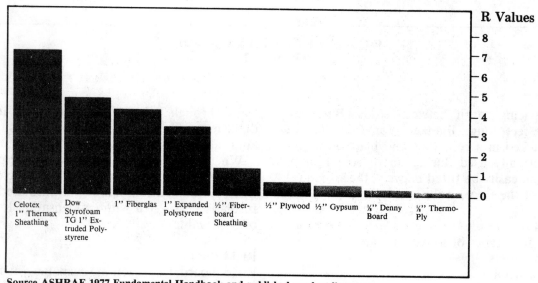

Source ASHRAE 1977 Fundamental Handbook and published product literature

Thermo-Ply—Reg. trademark of Simplex Industries, Adrian, MI
Denny Board—Reg. trademark of Denny-Corp., Calwell, OH
Expanded Polystyrene is sold by many different manufacturers
Styrofoam—Reg. trademark of Dow Chemical manufacturers
Fiberglas—Reg. trademark of Owens-Corning Fiberglas Corporation, Toledo, OH

**Comparative R Values (at 70° mean temperature)
of Sheathing in Available Thicknesses
Figure 12-3**

areas surrounded by fireproof materials. Ureatripolymer and low-flame-spread styrofoam are considered fire-resistant.

Reflective insulation is made from reflective foil such as aluminum, or polished metallic flake glued to reinforced paper. It retards the flow of infrared heat rays passing across an air space. Reflective insulation is available in single sheets, strips formed to create 3 or 4 separate air spaces of 3/4 inch each, and as a combination vapor barrier and reflective surface attached to batts or blankets.

To ensure the efficiency of reflective insulation, install it so the reflective surface always faces an air space of 3/4 inch or more. If the reflective surface

the floor or in the attic isn't important because there's usually plenty of space. The thickness for wall insulation has to be considered; you have only a 3½-inch space to work with. You can add a sheathing-type insulation to the wall or 2 x 6 studs if more insulation is needed.

Let's look at some comparative R values for different sheathing materials. (See Figure 12-3.)

Thermax sheathing is an insulation board consisting of a glass fiber-reinforced polyisocyanurate foam plastic core with aluminum foil facers. It has a uniform closed-cell structure which is exceptionally resistant to the flow of heat. It's available in 4 x 8- and 4 x 9-foot sizes from 1/2 to 3 inches thick. Standard cavity wall sizes of 8- and 9-foot lengths with 16- and 24-inch widths are available.

The sheathing may be installed horizontally on frame walls before the walls are placed into position. (See Figure 12-4.) Or it can be installed after the framing is completed, as shown in Figure 12-5.

Small tears can be repaired with aluminized tape if the sheathing is accidentally damaged. Larger holes or breaks should be "plugged" or patched with appropriately sized pieces of Thermax sheathing held in place with aluminized tape. (See Figure 12-6.)

The R20 Wall

In colder climates you need an insulating value of close to R20. This can be done with conventional 2 x 4 framing by substituting Thermax sheathing for ordinary exterior sheathing material. No major changes in house design, framing lumber or trim are necessary.

Let's use an example. The basic components of an R20 wall system, starting from the interior of the wall, might be as in Figure 12-7.

- 1/2-inch gypsum wallboard
- 6-mil polyethylene vapor barrier
- Nominal 2 x 4-inch framing

Sheathing installed on frame before lifting into position
Figure 12-4

Sheathing installed after framing
Figure 12-5

Repairing small tears
Figure 12-6

R-20 Wall
Figure 12-7

The variations of the wall design are:
- Stud cavity insulation
- Thermax sheathing thickness
- Exterior siding materials

A similar calculation can be made for other interior finishes.

In areas which exceed 4,000 degree days, the system may also include plastic vent strips which are installed horizontally along the exterior top plate prior to the installation of the Thermax.

Based on climate, the following are recommended for vapor barrier/vent use (see Figure 12-8):

Area I (4,000 winter degree days or less)—Use properly installed vapor barrier—6 mil-polyethylene on the interior side of the wall.

Area II (Above 4,000 winter degree days)—Use properly installed 6-mil polyethylene vapor barrier on the interior side of the wall.

Area III (8,000 winter degree days or more)—Use properly installed 6-mil polyethylene vapor barrier and vent strips.

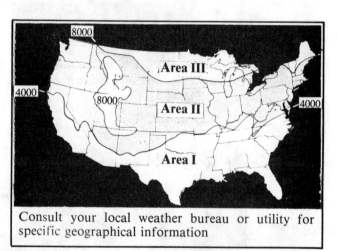

Consult your local weather bureau or utility for specific geographical information

Vapor Barrier and Vent Strip Areas
Figure 12-8

Masonry Systems

Thermax sheathing is applied on the interior side of masonry or concrete wall construction (Figure 12-9). The typical detail for such a system would include the following:

1/2-inch gypsum board
"Hat channel" or wood furring
1-inch Thermax sheathing
8-inch concrete block
Air space
Exterior finish (brick for this example)

The R value for such a wall system would be calculated as follows (Hat channel 14 inches o.c., 15 percent framing factor):

	R Value Thru Furring	R Value Between Furring
Inside surface film	0.68	0.68
½ inch gypsum wallboard	0.45	0.45
Reflective air space (Hat channel)	2.77	---
Reflective air space	---	2.77
1 inch (nominal thickness) Thermax sheathing	7.20	7.20
8-inch concrete block	1.11	1.11
Non-reflective air space	0.94	0.94
4-inch face brick	0.44	0.44
Outside surface film	0.17	0.17
R's at sections	13.76	13.76

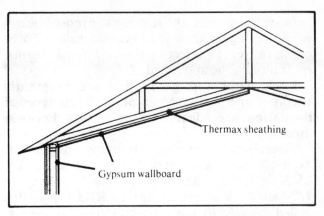

Thermax sheathing scissor truss system
Figure 12-10

Metal-furred underlayment system
Figure 12-9

Scissor Truss Construction

Scissor truss construction, low-pitched roofs and floored attics are examples of assemblies where limited depth of attic space may prevent the installation of adequate thickness of loose fill or batt insulation. It is sometimes desirable to install insulation on the room side of a ceiling because the depth of the attic space does not permit adequate insulation for today's energy efficient design. (See Figure 12-10.) Sheathing having the insulation qualities of Thermax is recommended in these cases.

"A" Frame Roof Construction with Exposed Wood Deck

Where the "A" frame deck is 2-inch tongue-and-groove wood or suitable plywood, the following procedures apply:

• Structural deck should be smooth, dry and free of debris.

• At eave and rake install a wood strip at least 2½ inches wide and flush with the top of insulation sheathing.

• Install Thermax with enough nails to hold in place. Fit snug at perimeter and butt joints.

• Install exterior grade plywood (minimum 3/8-inch thickness) over the sheathing and eave and rake strips. Stagger plywood joints so they do not coincide with joints of insulation. Nail 12 inches o.c. each direction with 3/16-inch-diameter head galvanized nails long enough to penetrate the structural deck at least 1 inch.

• Install the roof material.

When the sheathing is used over a wood deck and a drop ceiling is installed, adequate air circulation to the interior of the structure must be provided to eliminate condensation from the attic space created.

The comparative R values of this system would be as follows:

	R Values
Asphalt shingles and felt	0.50
⅜-inch exterior plywood	0.47
2-inch wood deck	1.89
2-inch Thermax sheathing	14.4
Inside surface film	0.61
Total R Value	17.88

In such a system, the R value at exposed beams would be higher than the indicated value. To be conservative, the effect of this additional thermal resistance has been omitted.

Rock wool and glass fiber are inherently fireproof and nearly vermin-proof. These types of insulation are ideal for application between members.

Fiberboard insulating sheathing panels are available in 4 x 8- and 4 x 9-foot sizes. The 25/32-inch-thick sheathing has an R2.1 value. The 1/2-inch-thick sheathing has an R1.3 value. It is nailed directly to the exterior side of the studs. Unless required by code, it's not necessary to cover the sheathing with building paper. Wood, hardboard and most other siding material can be applied directly over fiberboard sheathing. Where masonry veneer is applied, allow not less than 1/2 inch between the face of sheathing and the back of the veneer. When applying the sheathing, use 1¼-inch-gauge galvanized roofing nails with 7/16-inch head or 8d nails. Figure 12-11 A and B shows details.

Insulating Ceilings and Walls

Most houses have an accessible attic with exposed ceiling framing so that any type of insulation can be applied. The member spacing is usually 16 or 24 inches o.c., so batt or blanket insulation is ideal. Loose fill can also be used by dumping it between the joists and leveling it off.

Be careful to cover all those areas where air infiltration can add to the heating and cooling load of the house. Weatherstrip attic access doors. Also, insulation board can be cut to size and applied to the back of the door. Panels over attic scuttle hole covers should have the insulation fastened to them.

Extend the insulation over the top of the top plate. Install mineral fiber blankets in ceilings by stapling vapor barrier flanges from below, installing unfaced pressure-fit blankets, or laying in blankets from above after the ceiling is in place. The vapor barrier side faces the interior of the house. Don't forget to run the insulation on the exterior wall side of pipes and vents. Repair rips or tears in the vapor barrier with suitable tape. Figure 12-12 A and B illustrates the proper installation of vapor barrier and insulation.

When adding batts over existing insulation, it's best to use unfaced batts. These prevent moisture from condensing in the existing insulation. If faced batts are used, the facing should be stripped off or split at 2-foot intervals to allow free passage of moisture.

Where attic space is not easily accessible, or if flooring is present, loose fill insulation might be the best method. You may have to remove several

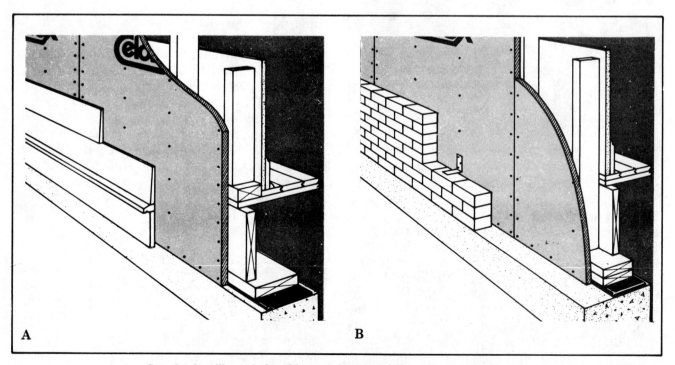

Standard nailing of sheathing used under siding (A) and veneer (B)
Figure 12-11

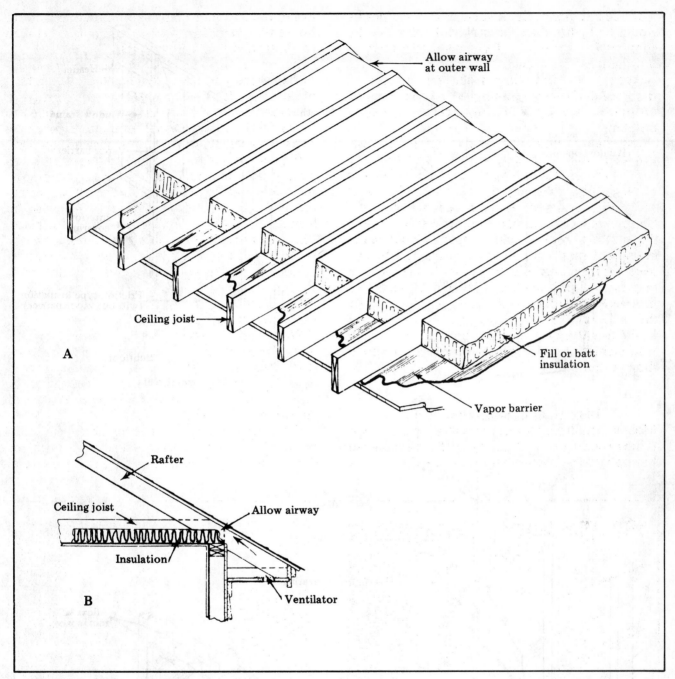

Installing Ceiling Insulation and Vapor Barrier: A, vapor barrier and insulation; B, airway for ventilation
Figure 12-12

floor boards to blow the insulation into place.

Rock wool can be poured between the ceiling joists. Recommended thicknesses and coverage are listed on the bag. Level the insulation out for uniform coverage, and be sure that eave ventilation openings are not blocked. Do not cover recessed lighting fixtures or exhaust fan motors that protrude into the ceiling.

In new construction, glass fiber blankets for the floor, wall and attic are a favorite insulation material. It can be cut to length and applied with a staple gun. 1/4-inch staples work well. The blanket is fitted between the studs or joists, and the tabs are lapped over the interior surface of the member and stapled. (See Figure 12-13 A).

In areas where condensation is a problem and

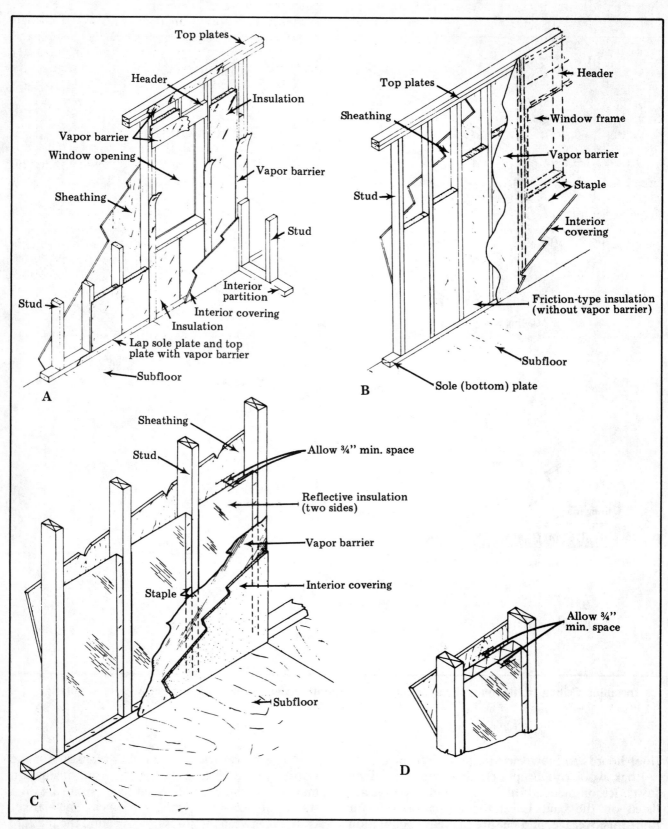

Installing Insulation: A, blanket insulation and vapor barriers in exterior wall; B, vapor barrier over friction-type insulation (enveloping); C, single sheet two side reflective insulation; D, multiple sheet reflective insulation
Figure 12-13

the tabs of the insulation are fastened to the inner surface of the stud/joist members, a vapor barrier should be applied over the insulation. (See Figure 12-13 B.)

At the junction of interior partitions with exterior walls, a vapor barrier should be applied.

Reflective insulation used in walls ordinarily consists of either a kraft sheet faced on two sides with aluminum foil, as shown in Figure 12-13 C, or the multiple-reflective "accordian type," as in Figure 12-13D. There must be a ¾-inch space between the reflective surface and the interior finish material for the insulation to be effective. When using reflective insulation, apply a vapor barrier over the studs of joists before applying the interior finish wall.

A vapor barrier is always required, regardless of the type or location of the insulation in those areas where the average temperature for January is 35 degrees or lower.

Figure 12-9 shows an excellent way to insulate masonry walls with a minimum of lost space. Another method is to frame a wall against the masonry wall and apply insulation in one of the ways previously discussed. However, you lose a little space with this method.

Insulating Floors

Batts or blankets, with the vapor barrier side placed toward the interior of the room, are applied between the joists. If placed before the subfloor is installed, the tabs can be stapled over the top of the joists. The problem with this method is that the insulation has a good chance of becoming rain-soaked before the house is dried in. So you might prefer to wait and insulate the floor after the house is dried in. Figure 12-14 offers three other methods of insulating the floor.

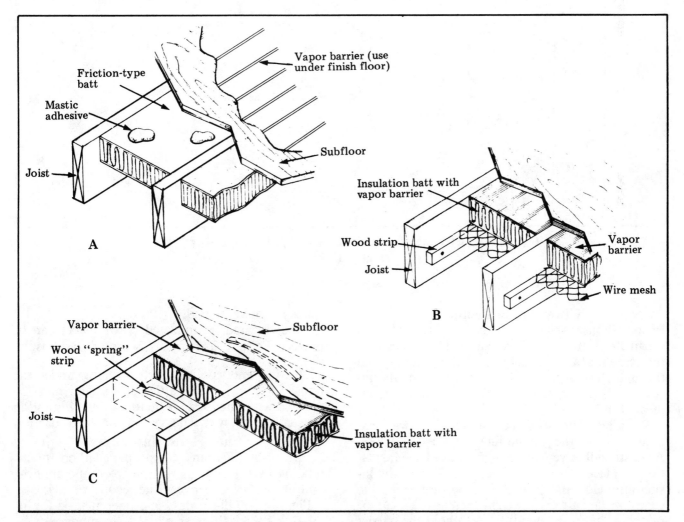

Installation of Vapor Barriers and Insulation in Floor (Unheated Crawl Space): A, friction-type batts; B, wire mesh support; C, wood strip support
Figure 12-14

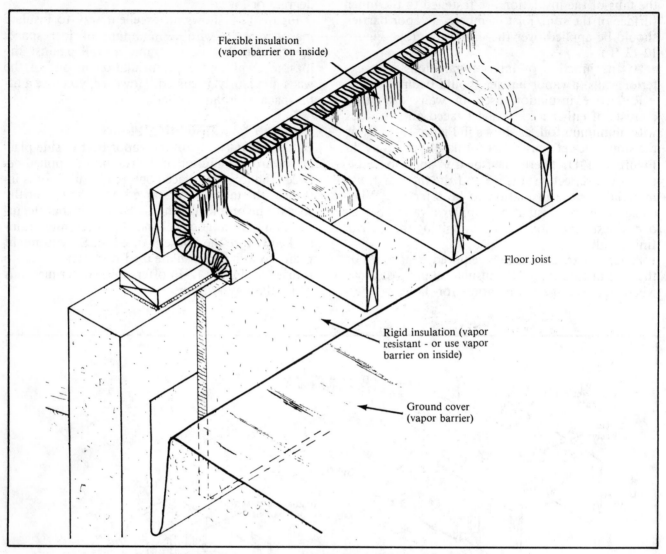

Flexible insulation
(vapor barrier on inside)

Floor joist

Rigid insulation (vapor
resistant - or use vapor
barrier on inside)

Ground cover
(vapor barrier)

Installation of Vapor Barrier and Insulation in Heated Crawl Space
Figure 12-15

Crawl Space Plenum

In some houses the crawl space is used as a forced warm air plenum chamber. Warm air is forced into the crawl space (usually a shallow area) and through floor registers. In such a system, the insulation is placed along the perimeter walls as indicated in Figure 12-15.

Rigid insulation is placed inside of the wall, extending below the ground line to reduce heat loss. A mastic adhesive can be used to hold the insulation in place. A vapor barrier is used over the insulation below the ground line. Be sure to lap the edges of the barrier. A weight such as brick can be placed at intervals over the lap to hold the barrier in place on the ground under the house. The crawl

space is not ventilated.

Pipes located in unheated crawl spaces can be protected from freezing by wrap-around insulation.

Heating and cooling ducts running under the house in crawl spaces, attics or garages, must be covered with duct wrap insulation. Duct insulation is usually faced with a heavy foil as a vapor barrier.

Other House Insulation

Figure 12-16 illustrates the proper method for insulating two-story construction. Note the application of insulation between the second floor joists.

Figure 12-17 illustrates the proper method for insulating knee wall areas of a 1½-story house.

Knee walls are partial walls which extend from

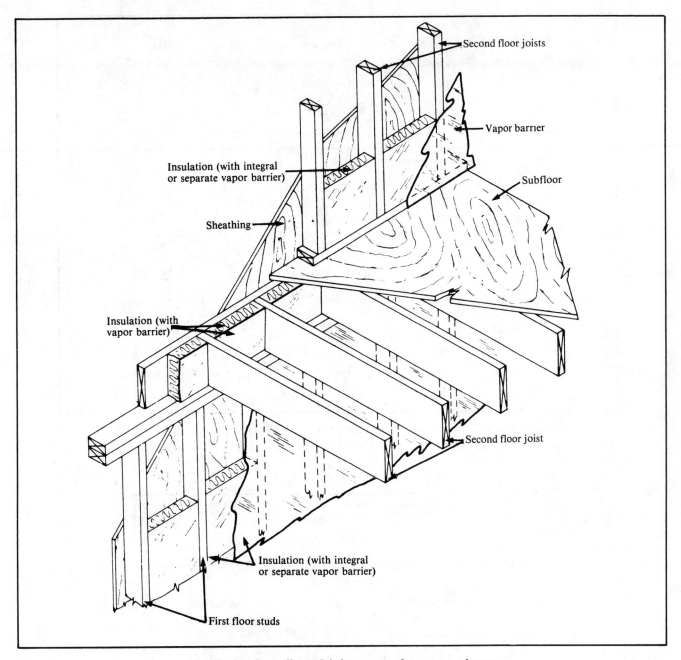

Insulation in walls and joist space of two-story house
Figure 12-16

the floor to the rafters. Place the batts, with the vapor barrier facing down, between joists from the outside wall plate to the knee wall. An airway for attic ventilation at the junction of the rafter and exterior wall is essential.

Use batt or blanket insulation between the rafters at the sloping portion of the heated room. Be sure to provide an airway between the top of the insulation and the roof sheathing around obstructions. This allows for the movement of air from the knee wall area to the attic area above the second floor rooms.

Figure 12-18 illustrates the proper method of insulating finished basement rooms. The same principles apply to basement insulation as framed construction. (Figure 12-9 offers one procedure for insulating a basement.)

When a vapor barrier has not been used under the concrete slab, apply a barrier on top of the slab before installing the sleepers. Remember to place

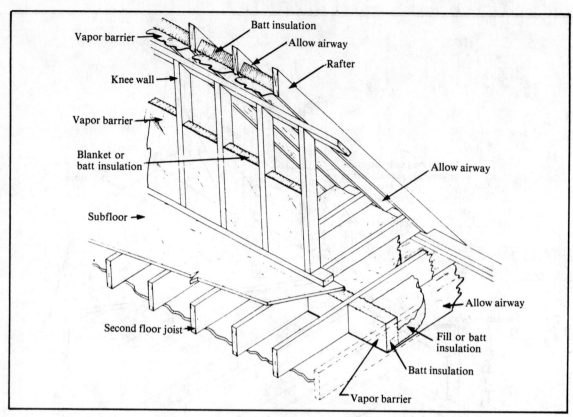

Installing Vapor Barrier and Insulation in Knee-wall Areas of 1½ Story House
Figure 12-17

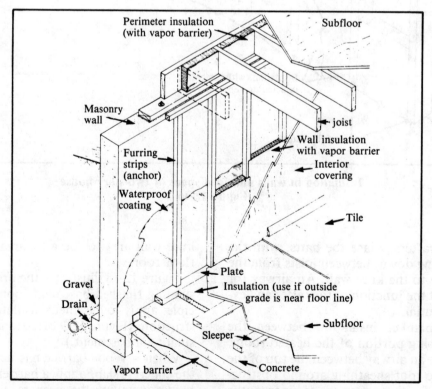

Installing Vapor Barrier in Floor and Wall of Finished Basement
Figure 12-18

insulation between the joists or along stringer joists around the perimeter of the floor framing above the foundation walls. The barrier faces the interior of the basement.

Figure 12-19 reflects the insulation R values recommended for different areas of the country. The figures are the recommendation of Owens-Corning Fiberglass and take into account national weather data, energy costs and projected increases. Figure 12-20 shows how much fiber, wood, or mineral type insulation is required to reach the desired R value.

Weatherstripping and Caulking

Air leaks around doors and windows have to be sealed if the house is going to be energy efficient. It does little good to insulate a house and then leave the door open, winter and summer. And that's what it amounts to if you don't caulk and weatherstrip those areas where air infiltration is likely.

There are many good weatherstripping materials available. These include sponge rubber with adhesive backing, plastic and rubber gaskets, spring bronze strips, interlocking bronze flanges, and magnetic strips. Non-hardening caulk is recommended where caulk is required. Use a top-grade caulk; inferior caulks don't last long. The quality material doesn't cost that much more and it'll last a while. Elastomeric caulks are the most durable and most expensive. They include silicones, polysulfides and polyurethanes. Latex, butyl or polyvinyl based caulks are good. They will bond to most surfaces.

Fill extra wide cracks with oakum, caulking cotton, sponge rubber, or glass fiber. It's a good backup for elastomeric caulks.

Caulk such places as:

• Joints between window and door frames and sidings (brick, plywood, hardboard, etc.).

• Between window sills and siding.

• Between drip caps and siding (windows and doors).

• At siding corners.

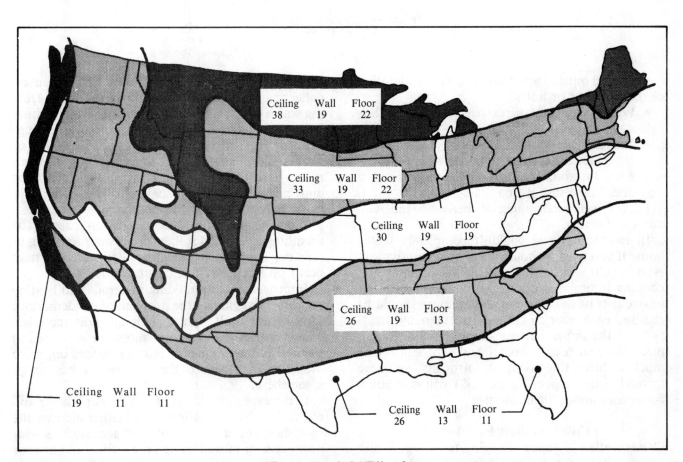

Recommended "R" values
Figure 12-19

Ceiling, double layers of batts
R-38, Two layers of R-19 (6'') mineral fiber
R-33, One layer of R-22 (6½'') and one layer of R-11 (3½'') mineral fiber
R-30, One layer of R-19 (6'') and one layer of R-11 (3½'') mineral fiber
R-26, Two layers of R-13 (3⅝'') mineral fiber

Ceiling, loose fill mineral and batts
R-38, R-19 (6'') mineral fiber and 20 bags of wool per 1,000 S.F. (8¾'')
R-33, R-22 (6½'') mineral fiber and 11 bags of wool per 1,000 S.F. (5'')
R-30, R-19 (6'') mineral fiber and 11 bags of wool per 1,000 S.F. (5'')
R-26, R-19 (6'') mineral fiber and 8 bags of wool per 1,000 S.F. (3¼'')

Walls, using 2'' x 6'' framing
R-19, R-19 (6'') mineral fiber batts

Walls, using 2'' x 4'' framing
R-19, R-11 (3½'') mineral fiber batts and 1'' polystyrene sheathing
R-13, R-13 (3⅝'') mineral fiber batts
R-11, R-11 (3½'') mineral fiber batts
R-20 (see page 215)

Floors
R-22, R-22 (6½'') mineral fiber
R-19, R-19 (6'') mineral fiber
R-13, R-13 (3⅝'') mineral fiber
R-11, R-11 (3½'') mineral fiber

Insulation recommendations
Figure 12-20

• Around outside water faucets, electrical outlets or other breaks or holes.

• Where pipe and wires penetrate the ceiling below and enter an unheated attic.

• Where the chimney or masonry meets siding.

A lot of air can penetrate under exterior doors where the threshold is worn or ill-fitted. Install a new one. Make sure it fits.

Fit the weatherstripping properly around windows and doors.

In most cases you couldn't completely seal a house if you tried. A house needs some ventilation, even in cold weather, to release built-up moisture, cooking fumes, and odors. Normally, one quarter to one half of the air volume in a house should be changed each hour to keep it from seeming "stuffy." At the most, a house should have one complete air change per hour. And it doesn't take much to bring this about. A lot of air can move through a tiny space. Figure 12-21 will give you a better idea about air infiltration.

Condensation Problems

Condensation is vapor changing to liquid. Water vapor within the house, if unrestricted, moves through the walls or ceiling during the heating season. When it meets a cold surface, it condenses, collecting generally in the form of frost or ice. During warm periods, it melts. The water from melting ice in unvented attics may drip to the ceiling. Moisture can also soak into the roof sheathing or rafters and cause decay. In walls, water from melting frost may run out between the siding laps and cause staining, or soak into the siding and cause paint blistering and peeling.

Wood and wood-base materials used for sheathing and panel siding often swell, resulting in bowing, cupping, or buckling. Efflorescence may occur on brick or stone of an exterior wall.

Improvements in building materials and heating equipment aggravate the problem of condensation. Houses are built tighter with air-tight materials which restrict the escape of moisture generated in the house. Every time the faucet is turned on, more moisture is added to the interior. It has to go somewhere.

If the excess water vapor is not properly removed, such as by ventilation, it will either increase the humidity, condense on cold surfaces such as window glass, or move through the walls, floor or roof cavities. Heating systems with humidifiers will increase the humidity and add to the problem. The

Building Component	"Standard" One Air Change Per Hour	"Below Standard" Two Air Changes Per Hour	"Poor" Three Air Changes Per Hour
Building with cellar OR	Tight, no cracks, caulked sills, sealed cellar windows, no grade entrance leaks.	Some foundation cracks, no weatherstripping on cellar windows, grade entrance not tight.	Stone foundation, considerable leakage area, poor seal around grade entrance.
Building with crawl space or on posts.	Plywood floor, no trap door leaks, no leaks around water, sewer and electrical openings.	Tongue-and-groove board floor, reasonable fit on trap doors, around pipes, etc.	Board floor, loose fit around pipes, etc.
Windows	Storm windows with good fit.	No storm windows, good fit on regular windows.	No storm windows, loose fit on regular windows.
Doors	Good fit on storm doors.	Loose storm doors, poor fit on inside door.	No storm doors, loose fit on inside door.
Walls	Caulked windows and doors building paper used under siding.	Caulking in poor repair, building needs paint.	No indication of building paper, evident cracks around door and window frame.

Usually building components are not all in the same infiltration category. You can estimate the approximate rate by considering how many of the components are in each category. For example, if two components are in the three air change category and two are in the two air change category, the overall infiltration would be 2 ½ air changes per hour.

Air infiltration categories
Figure 12-21

more efficient the insulation is in retarding heat transfer, the colder the outside surfaces become and the greater the potential for condensation. Moisture migrates toward cold surfaces and will condense or form as frost or ice on these surfaces.

When you have the walls open it's easy to prevent condensation problems. Envelop the walls and ceilings behind the finish materials in a polyethylene film, ventilate them properly and you've got it. It's like covering the baby's crib mattress with a protective covering.

Water vapor from the soil of crawl-space houses does not normally affect the interior. However, it can cause problems in exterior walls over the area and within the crawl space. Figure 12-22 illustrates one way to solve this problem.

Concrete slabs without radiant heat are sometimes subjected to surface condensation. Figure 12-23 illustrates one way to correct the problem before it occurs.

Figures 12-24 through 12-26 show the proper locations of vapor barriers and insulation.

Where loose fill insulation has been used in walls and ceilings and no new interior covering is planned, a vapor-resistant coating should be applied to the inside surface. One method for applying such a coating is to paint the interior surfaces of all outside walls with two coats of aluminum primer. This is then covered with the finish paint. This is not as effective as polyethylene but sometimes it's the best you can do.

If the exterior wall covering is permeable enough to allow the moisture to escape from the wall, a vapor-resistant coating on the inside should be adequate. Moisture will, however, continue to collect in the insulation. A material or coating of low permeability on the outside can retard the escape of moisture that has been forced into the wall from the inside. As an alternative finish, penetrating stains may be used on wood surfaces. Staining will not form a coating on the wood surface and so will not retard the movement of moisture, and it doesn't peel or blister.

Where the older house has a paint-peeling problem from condensation, paint the siding with a porous latex paint. White paint does not fade, keeps its appearance, and is less obvious when touch-up is required because of subsequent peeling.

Ventilation

Adequate ventilation in attics and crawl spaces is essential in all houses. The free movement of air through these areas is necessary in hot areas as well as in cold areas. It helps prevent condensation and heat build-up, and discourages termites.

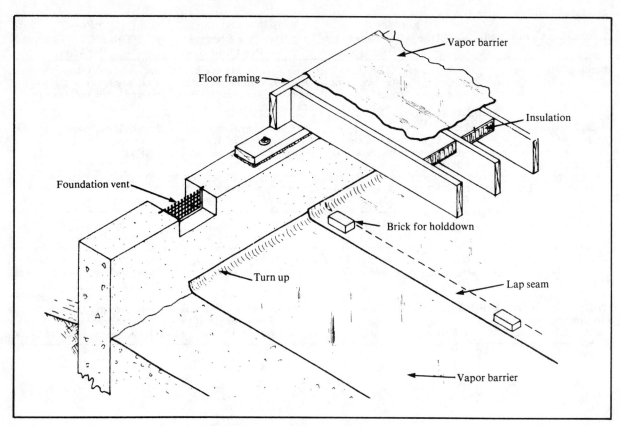

Vapor barrier for crawl space (ground cover)
Figure 12-22

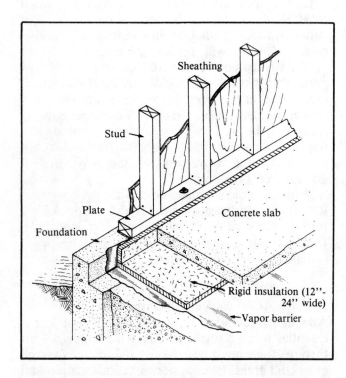

Installation of vapor barrier under concrete slab
Figure 12-23

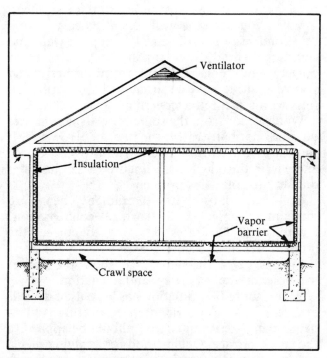

Location of vapor barriers and insulation in crawl
space of another one-story house
Figure 12-24

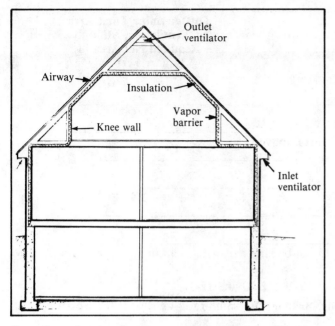

Location of vapor barriers and insulation in 1½ story house with basement
Figure 12-25

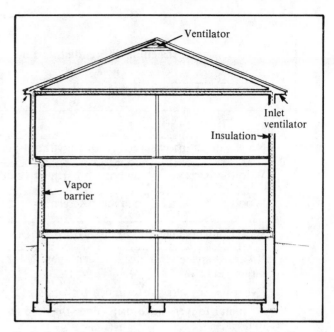

Location of vapor barriers and insulation in full two-story house with basement
Figure 12-26

Foundation vents are available for crawl space construction. These are generally placed near the top of the masonry wall. In concrete block foundations, the ventilator is usually the size of an 8 x 16-inch block. In heated crawl spaces the ventilator is not used. A vapor barrier and insulation is used as previously discussed.

The amount of ventilation required for crawl spaces is based on the total square foot area of the house and the presence of a vapor barrier soil cover. (See Figure 12-27.)

Air flow through the vent is restricted by the screen and louvers on the vent. The type of mesh or screen and louvers must be considered when determining the size vent to use. Figure 12-28 will help you with this.

Attic ventilation can be provided by various types of ventilators. The minimum amount of attic or roof space ventilation required is determined by the total ceiling area. Figures 12-29 through 12-31 show various roofs and ventilation systems.

Determine the total net area of ventilators by applying the rules in Figure 12-28. Divide the total area by the number of ventilators used to find the

Crawl Space	Ratio of Total Net Ventilating Area To Floor Area[1]	Minimum Number of Ventilators[2]
Without vapor barrier	1/150	4
With vapor barrier	1/1500	2

[1]The actual area of the ventilators depends on the type of louvers and size of screen used—see Figure 12-28.

[2]Foundation ventilators should be distributed around foundation to provide best air movement. When two are used, place one toward the side of prevailing wind and the other on opposite side.

Crawl space ventilation
Figure 12-27

Obstructions in Ventilators-Louvers and Screens	To Determine Total Area of Ventilators, Multiply Required Net Area in Square Feet by[2]
¼ inch mesh hardware cloth	1
⅛ inch mesh screen	1¼
No. 16 mesh insect screen (with or without plain metal louvers)	2
Wood louvers and ¼ inch mesh hardware cloth[3]	2
Wood louvers and ⅛ inch mesh screen[3]	2¼
Wood louvers and No. 16 mesh insect screen[3]	3

[1]In crawl-space ventilators, screen openings should not be larger than ¼ inch; in attic spaces no larger than ⅛.

[2]Net area for attics determined by ratios in Figures 12-29, 12-30 and 12-31.

[3]If metal louvers have drip edges that reduce the opening, use same ratio as shown for wood louvers.

Ventilating Area Increase Required If Louvers and Screening are used in Crawl Spaces and Attics
Figure 12-28

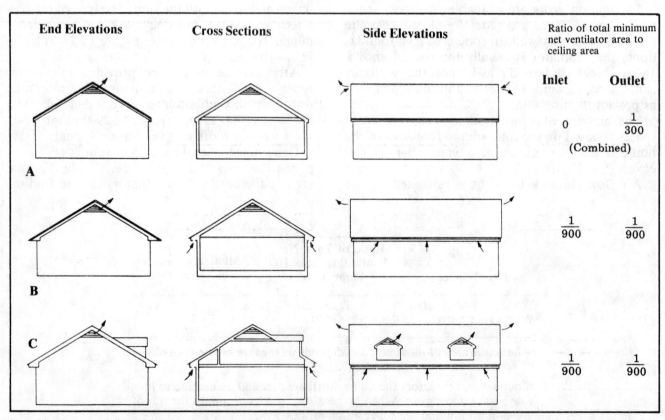

Ventilating Area of Gable Roofs: A, louvers in ends walls; B, louvers in end walls with additional openings at eaves; C, louvers at end walls with additional openings at eaves and dormers. Cross section of C shows free opening for air movement between roof boards and ceiling insulation of attic room
Figure 12-29

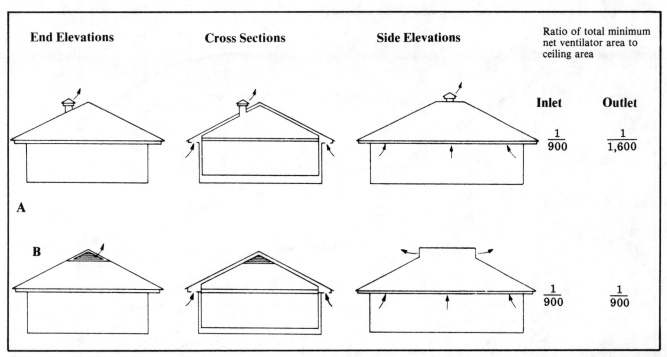

Ventilating Areas of Hip Roofs: A, inlet openings beneath eaves and outlet vent near peak; B, inlet openings beneath eaves and ridge outlets
Figure 12-30

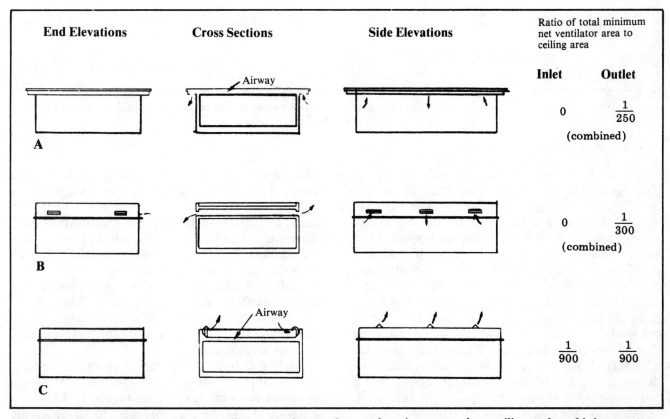

Ventilating Area of Flat Roofs: A, ventilator openings under overhanging eaves where ceiling and roof joists are combined; B, for roof with a parapet where roof and ceiling joists are separate; C, for roof with a parapet where roof and ceiling joists are combined
Figure 12-31

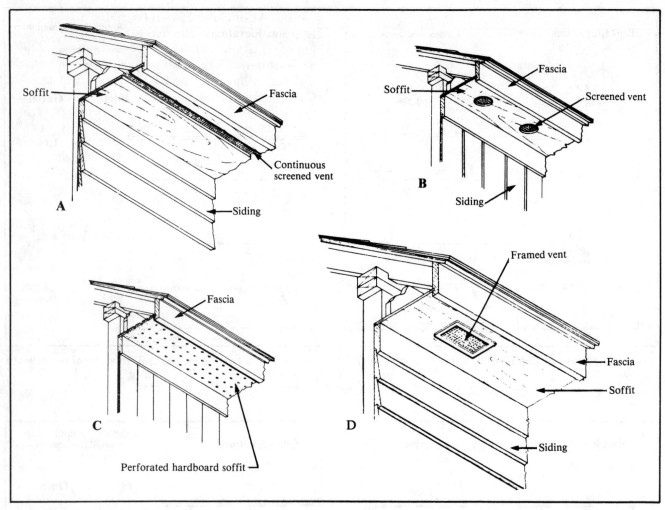

Inlet Ventilators in Soffits: A, continuous vent; B, round vents; C, perforated; D, single ventilator
Figure 12-32

recommended square foot area of each. For example, a gable roof similar to (B) in Figure 12-29 with inlet and outlet ventilators, has a minimum required total inlet and outlet ratio of 1/900 of the ceiling area. If the ceiling area of the house is 1350 square feet, each inlet and outlet ventilator area should be 1350 divided by 900, or 1½ square feet.

If ventilators have a No. 16 mesh screen and plain metal louvers (Figure 12-28) the minimum gross area must be 2 x 1½, or 3 square feet. When one outlet ventilator is used at each gable end, each should have a gross area of 1½ square feet (3 divided by 2). When there are three soffit inlet ventilators on each side, each ventilator should have a gross area of 0.5 square feet. For long houses, use 6 or more on each side.

When it comes to installing attic vents, I use a rule of thumb: one square inch for each square foot of attic space, and forget about the screen mesh and louver sizes. In an attic with a 1350-square-foot ceiling I would use a vent with about 672 square inches (24 x 28-inch vent) on each gable end. I like to make sure there's plenty of attic ventilation. Throw in a few soffit inlet vents on each side and you're in good shape.

Cathedral ceilings require the same type of ventilation as flat roofs. A continuous ridge vent is also desirable. Even with holes in the ridge plate, air travel through the rafter space is sluggish without a ridge vent.

Inlet vents in the soffit can be of several designs. The important thing is to distribute them to eliminate dead air space. (See Figure 12-32.)

Figure 12-33 shows one method of installing vents in an open cornice design.

Outlet vents should be installed at the highest

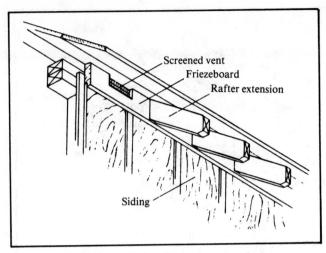

Frieze ventilator (for open cornice)
Figure 12-33

point feasible in a gable roof, as shown in Figure 12-34.

Attic exhaust fans are another way to move air quickly through the attic. They can be thermostatically activated to improve uniform ventilation.

Additional Protective Measures

In snow belt areas water leakage into walls and ceiling areas can be caused by the "freeze and thaw" cycle of snow, which causes ice dams on the roof and backs up water under the shingles and down through the sheathing. This can happen after a heavy snowfall when there is sufficient heat loss from the interior to melt the snow on the roof. The water moves down the roof to the colder overhang of the roof, where it freezes. This produces a ledge of ice and backs up water, which can enter the wall or drip down onto the ceiling finish. (See Figure 12-35 A.)

To minimize or eliminate this problem, reduce the attic temperature in the winter so that it is only slightly above outdoor temperatures. Add insulation in the attic to retard heat movement from the interior; add inlet and outlet vents; or apply a 36-inch-wide, 50-pound weight flashing strip of roll roofing paper along the eave line before reshingling (Figure 12-35 B). This keeps water seepage underneath the shingles to a minimum.

Where "freeze and thaw" creates dams in the valleys, it might be necessary to lay electric-thermal wire in the area in a zig-zag pattern. The wire is

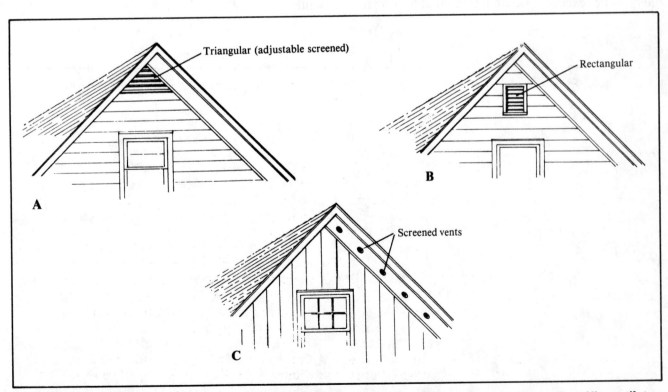

Gable Outlet Ventilators: A, triangular gable end ventilator; B, rectangular gable end ventilator; C, soffit ventilators
Figure 12-34

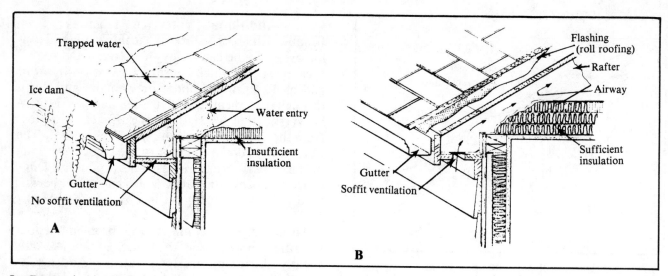

Ice Dams: A, insufficient insulation and ventilation can cause ice dams and water damage; B, good ventilation, insulation, and roof flashing
Figure 12-35

connected during snowfall to keep the ice dam from forming.

Summary

A properly insulated house with storm windows and doors reduces heating and cooling costs. You have some excellent materials available for in-sulating any house. It's your job to select the right material and install it correctly. Your customer and future owners will appreciate the value you build into the work you do.

Insulation will pay for itself over a reasonable period of time. It is one of the very few things that will!

13

Asphalt Roofing

More than 80 percent of all residential roofing applied in the U.S. is asphalt, either shingles, roll roofing or saturated felt.

Shingles and roll roofing are outer roof coverings. They are exposed to the weather and designed to withstand the elements. Saturated felts are inner roof coverings. They provide underlayment protection for the exposed roofing materials.

As outer roof coverings, shingles and roll roofing have three basic components that provide the protection and long-term wear:

1. A base material made of an organic felt or fiberglass mat. This base supports the other components and gives the product the strength to withstand handling, installation and service conditions.

2. A specially formulated asphalt coat which resists weathering and remains stable under severe temperature extremes.

3. A surfacing of ceramic-coated mineral granules which shields the asphalt coating against the sun's rays, adds color and provides fire resistance.

Shingles

Over the last 10 years the trend in asphalt roofing has been to use fiberglass base shingles rather than conventional organic or paper base.

Asphalt shingles are the most common roofing material used today. They are manufactured as strip shingles, interlocking shingles and large individual shingles.

Strip shingles are rectangular, measuring approximately 12 inches wide by 36 inches long, and may have as many as five cutouts. Cutouts separate the shingle's tabs so the finished roof looks like it is made of many smaller units. Strip shingles are also made without cutouts to produce a much different effect.

Many of the shingles are treated with a factory-applied, self-sealing adhesive—generally a thermoplastic material—which is activated by the heat of the sun after the shingle is in place. Exposure to the sun's heat bonds each shingle securely to the one below for greater wind resistance. In the spring, summer and fall this self-sealing action takes just a few days. In winter, it can take much longer, depending on the temperature, roof slope and orientation toward the sun.

Strip shingles with a surface of factory-applied ceramic-coated mineral granules add color and protection to the roof. Colors range from white and black to hues of red, brown and green.

Strip shingles are available with trimmed buttlines to give a deeply textured look to the finished roof. They also come embossed or built up from a number of laminations of base material to give a three-dimensional effect. Each of these shingle characteristics—staggered buttlines, embossing and laminations—can be combined in various ways to create textures on the finished roof surface that resemble wood, slate or tile.

Interlocking shingles are designed to provide resistance to strong winds. The shingles come in various shapes and with various types of locking devices that provide a mechanical interlock of the roof. Large individual shingles are generally rectangular or hexagonal. Typical strip and interlocking shingles are illustrated and described in Figure 13-1.

Roll Roofing

Roll roofing comes in rolls approximately 36 inches wide and 36 to 38 feet long. It weighs from 40 to 90 pounds per roll and is available with either a smooth surface or a surface embedded with mineral granuales.

Mineral-surfaced roll roofing is available with a granule-free selvage edge that shows the amount each succeeding course should overlap the preceding course. Be sure to follow the manufacturer's recommendations for top lap as well as side and end lap. The amount of overlap determines how much of the material is exposed to the weather and the extent of coverage to the roof surface.

Roll roofing is also used as a flashing material.

Saturated Felts

This material has dry felt impregnated with asphalt. It is used primarily as an underlayment for asphalt shingles, roll roofing and other types of roofing materials and as sheathing paper. It's available in different weights, the most common being No. 15, weighing about 15 pounds per 100 square feet ("square"), and No. 30, which weighs about 30 pounds per square.

Selecting the Right Asphalt Roofing Product

The "right" roof for your job depends on roof slope, coverage, local wind conditions and esthetics. One of the most critical is the slope of the roof because it affects surface drainage. Some types of roofing materials can't be used on steep or low pitch roofs. Good drainage is essential to every asphalt roof and can make the difference between a weathertight roof and one that leaks. I'm sure that you've seen a low-pitched roof covered with shingles leaking like a rusted-out tub. Apparently some do-it-yourselfers (and some builders, too) think that any roofing material will do the job on most any roof. Unfortunately, it just isn't so.

The slope of the roof is usually set by the style of the house. A number of common styles are shown in Figure 13-2.

Generally, asphalt shingles may be used on roof slopes between 4 inches and 21 inches per foot, using standard application methods. Beyond this maximum slope, special steep slope application procedures must be followed. Square-tab strip shingles may be used on slopes between 2 inches

and 4 inches per foot if special low slope application procedures are followed.

The minimum slope for roll roofing depends on the application method and the type of roll roofing used. As a general rule, roll roofing may be used on roof slopes down to 2 inches per foot if the roofing is applied by the exposed nail method. With the concealed nail method and at least 3 inches of top lap, roll roofing may be used on slopes as low as 1 inch per foot. Don't use roll roofing on slopes of less than 1 inch per foot unless specified by the manufacturer.

Slope limitations for asphalt roofing materials are shown in Figure 13-3.

Coverage and Exposure

Exposure is that portion of the roofing exposed to the weather after the roofing is installed. Exposure for asphalt roofing products are set by the manufacturer. (See Figure 13-1.)

Coverage is an indication of the amount of weather protection the roofing provides. Depending on the number of plies or layers of material that lie between the exposed surface of the roofing and the deck, the material is designated single, double, or triple coverage. Where the number of plies varies, coverage is usually considered to be that which exists over most of the roof area. For example, where no significant roof area has less than two thicknesses of material, the installation would be considered double coverage.

Asphalt roll roofings are generally considered single coverage products because they provide a single layer of material over the greater part of the roof area. An exception is roll roofing applied with a 19-inch overlap and 17-inch exposure. This is considered a double coverage material. Asphalt strip shingles are also considered double coverage materials because their top layer is 2 inches or more greater than their exposure.

Esthetics

Asphalt roofing shingles give you, the builder, lots of features to sell. Many colors are now available in addition to such standard colors as white, black and light pastels. Asphalt roofing is available in blends of red, brown, and green. If your client wants to create a roof that relates to the natural environment, consider greens or autumn reds. Color can be used to complement and reinforce the natural colors of other building elements such as brick walls or wood siding. Figure 13-4 is a guide to choosing shingle color for compatibility with colors of siding, trim, shutters, and doors.

PRODUCT	Configuration	Per Square			Size			Underwriters Laboratories Listing
		Approximate Shipping Weight	Shingles	Bundles	Width	Length	Exposure	
Self-sealing random-tab strip shingle 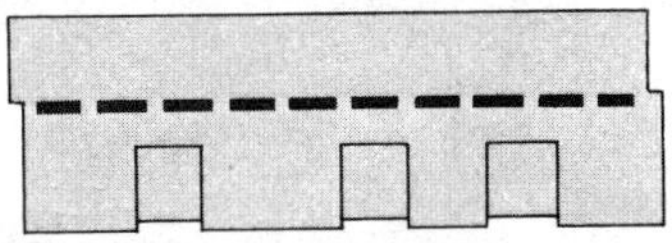 Multi-thickness	Various edge, surface texture and application treatments	285# to 390#	66 to 90	4 or 5	11½" to 14"	36" to 40"	4" to 6"	A or C - Many wind resistant
Self-sealing random-tab strip shingle 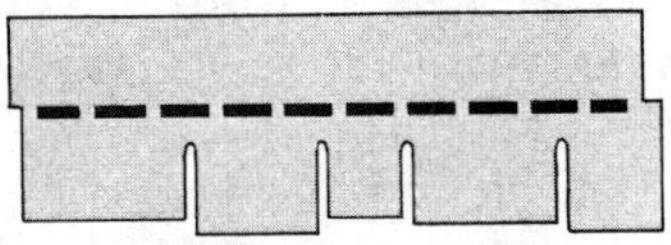 Single-thickness	Various edge, surface texture and application treatments	250# to 300#	66 to 80	3 or 4	12" to 13¼"	36" to 40"	5" to 5⅝"	A or C - Many wind resistant
Self-sealing square-tab strip shingle 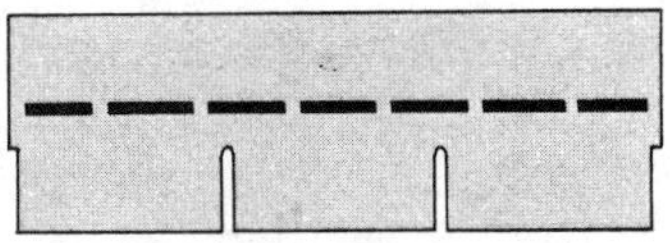 Three-tab	Two-tab or Four-tab	215# to 325#	66 to 80	3 or 4	12" to 13¼"	36" to 40"	5" to 5⅝"	A or C - All wind resistant
	Three-tab	215# to 300#	66 to 80	3 or 4	12" to 13¼"	36" to 40"	5" to 5⅝"	
Self-sealing square-tab strip shingle 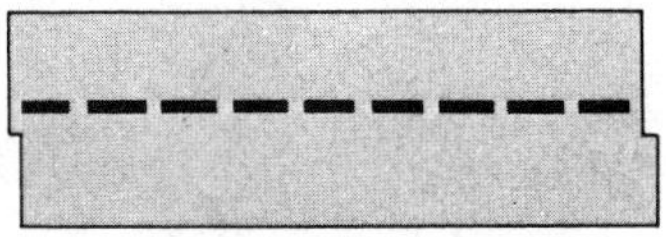 No-cutout	Various edge and surface texture treatments	215# to 290#	66 to 81	3 or 4	12" to 13¼"	36" to 40"	5" to 5⅝"	A or C - All wind resistant
Individual interlocking shingle 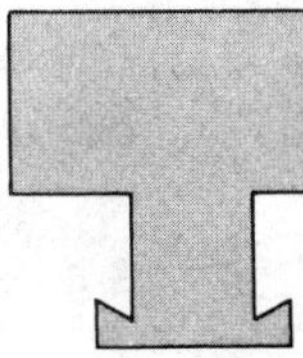 Basic design	Several design variations	180# to 250#	72 to 120	3 or 4	18" to 22¼"	20" to 22½"	—	C - Many wind resistant

Typical Asphalt Shingle
Figure 13-1

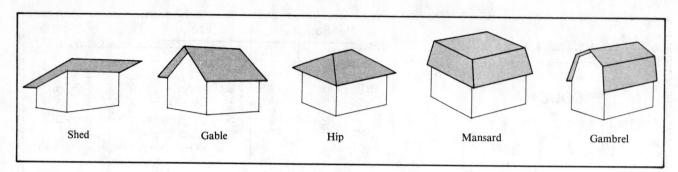

Common Roof Styles
Figure 13-2

Roof color can also be an important design element. A small house may be given added dimension by a light colored roof that will direct the eye upward and help create a sense of spaciousness. Dark colors on a steep roof create the opposite effect, bringing the structure down to scale visually.

Laminated asphalt shingles can complement the overall architectural effect of a building. Many of these shingles offer the "look" of wood but with the long-term wear and fire safety of asphalt shingles. Others offer the "look" of slate and tile. All create interesting visual effects of light and shadow over the roof area because of their three dimensional construction.

Roofing manufacturers have been working to make you look good when selling a prospective customer. Learn to use these little tricks that are available not only in asphalt shingles but in most other building materials as well. It will put you a mile ahead of the builder who doesn't have this knowledge.

Building is an art. You pick up a tip here and a tip there. The more you learn, the more you can apply to the art. No one just builds anymore. Peo-

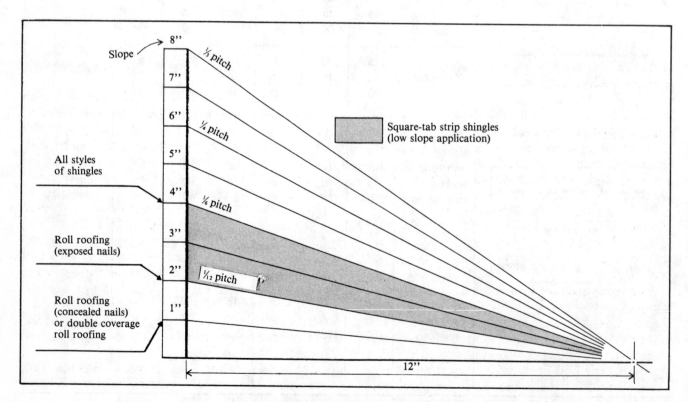

Minimum Pitch and Slope Requirements For Various Asphalt Roofing Products
Figure 13-3

Roof Shingles	Siding	Trim	Shutters and/or Doors
White	White	White	Deep Gold, Maroon
	White	Gray	Charcoal
	Green	White	Dark Brown, Dark Green
Black	White	White	Black, Maroon
	Yellow	White	Black, Deep Olive Green
	Gold	White	Black, Deep Olive Green
Gray	Red	White	Black, White
	Yellow	White	Gray, Charcoal, Green
	Coral Pink	Light Gray	Charcoal
Red	White	Gray	Charcoal
	White	White	Red
	Beige	White	Dark Brown
Brown	White	White	Dark Brown, Terra Cotta
	Green	White	Dark Brown, Dark Green
	Yellow	White	Dark Brown, White
Green	White	White	Dark Green, Black
	Yellow	White	Dark Green
	Light Green	White	Dark Green, Terra Cotta

Asphalt Shingle Color Guide
Figure 13-4

ple look to builders for creativity. That's why most manufacturers hired specialists to come up with interesting products. So now, even barns are works of art.

Estimating How Much Roofing Is Required

Every roofing job you have will include accessories like starter strips, drip edges, valley flashings and hip and ridge shingles. Before the job begins, estimate the quantity of each material. These estimates are based on calculations made from the dimensions of the roof.

The calculations are fairly simple. Some handy rules of thumb can be used to simplify the process even further.

Estimating area for simple roofs: Roofs come in many shapes and styles, but virtually every kind of roof consists of plane surfaces that can be divided into simple geometric shapes—squares, rectangles, trapezoids and triangles. Roofing area calculations are simply the measurement of these basic shapes.

The simplest type of roof has no projecting dormers or intersecting wings. Examples of this type are shown in Figure 13-5. Each of the illustrated roofs comprises one or more rectangles. The area of the entire roof in each case is the sum of the areas of each rectangle.

For the shed roof which has only one rectangle, the area is found by simply multiplying the rake line by the eaves line, or A x B. The gable roof comprises two rectangular planes, and its area is found by multiplying the sum of the rake lines by the eaves line, or A(B + C). For the gambrel roof, four rake lines are involved. The total area is calculated by multiplying the sum of the rake lines by the eaves line, or A(B + C + D + E).

Estimating area for complex roofs: The more complex roofs include those with intersecting wings or dormer projections through the various roof planes. Area calculations for these roofs use the same basic approach taken for simple roofs but involve a number of subdivisions of the roof surface that are calculated separately, then added together to get the total roof area.

Use building plans, if available, to find the roof dimensions. Otherwise, you may have to measure the roof yourself.

However, there is an alternative that lets you

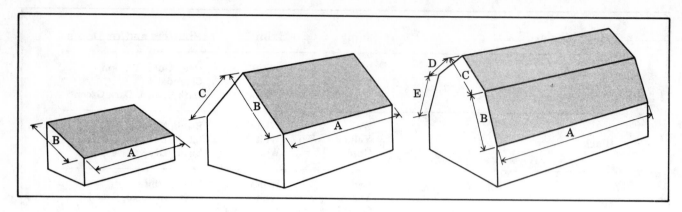

Simple Roofs
Figure 13-5

make the measurements without actually taping off the various lengths. It involves calculating the projected horizontal areas of the roof, then combining these with the slope or slopes to find the true areas. Both the roof slope and the horizontal projection of the various roof surfaces are determined indirectly as described below. The tables in the following sections convert indirect measurements and calculations to actual lengths and roof areas.

Roof pitch and slopes: The degree of incline of a roof is usually expressed as its "pitch" or "slope." Pitch is the ratio of the rise of the roof to the span of the roof. (See Figure 13-6.) Slope is the ratio of rise in inches to horizontal run in feet (run equals half the span). For example, if the span of a roof is 24 feet and the rise is 8 feet, the pitch is 8/24, or 1/3. Expressed as slope, the same roof is said to rise 8 inches per foot of horizontal run. If the rise of the same roof span were 6 feet, the pitch would be 1/4 and its slope would be 6 inches per foot of

run. Whether a particular roof incline is expressed in pitch or slope, the results of the area calculations will be the same.

It isn't necessary to go onto a roof to measure pitch or slope. You can estimate the slope from the ground with a carpenter's folding rule: Stand away from the building and form the rule into a triangle, with the 6-inch joint at the apex and the 12-inch joint at one side of the horizontal base line. Holding the rule at arm's length, line up the sides of the triangle with the roof as shown in Figure 13-7, being sure to keep the base of the triangle horizontal.

Then, with the zero point of the rule aligned with the center of the base, read the intersection of the zero point with the base. In the example shown in Figure 13-7, this occurs at the 22-inch mark. Next, locate the "Rule Reading" in Figure 13-8 nearest to the one read in the field. Directly under it you'll find the pitch and slope of the roof. For the example, the pitch is shown as 1/3, the slope as 8 inches per foot.

Projected horizontal area: No matter how complicated a roof may be, its projection onto a horizontal plane will easily define the total horizontal surface the roof covers. Figure 13-9 illustrates a typical roof complicated by valleys, dormers and ridges at different elevations. The lower half of the figure shows the projection of the roof onto a horizontal plane. In the projection, inclined surfaces appear flat and intersecting surfaces appear as lines.

Measurements for the horizontal projection of the roof can be made from the plans, from the ground, or from inside the attic. Once the measurements are made, the horizontal area

Pitch and Slope Relationships
Figure 13-6

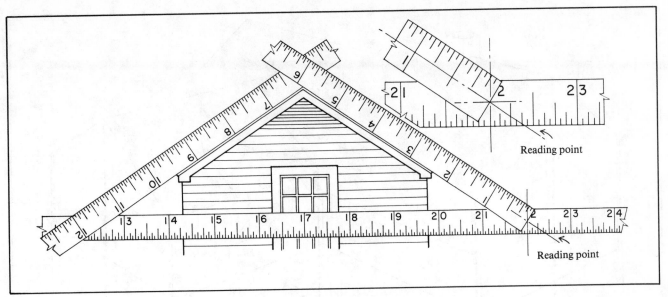

Use of Carpenter's Folding Rule to Determine Pitch and Slope
Figure 13-7

covered by the roof can be drawn to scale and calculated.

Sample calculations are illustrated below for the roof in Figure 13-9, using the dimensions and slopes indicated.

Because the actual area is a function of the slope, calculations must be grouped in terms of roof slope. Those of different slopes are not combined until the true roof areas have been found.

The horizontal area under the 9-inch slope roof is:

26 x 30 = 780
19 x 30 = 570
_____
Total - 1350 square feet

From this gross figure, deductions must be made for the area of the chimney and for the triangular area of the ell roof that overlaps and is sloped differently from the main roof:

Chimney 4 x 4 = 16
Ell roof   ½ (16 x 5) = 40 (triangular area)
_____
56 square feet

The net projected area of the main roof is:

1,350-56 = 1,294 square feet

The horizontal area under the 6-inch slope roof is:

20 x 30 = 600
½ (16 x 5) = 40
_____
Total = 640 square feet

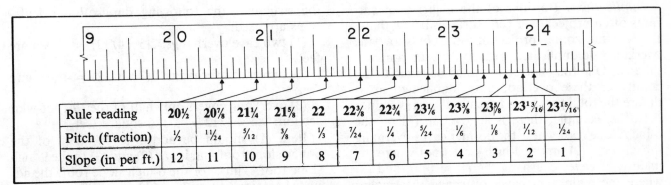

Rule reading	20½	20⅞	21¼	21⅝	22	22⅜	22¾	23⅛	23⅜	23⅝	23¹³⁄₁₆	23¹⁵⁄₁₆
Pitch (fraction)	½	¹¹⁄₂₄	⁵⁄₁₂	⅜	⅓	⁷⁄₂₄	¼	⁵⁄₂₄	⅙	⅛	¹⁄₁₂	¹⁄₂₄
Slope (in per ft.)	12	11	10	9	8	7	6	5	4	3	2	1

Reading Point Conversions to Pitch and Slope
Figure 13-8

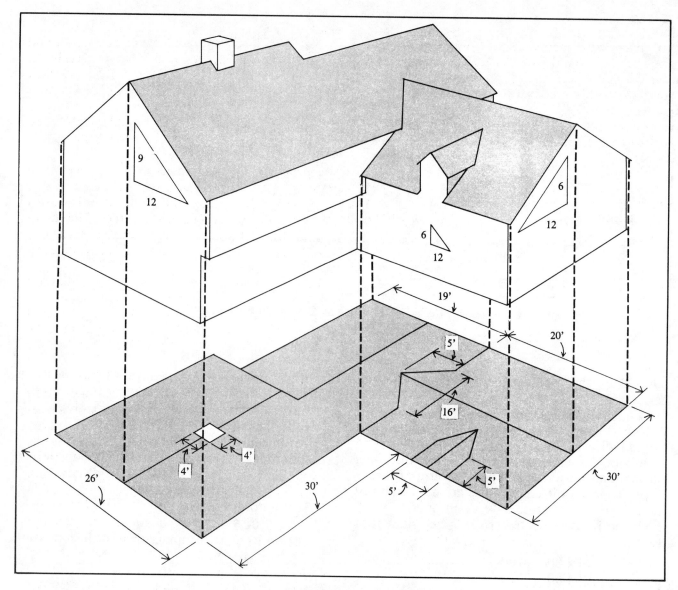

Horizontal Projection of Complex Roof
Figure 13-9

Duplications: Portions of the higher roof surfaces often project over roof surfaces below them. But the horizontal projections do not show the overlap. Add duplicated areas to the total horizontal area. One final correction must be made to account for these overlapped or duplicated areas before the total projected horizontal area is found.

In the example, there is an overlap (1) on the 6-inch slope roof where the dormer eaves overhang the ell roof, (2) on the 9-inch slope roof where the main roof eaves overhang the ell section, and (3) where the main roof eaves overhang the smaller section of the main roof in the rear of the building. In each case, if the eaves extend 4 inches beyond

the structure, the following duplication calculations are used:

(1) two eave overhangs: 2(5 x 4/12) = 3⅓ square feet.

(2) two eave overhangs: 2(7 x 4/12) = 4⅔ square feet.

(3) overhang covers only half of the 19-foot-wide section: 9.5 x 4/12 = 3-1/6 square feet.

Item (1) should be added to the area of the 6-inch slope roof and items (2) and (3) to the 9-inch slope roof. Thus, for the 6-inch slope roof, the adjusted total is 640 + 3 = 643 square feet, and for the 9-inch slope roof, 1294 + 8 = 1302 square feet. (Fractions are rounded off to the nearest foot.)

Conversion to actual area: Now that the total projected horizontal areas for each roof slope have been calculated, the results can be converted to actual areas with the aid of Figure 13-10.

To use the table, simply multiply the projected horizontal area by the conversion factor for the appropriate roof slope. The result is the actual area of the roof.

Slope (Inches Per Foot)	Area/Factor Factor
4	1.054
5	1.083
6	1.118
7	1.157
8	1.202
9	1.250
10	1.302
11	1.356
12	1.414

Area/Rake Conversion Table
Figure 13-10

For example, for the 9-inch slope roof:

Horizontal area	x	Conversion factor	=	Actual area
1,302 S.F.	x	1.250	=	1,627.5 S.F.

For the 6-inch slope roof:

| 643 S.F. | x | 1.118 | = | 718.8 S.F. |

Horizontal length	x	Conversion factor	=	Actual length
61.5 feet	x	1.250	=	76.9 feet

If this seems a little confusing, try it again. With a little practice you'll almost be able to determine how many shingles it will take to cover a roof by just looking at it.

After converting the horizontal areas to actual areas, add the results to get the total area of roof to be covered: 1628 + 719 = 2347 square feet.

For an actual job estimate, you'll have to make an allowance for waste. In this case, assume a 10 percent waste allowance. Thus, the total area of roofing material required is: 2347 + 235 = 2582 square feet.

Note: The same horizontal area projection and roof slope will always result in the same actual area regardless of roof style. In other words, if a shed roof, gable roof or hip roof, with or without dormers, each covered the same horizontal area and had the same slope, they would each require the same amount of roofing to cover them.

Additional material estimates: To complete the estimate, the required quantity of starter strips, drip edges, hip and ridge shingles and valley strips must be determined. Each of these estimates depends on the length of the eaves, rakes, hips, ridges and valleys on which the material will be applied.

Because eaves and ridges are horizontal, their lengths may be determined directly from the horizontal projection drawing. Rakes, hips, and valleys are sloped. Thus, their lengths must be calculated following a procedure similar to that for calculating sloped areas.

To determine the true length of a rake, first measure its projected horizontal distance. The table in Figure 13-10 that you used for horizontal-to-actual area conversions is used to convert horizontal-to-actual lengths. To use the table, multiply the rake's projected horizontal length by the conversion factor for the appropriate roof slope. The result is the actual length of the rake.

For the house in Figure 13-9, the rakes at the ends of the main house have horizontal distances of 26 feet and 19 feet. There is another rake in the middle of the main house where the higher roof section meets the lower. Its horizontal distance is 13 plus 3.5, making 16.5 feet. Adding all these horizontal distances gives a total of 61.5 feet. From Figure 13-10 for the 9-inch slope roof:

Horizontal length	x	Conversion factor	=	Actual length
61.5 feet	x	1.250	=	76.9 feet

Following the same procedure for the ell section with its 6-inch slope roof and dormer, the total length of rakes is found to be 39.1 feet.

These rake lengths can now be added to the total length of the eaves (actual horizontal distances, no conversion is necessary) to estimate the quantity of drip edges required for the job.

The quantity of ridge shingles required is estimated directly from the drawings, since ridge lines are true horizontal distances.

Hips and valleys again involve sloped distances. As a result, their projected horizontal lengths must be converted to actual lengths with the aid of Figure 13-11A.

To use this table, measure the length of the hip or valley on the horizontal projection drawing and multiply it by the conversion factor for the appropriate roof slope. The result is the actual length of the hip or valley.

Slope (Inches Per Foot)	Hip/Valley Factor
4	1.452
5	1.474
6	1.500
7	1.524
8	1.564
9	1.600
10	1.642
11	1.684
12	1.732

Hip/Valley Conversion Table
Figure 13-11A

In the following calculations, we'll determine the total length of valleys for the house in Figure 13-9 and estimate the valley flashing material required.

There is a valley formed on both sides of the ell roof intersection with the main roof. The total measured distance of these valleys on the horizontal projection is 16 feet.

The fact that two different slopes are involved complicates the procedure somewhat. If there were only one roof slope, the true length could be calculated directly from Figure 13-11A. But in this case, calculations for each slope must be made and then averaged to find the true length of the valleys. Thus:

Horizontal Length	X	Conversion Factor	=	Actual Length
16 feet	X	1.600 (for 9-inch	=	25.6 ft. slope)
16 feet	X	1.500 (for 6-inch	=	24.0 ft. slope)
Average:		(24.0 + 25.6)/2		= 24.8 feet

Figure 13-11B

The approximate length of the two valleys is 24.8 feet, or 12.4 feet each.

The total projected horizontal length of the dormer valleys in Figure 13-9 is 5 feet. From Figure 13-11A, with the slope of the ell roof and the dormer both 6 inches, the actual length of the valleys is calculated to be 7.5 feet.

The total length of valleys for the house is 25 plus 7, equaling 32 feet. Based on this true length, the amount of valley flashing material can now be estimated.

Fasteners

Nails and staples are the two types of fasteners generally used to secure asphalt roofings to the deck. Nails are recommended for a wide range of applications, while staples are an acceptable fastener only on new construction if they meet the requirements discussed below.

Nails: Roofing nails can be steel or aluminum. Steel roofing nails should be zinc-coated for corrosion protection.

Roofing nails should have barbed or deformed shanks. They should be 11- or 12-gauge and have large heads of 3/8 to 7/16 inch in diameter. The nails should be long enough to penetrate through the roofing material and at least 3/4 inch into the deck lumber or through a plywood deck. Figure 13-12 gives recommended lengths for various applications.

Application	Nail Length (Inch)
Roll roofing on new deck	1
Strip or individual shingles on new deck	1¼
Reroofing over old asphalt roofing	1½ to 2
Reroofing over old wood shingles	2

Recommended Nail Lengths
Figure 13-12

The number and location of nails required for various asphalt roofing products is covered in the following sections. As a general rule, roofing manufacturers specify this information in the application instructions that accompany each type of product. Follow the individual manufacturer's specifications.

Staples: The use of staples as fasteners for asphalt roofing is limited to wind resistant shingles with factory-applied adhesives on new construction. Staples are not recommended for reroofing unless the old roofing has been removed.

In general, staples should be zinc-coated for corrosion protection and 16-gauge minimum. They should have a minimum crown width of 15/16 inch and be long enough to penetrate 3/4 inch into deck lumber or through plywood decks.

The staples should be driven with an accurately adjusted pneumatic staple gun so that the entire crown bears tightly against the shingle but does not cut the shingle surface. An improperly adjusted staple gun or poor placement of staples will cause sealing failure, raised tabs, buckling, leaks and blow-offs. The number and location of staples to be used in recommended applications is discussed in the following section.

Asphalt roofing materials must not become waterlogged before installation. Protect shingles against the weather.

Don't stack asphalt roofing material more than 4 feet high. Don't place it in direct contact with the ground or in the hot sun.

Applying Strip Shingles on New Construction

No roof is better than its installer. Asphalt roofing materials are no exception. They are designed to give years of service when applied carefully and correctly.

Asphalt roofing is easy to install. Even so, there are things you must do to put down a good asphalt roof. These include proper deck preparation, flashing details, underlayment, alignment and starter strip applications.

Deck preparation: A number of roofing problems are caused by poor deck preparation. The support for the roof deck itself must provide a rigid deck surface. It should not sag, shift or deflect under the weight of the roofing and your crew, or under snow loads the roof may have to support. An unstable roof deck is the beginning of trouble.

The deck should be tightly built of good quality, well-seasoned lumber or good decking grade plywood. Never use "green" or poorly-seasoned deck lumber. It will eventually dry out and warp. This could buckle overlying shingles or wrinkle and buckle roll roofing.

Warping as well as other problems may result even with well-seasoned lumber if the attic space under the roof deck or the spaces between the rafters and beams in a cathedral ceiling or mansard roof are improperly ventilated. Poor ventilation results in an accumulation of moisture that eventually condenses on the underside of the roof deck.

When using sheathing boards, don't use boards more than 6 inches wide. They will swell and shrink after installation. Wider lumber is likely to swell and shrink enough to buckle asphalt roofing materials.

Inspect the lumber before installation. Don't use boards that are badly warped or have loose knots or excessive pitch areas. Wood resin will react with the asphalt roofing materials.

If the sheathing is already in place, check for pitch and loose knots. Cover each of these areas with sheet metal patches before the roofing is applied.

Your local building code controls the type, grade, thickness and installation of plywood used for decks. Figure 13-13 illustrates a typical plywood deck.

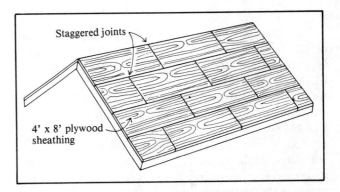

Typical Plywood Deck
Figure 13-13

Drip edges: Drip edges provide efficient water shedding at the rakes and eaves and protect the underlying wood from rotting. Use corrosion-resistant material that extends approximately 3 inches back from the roof edge. Apply drip edges directly to the deck along the eaves and over the underlayment along the rakes. Figure 13-14 shows drip edge details.

Underlayment: After the deck has been properly prepared, cover it with an asphalt-saturated felt underlayment. Apply the felt when the deck is dry.

On decks with a slope of 4 inches per foot or more, the underlayment should consist of one layer of non-perforated No. 15 asphalt-saturated felt. Some local codes may require a No. 30 asphalt-saturated felt. Don't use coated felts, tar-saturated materials, or laminated waterproof papers. These could act as a vapor barrier to trap moisture or frost between the covering and the roof deck.

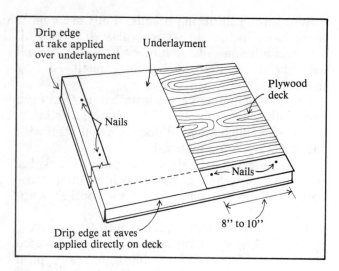

Application of Drip Edge at Rake and Eaves
Figure 13-14

Lay the felt parallel to the eaves, lapping each course at least 2 inches over the underlying course. Nail the felt using only enough nails to hold it in place. (See Figure 13-15.) If two or more pieces are required to continue a course, lap the ends at least 4 inches. End laps in a succeeding course should be located at least 6 feet from the end laps in the preceding course. Lap the felt 6 inches from both sides over all hips and ridges. Where the roof meets a vertical surface, carry the underlayment 3 to 4 inches up the surface.

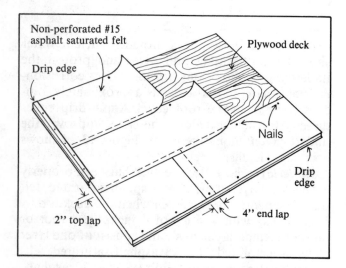

Application of Underlayment
Figure 13-15

An asphalt-saturated felt underlayment should always be used in new construction because it serves two important functions. It keeps the deck dry until shingles are applied. That eliminates any problems that would result if shingles were placed on wet deck lumber or if the plywood delaminated from moisture absorption. And, if shingles should be lifted, damaged or torn by winds after application, it provides secondary protection by shielding the deck from wind-driven rain and preventing water from reaching the structure.

Chalk lines: Even shingles taken from the same package will vary in size. The variation is small. But on a large roof, many small differences can cause irregular butt lines or misaligned cutouts. Whether you're working on a large roof or a small one, use chalk lines as visible guides to guarantee correct horizontal and vertical alignment and the right exposure.

Begin with a horizontal chalk line parallel to the eaves. Measure the appropriate distance on the roof and mark it at three locations—at each end and in the middle—as a check against possible measurement errors. Put a nail on the mark at each end. Stretch the chalk line between them and pull it taut. Check the alignment of the middle mark, then snap the line from the center.

Vertical chalk lines are important for aligning cutouts from eaves to ridge. They are also important for aligning shingles on each side of a dormer so that the shingles and cutouts meet above the dormer in proper alignment without any gaps or overlaps.

On long runs, snap a vertical chalk line in the center of the run and apply shingles to the left and right of the line. Check horizontal chalk lines as the shingle application approaches the ridge so that the upper courses will be parallel to them.

Eave flashing: Ice dams formed by the thawing and freezing of snow can force water under roofing and damage a home's ceilings, walls and insulation. Eave flashing is the best way to prevent such leakage.

The flashing should be installed wherever there is a possibility of icing along the eaves. The flashing material and the width of the flashing strip depend on the severity of icing you expect.

On roofs with a slope of 4 inches per foot or more, install a course of smooth, coated roll roofing of not less than 50 pounds parallel to the eaves and overhanging the drip edge by 1/4 to 3/8 inch. Apply the roll roofing flashing strip up the roof to a point at least 12 inches beyond the interior wall

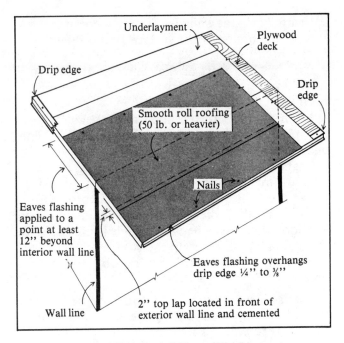

Application of Eaves Flashing
Figure 13-16

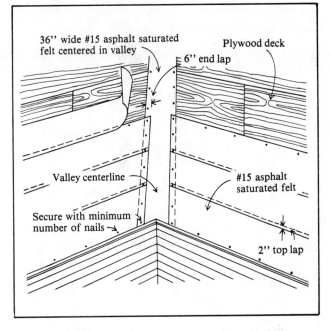

Application of Underlayment in a Valley
Figure 13-17

line. If a second flashing strip is needed to reach that point, locate the lap in front of the exterior wall line. Overlap the flashings at least 2 inches and cement the horizontal joint over its entire length. End laps should be 12 inches and cemented. (See Figure 13-16.)

In areas of severe icing, extend the roll roofing flashing strip at least 24 inches beyond the interior wall and cement it to the underlayment. Apply asphaltic plastic cement at the rate of two gallons per 100 square feet. Press the flashing strip firmly into the cemented area.

Valley flashing: Valleys are formed where the sloping roof planes meet at an angle. The sloping roof areas direct water toward the valley, concentrating water along the joint and making it especially prone to leaks. One of the most important installation details for good roofing is good valley flashing.

Begin by centering a 36-inch-wide strip of No. 15 asphalt-saturated felt in the valley. Secure it with only enough nails to hold it in place. Then trim the horizontal courses of felt underlayment applied on the roof to overlap the valley strip by at least 6 inches. (See Figure 13-17.)

There are three types of valleys: open, woven and closed cut. Open valleys may be used with most types of shingles and roll roofing. Woven and closed cut valleys are limited to strip shingles and

are the preferred valley treatment for this type of roofing. Regardless of the type of valley used, it must be smooth, even, big enough to carry water away rapidly and capable of surviving an occasional backing up of water.

Open valleys—The recommended flashing material is 90-pound mineral-surfaced roll roofing. Select a color that is neutral or that matches the roofing shingles being installed. Valley flashing is applied in two layers. (See Figure 13-18.)

Center the first layer, 18 inches wide, in the valley with the mineral surface facing down. Trim the lower edge flush with the eaves' drip edge. Install the first layer up the entire length of the valley. If two or more strips of roll roofing are required, lap the upper piece over the lower so that water is carried over the joint, not into it. The overlap should be 12 inches and fully bonded with asphalt plastic cement. Use only enough nails to hold the strip in place. Nail along a line 1 inch from each edge. Start at one edge and work all the way up. Then return to nail the other side, pressing the flashing strip firmly into the valley at the same time.

After the 18-inch strip has been secured, center a second strip, 36 inches wide, in the valley and lay it over the first strip, this time with the mineral surface facing up. Nail the strip in place in the same manner as the underlying strip. Overlaps should be

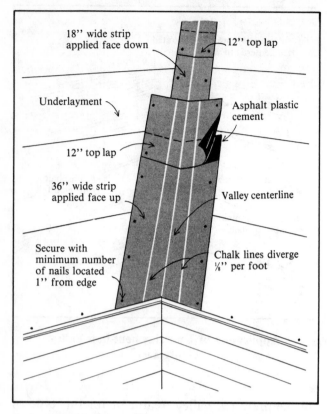

**Application of 90-Pound Roll Roofing
as Flashing for an Open Valley
Figure 13-18**

12 inches and cemented. The valley will be completed as the shingles are installed.

Woven and closed-cut valleys—Cover both these types of valleys with 36-inch-wide mineral- or smooth-surfaced roll roofing, 50 pounds or heavier. Center the strip in the valley. Drive only enough nails to hold it in place. Nail the strip along a line 1 inch from the edges, first on one edge all the way up, then on the other, while pressing the flashing strip firmly and smoothly into the valley. Laps should be 12 inches and cemented. The valley will be completed as the shingles are installed.

Shingle application: The general application procedures for all asphalt strip shingles are about the same. But there are differences in applying the first shingle in each course and in fastening shingles. You should read the instructions supplied with each pack of shingles. The directions that follow will apply on most shingle jobs.

Before beginning, check that all chimneys are completed and all vent pipes, soil stacks and ventilators are in place. Plan for the additional flashing that will be required as the job is done.

If a roof surface is broken by a dormer or valley,

start applying the shingles from a rake and work toward the break. If the surface is unbroken, start at the rake that is most visible. If both rakes are equally visible, start at the center and work both ways. On hip roofs, start at the center and work both ways.

No matter where the application begins, apply the shingles across and diagonally up the roof. This guarantees that each shingle is nailed properly. Straight-up application or "racking" requires that part of the shingles in some courses be placed under those already applied in the course above. Because part of the shingle is hidden, you may overlook that area when the shingle is nailed. With a diagonal application up the roof, each shingle is completely visible until covered by the course above.

Starter strip: The starter strip may be either a row of shingles trimmed to the shingle manufacturer's recommendations or a strip of mineral-surfaced roll roofing at least 7 inches wide. The starter strip protects the roof by filling in the spaces under the cutouts and the joints of the first course of shingles. It should overhang the eaves and rake edges by 1/4 to 3/8 inch.

If self-sealing shingles are used for the starter strip, remove the tab portion of each shingle and position the remaining strip with the factory-applied adhesive face up along the eaves. Trim at least 3 inches from the end of the first shingle in the starter strip. This ensures that the cutouts of the first course of shingles are not placed over the starter strip joints. Nail starter strips parallel to the eaves along a line 3 to 4 inches above the eaves. Position the nails so that they will not be exposed under the cutouts in the first course. (See Figure 13-19.)

If shingles without a self-sealing adhesive are used for the starter strip, remove the tab portion of each shingle and position the remaining strip along the eaves.

If roll roofing is used for the starter strip, nail along a line 3 to 4 inches above the eaves. Space the nails 12 inches apart. If more than one piece of roll roofing must be used, lap the end joint 2 inches and cement it.

First and succeeding courses: The first course is the most critical. Be sure it's laid perfectly straight, checking it regularly during application against a horizontal chalk line. A few vertical chalk lines aligned with the ends of shingles in the first course make alignment of cutouts easier.

If you use free-tab shingles or roll roofing for the starter strip, bond the tabs of each shingle in the

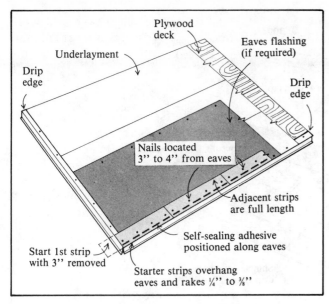

Application of Starter Strip
Figure 13-19

The first course starts with a full shingle. Later courses start with portions removed according to the style of shingle being applied and the pattern desired. Don't discard the pieces cut from the first shingle in each course. If they are full tabs, use them for finishing the opposite end of the course and for hip and ridge shingles.

Here's how to get the correct exposure for square-tab strip shingles. Align the butts with the top of the cutouts in the course below. For no-cutout shingles and shingles with variable buttlines, follow the manufacturer's directions.

There are three methods of applying three-tab strip shingles: the 6-inch method, 5-inch method and 4-inch method. The methods correspond to the additional amount removed from the first shingle in each successive course to get the desired pattern. Removing a small amount from the first shingle keeps the cutouts in one course from lining up directly with those of the course below.

1. The 6-inch method. This method starts each succeeding course after the first and goes up to the sixth with a shingle from which an additional 6 inches has been removed. The first course starts with a full-length shingle. The second starts with a shingle that has 6 inches removed, the third course with a shingle that has 12 inches removed, and so on through the sixth course which starts with a

first course to the starter strip with a spot of asphalt plastic cement about the size of a quarter under each tab. Then press the tabs firmly into the cement. Don't use too much cement; it can cause blistering.

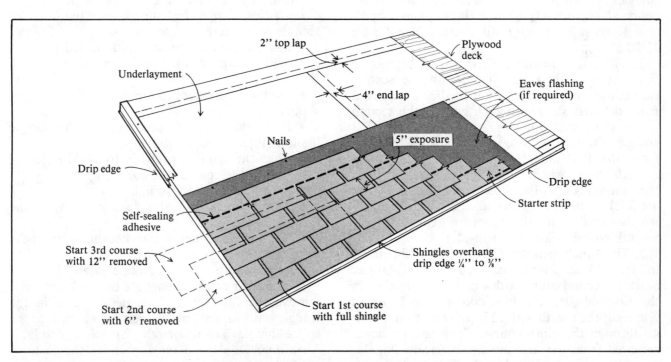

Application of Shingles Using the 6-Inch Method
Figure 13-20

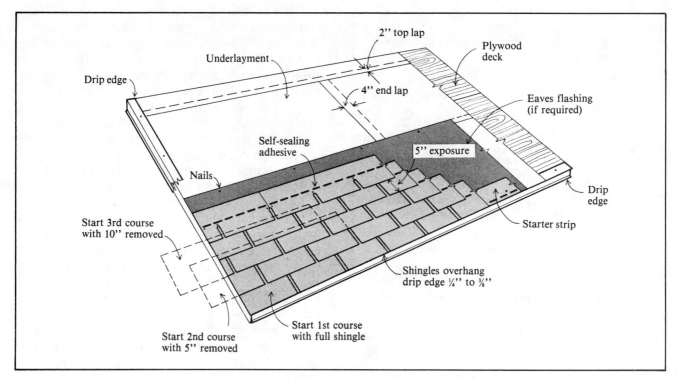

Application of Shingles Using the 5-Inch Method
Figure 13-21

shingle that has 30 inches removed. Adjacent shingles in each course are full length. The seventh course starts with a full-length shingle, and the pattern is repeated every sixth course. (See Figure 13-20.)

2. The 5-inch method. This method begins the first course with a full shingle. In the second through seventh courses a 5-inch slice is removed from the first shingle in each course: the second course starts with 5 inches removed from the first shingle, the third course with 10 inches removed from the first shingle and so on through the seventh course which has 30 inches removed from the first shingle. Adjacent shingles in each course are full length. The eighth course begins with a full-length shingle, and the pattern is repeated every seventh course. (See Figure 13-21.)

3. The 4-inch method. This method is illustrated in Figure 13-22. Start the course with a full shingle. Start the second course with 4 inches removed from the first shingle, the third course with 8 inches removed, the fourth with 12 inches removed and so on through the ninth course which has 32 inches removed from the first shingle. Adjacent shingles in each course are full length. The tenth course again begins with a full-length shingle, and the pattern is repeated every ninth course.

Never use an alignment system where shingle joints are closer than 4 inches to each other.

Fastening: Proper fastening is essential on every shingle roof. Here are some good rules to follow:
• Use the correct size and grade fastener.
• Use zinc-coated fasteners for corrosion protection.
• Use the recommended number of fasteners per shingle.
• Place the fasteners according to the shingle manufacturer's specifications.
• Align the shingles properly to avoid exposing fasteners in the course below.
• Drive the fasteners straight.
• Don't break the shingle surface with the fastener head.
• Don't drive fasteners into knotholes or cracks in the roof deck.
• Repair faulty fastening immediately.

If a fastener doesn't penetrate the deck properly, remove the fastener and repair the hole in the shingle with asphalt plastic cement or replace the entire shingle. Then place the fastener properly.

Nailing: Don't nail into or above the factory-applied adhesives. Align each shingle carefully. Wherever possible, make sure that no cutout or end joint is less than 2 inches from a nail in an

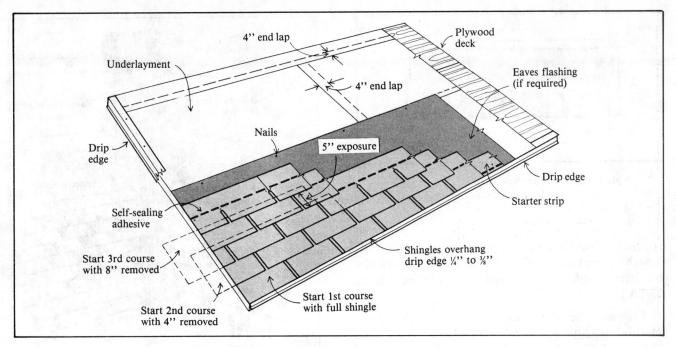

**Application of Shingles Using the 4-Inch Method
Figure 13-22**

underlying course.

Start nailing from the end nearest the last shingle laid and proceed across. This will prevent buckling. Don't try to realign a shingle by shifting the free end after two nails are in place; this may cause distortion. Drive nails straight so that the edge of the nail head doesn't cut into the shingle. Nail heads should be driven flush with the shingle surface, not sunken into it.

Three-tab strip shingles—These shingles require four nails each. When the shingles are applied with an exposure of 5 inches, the nails should be placed on a line 5/8 inch above the top of the cutouts and located 1 inch from each end and centered over each cutout. (See Figure 13-23.)

Two-tab strip shingles—Four nails are required per shingle. For a 5-inch exposure, place nails on a line 5/8 inch above the top of the cutout and 1 inch and 13 inches from each end. (See Figure 13-24.)

No cutout strip shingles—Each shingle requires four nails. For a 5-inch exposure, place nails on a line 5⅝ inches above the butt edge and 1 inch and 12 inches from each end. (See Figure 13-25.)

Stapling: Staples may be used instead of roofing nails on a new deck when installing wind-resistant shingles with factory-applied adhesives. If there's any doubt about the acceptability of staples in a particular application, consult the shingle manufacturer.

Staples should be placed in the same locations as

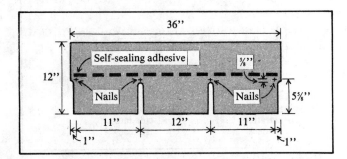

**Nail Locations for Three-Tab Strip Shingle
Figure 13-23**

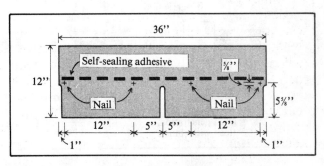

**Nail Locations for Two-Tab Strip Shingle
Figure 13-24**

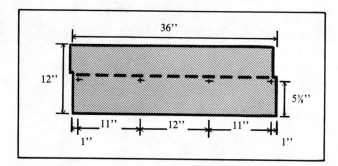

Nail Locations for No-cutout Strip Shingle
Figure 13-25

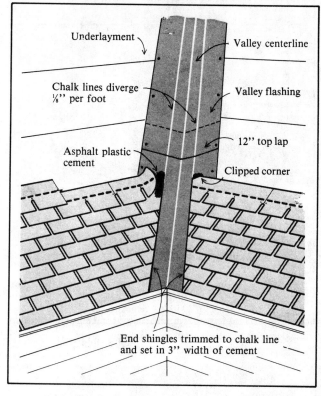

Application of Shingles in Open Valley
Figure 13-27

roofing nails. They should be driven parallel to the length of the shingle. Be sure the stapler is adjusted to drive staples to the right depth. (See Figure 13-26.)

Valleys: Use the open, woven, or closed-cut method to apply shingles to valleys. Woven or closed-cut valleys are preferred for strip shingles. For all the methods, valley flashing should be in place before applying shingles. An exception is open valleys around dormers where the valley flashing must overlap the top courses of shingles along the dormer sidewalls.

Open valley—Snap two chalk lines, one on each side of the valley centerline, over the full length of the valley flashing. Locate the upper ends of the chalk lines 6 inches apart at the ridge (3 inches to either side of the valley centerline). The lower ends should diverge at about 1/8 inch per foot. Thus, for an 8-foot-long valley, the chalk lines should be 7 inches apart at the eaves; for a 16-foot-long valley, they should be 8 inches apart. (See Figure 13-27.)

As you move toward the valley, trim the last shingle in each course to fit on the chalk line. Clip

an inch from the upper corner of the shingle on a 45-degree angle to direct water into the valley and prevent it from penetrating between courses. To form a tight seal, cement the shingle to the valley lining with a 3-inch width of asphalt plastic cement. There should be no exposed nails along the valley flashing.

Open valley for dormer roof—The valley between a dormer and main roof requires special treatment. Don't install valley flashing until the

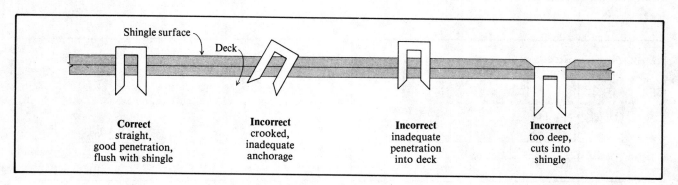

Application of Staples
Figure 13-26

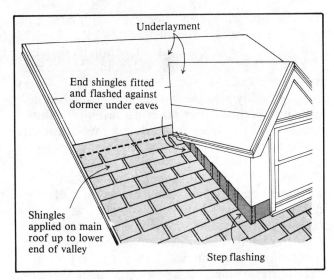

**Point at Which Installation of Open Valley
at Dormer Roof Begins
Figure 13-28**

shingle application reaches a point just above the lower end of the valley. (See Figure 13-28.)

Apply the first or bottom layer of valley flashing (18-inch-wide, 90-pound mineral-surfaced roll roofing) the same way as any open valley flashing. Trim the lower section of the flashing so that it extends 1/4 inch below the edge of the dormer deck. The lower section in contact with the main roof deck should project at least 2 inches below the point where the two roofs meet. Extend the upper section so that the portion on the main roof extends 18 inches above the point where the dormer intersects the roof. Trim the portion on the dormer at the ridge. (See Figure 13-29.)

Apply valley flashing on the other side of the dormer in the same way, extending the portion on the main roof up and over the portion from the first valley. Cement and nail the overlap. Lap the flashing on the dormer side over the ridge, then cement and nail it.

Trim the second or top layer of flashing (36-inch-wide, 90-pound mineral-surfaced roll roofing) on the dormer side to match the lower end of the underlying 18-inch strip. Trim the side that will lie on the main deck to overlap the nearest course of shingles. This overlap is the same as the normal lap of one shingle over another for the shingles being applied. For example, for 12-inch-wide, three-tab strip shingles, extend the flashing to the top of the cutouts.

Nail the top flashing strip over the bottom as in standard open valley construction. Work the

flashing into the valley joint so that it lies flat and smooth in both planes up to the edge of the dormer eaves. Trim the top layer horizontally on a line with the top of the dormer ridge where it intersects the main roof.

Apply the top layer in the valley on the other side of the dormer in the same manner except at the dormer ridge where it is cemented and nailed to the first valley flashing.

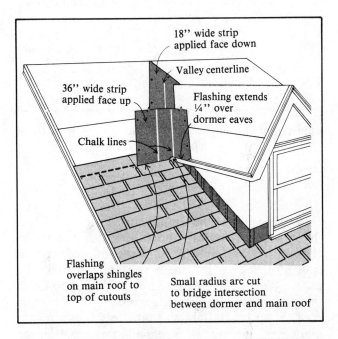

**Application of 90-Pound Roll Roofing as Flashing for
an Open Valley at a Dormer Roof
Figure 13-29**

Trim the lower end of the flashing on a small radius arc that bridges slightly over the point of intersection between the dormer and main roof. This shape forms a small canopy over the joint between the two decks.

Snap chalk lines on the valley flashing 3 inches on each side of the centerline at the top and diverging 1/8 inch per foot to the bottom of the valley. Continue applying shingles, trimming the end shingle in each course to the chalk lines. Clip the upper corner and embed the end shingle in a 3-inch-wide strip of asphalt plastic cement to seal it to the flashing. Complete the valley in the usual manner. (See Figure 13-30.)

After shingles have been applied to both sides of the dormer roof, apply the dormer ridge shingles. Start at the front of the dormer and work toward

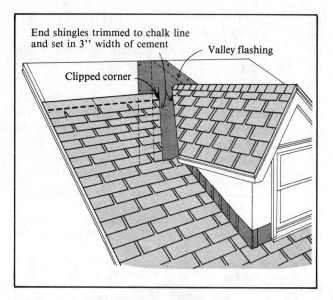

**Application of Shingles in Open Valley at Dormer Roof
Figure 13-30**

eaves of one of the intersecting roof planes and across the valley. Extend the end shingle at least 12 inches onto the adjoining roof. Apply succeeding courses in the same manner, extending them across the valley and onto the adjoining roof. Press shingles tightly into the valley. Use normal shingle fastening methods, but place no fastener within 6 inches of the valley centerline. Two fasteners should be placed at the end of each shingle crossing the valley.

Then apply shingles on the adjoining roof plane, starting along the eaves and crossing the valley onto the previously applied shingles. Trim the shingles being installed no less than 2 inches back from the valley centerline, following a chalk line snapped over the shingles to ensure a neat installation. Trim one inch on a 45-degree angle from the upper corner of each end shingle to direct water into the valley. Finally, embed each end shingle in a 3-inch-wide strip of asphalt plastic cement. (See Figure 13-32.)

the main roof. Apply the shingles as described later in this chapter under "Hips and Ridges." Apply the last ridge shingle so that it extends at least 4 inches onto the main roof. Slit the center of the portion attached to the main roof and nail it into place. Then apply the main roof courses to cover the portion of the last ridge shingle on the main roof. Snap chalk lines so that the shingles on the main roof continue the same alignment pattern on both sides of the dormer, as shown in Figure 13-31.

Closed-cut valley—With valley flashing already in place, apply the first course of shingles along the

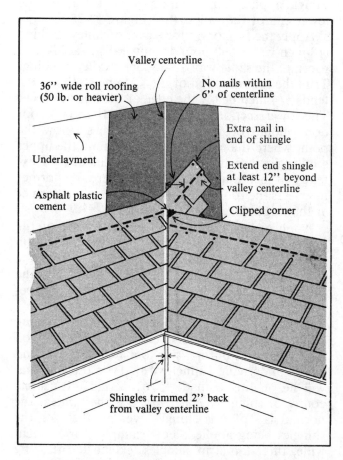

**Application of Shingles in Closed-Cut Valley
Figure 13-32**

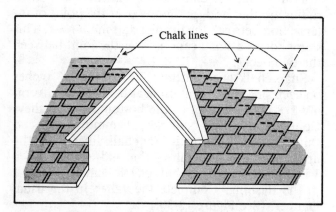

**Application of Chalk Lines for Proper Alignment of Shingles and Cutouts Above Dormer
Figure 13-31**

Woven valley—The valley flashing should already be in place as described earlier. Shingles on the intersecting roof surfaces may be applied toward the valley from both roof areas simultaneously or each roof area may be worked separately up to a point about 3 feet from the center of the valley and the gap closed later.

Regardless of the procedure you use, apply the first course along the eaves of one roof area up to and over the valley, with the last shingle extending at least 12 inches onto the intersecting roof. Then apply the first course on the intersecting roof along the eaves and extend it across the valley over the top of the shingles already crossing the valley and at least 12 inches onto the other roof surface. Apply successive courses alternately from the adjoining roof areas, weaving the valley shingles over each other as shown in Figure 13-33. Press each shingle tightly into the valley and follow the same nailing procedures as for the closed-cut valley.

be flashed carefully.

Flashing against vertical sidewall—Roof planes that butt against vertical walls at the end of shingle courses are best protected by metal *flashing shingles* placed over the end of each course. This method is called *step flashing*.

The metal flashing shingles are rectangular, 10 inches long and 2 inches wider than the exposed face of the roofing shingles. For example, when used with strip shingles with a 5-inch exposure, they are 10 x 7 inches over the roof deck and 5 inches up the wall surface. Each flashing unit is placed just up the roof from the exposed edge of the shingle that will overlap it so that it is not visible when the overlapping shingle is in place. (See Figure 13-34.)

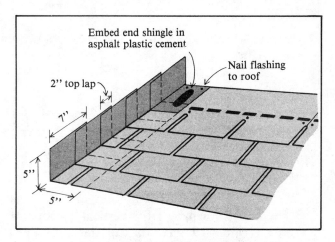

Application of Step Flashing
Figure 13-34

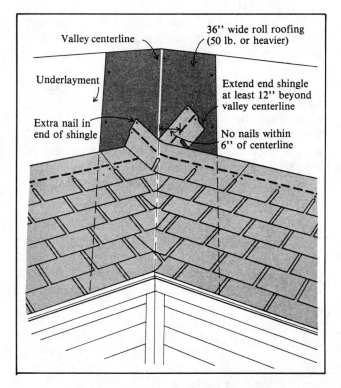

Application of Shingles in Woven Valley
Figure 13-33

Flashing: Intersections of the roof surfaces with one another or with vertical walls and projections through the roof surface such as chimneys or soil stacks are critical points. Each of these points must

To install step flashing, place the first flashing unit over the end of the starter strip and position it so that the tab of the end shingle in the first course covers it completely. Secure the horizontal arm to the roof with two nails. Don't nail flashing to the wall. Settling of the roof could damage the seal.

Apply the first course of shingles up to the wall. Position the second step flashing strip over the end shingle in the first course 5 inches up from the butt so that the tab of the end shingle in the second course covers it completely. Fasten the horizontal arm to the roof. The second course of shingles follows. The end is flashed as in preceding courses and so on to the top of the intersection. Because the metal strip is 7 inches wide and the roof shingles are laid with a 5-inch exposure, each flashing unit will overlap the course below by about 2 inches.

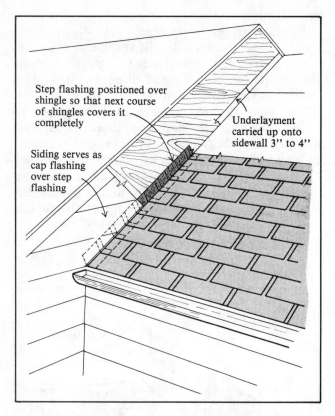

Application of Step Flashing Against Vertical Sidewall
Figure 13-35

Bring siding down over the vertical sections of the step flashing to serve as cap flashing. Keep wood siding far enough away from the roof shingles to allow for painting. (See Figure 13-35.)

Flashing against vertical front wall—Apply shingles up the roof until a course must be trimmed to fit at the base of the vertical wall. Adjust the exposure slightly in the previous two courses so that the last course is at least 8 inches wide. Apply a continuous piece of metal flashing over the last course of shingles by embedding it in asphalt plastic cement and nailing it to the roof. The metal flashing strip should be 26 gauge and bent to extend at least 5 inches up the vertical wall and at least 4 inches onto the last shingle course. Don't nail the strip to the wall. Apply an additional row of shingles over the metal flashing strip, trimmed to the width of the strip. (See Figure 13-36.)

Bring the siding down over the vertical flashing to serve as cap flashing. Keep wood siding far enough away from the roof shingles to allow for painting. Don't nail siding into the vertical flashing.

If the vertical front wall meets a sidewall, as in dormer construction, cut the flashing so that it extends at least 7 inches around the corner. Then continue up the sidewall with step flashing as described earlier.

Soil stacks and vent pipes—Practically every building has circular vent pipes or ventilators projecting through the roof. These require special flashing.

Apply shingles up to the vent pipe. Then cut a hole in a shingle to fit over the pipe. Set the shingle in asphalt plastic cement. (See Figure 13-37.) A preformed flashing flange that fits snugly over the pipe is then placed over the shingle and vent pipe and set in asphalt plastic cement. Place the flange over the pipe to lay flat on the roof. (See Figure 13-38.)

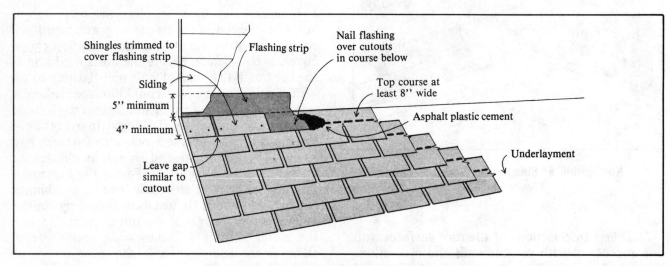

Application of Flashing Against Vertical Front Wall
Figure 13-36

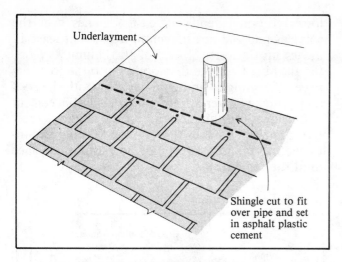

Application of Shingle Over Vent Pipe
Figure 13-37

After the flashing is in place, resume shingle application. Cut shingles in successive courses to fit around the pipe. Embed them in asphalt plastic cement where they overlap the flange. Too much cement may cause blistering. Don't drive fasteners close to the pipe. The completed installation should look like Figure 13-39, with the lower part of the flange overlapping the lower shingles and the side and upper shingles overlapping the flange.

Follow the same procedures where a ventilator or exhaust stack is located at the ridge. But there is one difference: Bring the shingles up to the pipe from both sides and bend the flange over the ridge

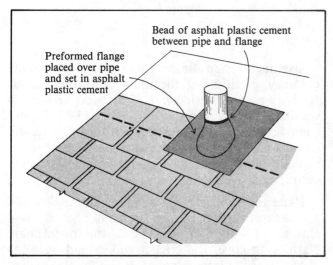

Application of Flashing over Vent Pipe
Figure 13-38

to lie in both roof planes, overlapping the roof shingles at all points. Ridge shingles are then positioned to cover the flange. Embed the ridge shingles in asphalt plastic cement where they overlap the flange.

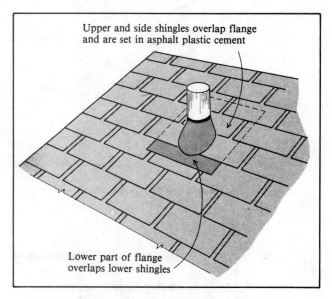

Application of Shingles Around Flashing
Figure 13-39

Flashing around chimneys—To prevent problems that uneven settling could cause, chimneys are usually built on a separate foundation from that of the main structure. This doesn't eliminate settling problems. It only frees the chimney to settle at its own rate.

Because the chimney moves separately, flashing at the point where the chimney projects through the roof calls for a water seal. For this, you need to apply base flashing secured to the masonry. If movement occurs, the cap flashing slides over the base flashing without affecting water runoff.

Chimneys projecting through the roof surface should have a cricket installed at the intersection of the back face of the chimney and the roof deck. The cricket or wood saddle protects the chimney flashing by preventing build-up of ice and snow at the rear of the chimney. It also diverts water runoff around the chimney. (See Figure 13-40.)

The cricket should be in place from the start because all roofing materials from the felt underlayment to the roofing shingles are carried over it. If it is not in place, build one as part of the

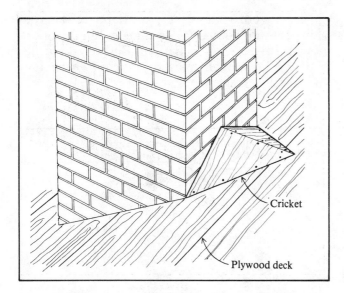

Location and Configuration of Chimney Cricket
Figure 13-40

deck work before you apply underlayment and shingles.

A cricket is made from two triangular sections of plywood joined to form a level ridge that extends from the centerline of the chimney back to the roof deck. Nail the sections to the deck and to each other as shown in Figure 13-40.

Apply shingles up to the front edge of the chimney before installing any flashings. Apply a coat of asphalt primer to the chimney's brickwork to seal the surface and to provide good adhesion at all points where asphalt plastic cement will later be applied.

Begin the flashing by installing 26-gauge corrosion-resistant metal base flashing between the chimney and the roof deck on all sides. Apply the base flashing to the front first, as shown in Figures 13-41 and 13-42. Bend the base flashing so that the

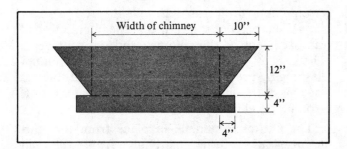

Pattern for Cutting Front Base Flashing
Figure 13-41

lower section extends at least 4 inches over the shingles and the upper section extends at least 12 inches up the vertical face of the chimney. Work the flashing firmly and smoothly into the joint between the shingles and chimney. Set both the roof and chimney overlaps in asphalt plastic cement placed over the shingles and on the chimney face. Drive one or two nails through the flashing and into the mortar joints to hold the flashing in place until the cement sets.

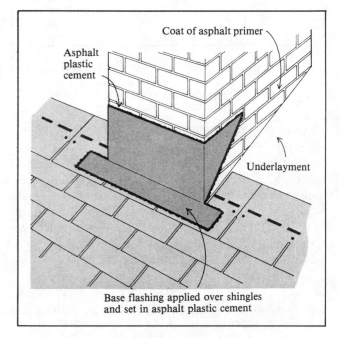

Base flashing applied over shingles
and set in asphalt plastic cement

Application of Base Flashing at Front of Chimney
Figure 13-42

Use metal step flashing for the sides of the chimney, positioning the units the same way as flashing a vertical sidewall. Cut, bend and apply the step flashing as shown in Figure 13-43. Secure each flashing to the masonry with asphalt plastic cement and to the deck with nails. Embed in asphalt plastic cement the end shingles in each course that overlaps the flashing.

Place the rear base flashing over the cricket and the back of the chimney as shown in Figures 13-44 through 13-46. Cut and bend the metal base flashing to cover the cricket and extend onto the roof surface at least 6 inches. It should also extend at least 6 inches up the brickwork and far enough laterally to lap the step flashing on the sides.

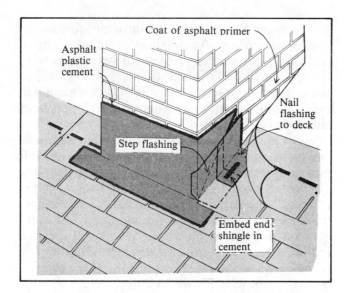

Application of Base Flashing at Side of Chimney
Figure 13-43

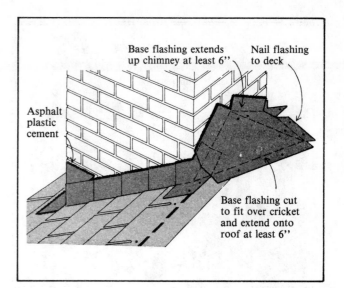

Application of Base Flashing Over Cricket
Figure 13-45

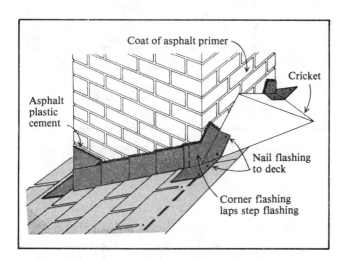

Application of Corner Base Flashing
at Rear of Chimney
Figure 13-44

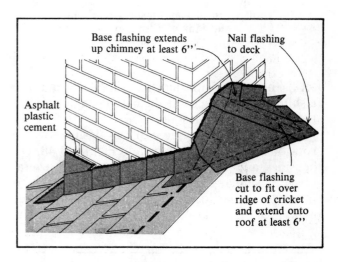

Application of Base Flashing over Ridge of Cricket
Figure 13-46

If large enough, the cricket may be covered with shingles. Otherwise, apply the rear base flashing, bring the end shingles in each course up to the cricket, and cement them in place.

Cap flashing must now be placed over all base flashings to keep water out of the joints. Begin by setting the metal cap flashing into the brickwork as shown in Figure 13-47. This is done by raking out a mortar joint to a depth of 1½ inches and inserting the bent edge of the flashing into the cleared joint. Once in place and under a slight amount of spring tension, the flashing can't be dislodged easily.

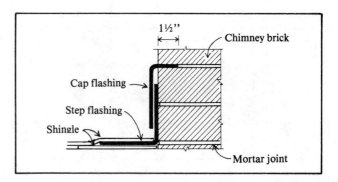

Application of Cap Flashing
Figure 13-47

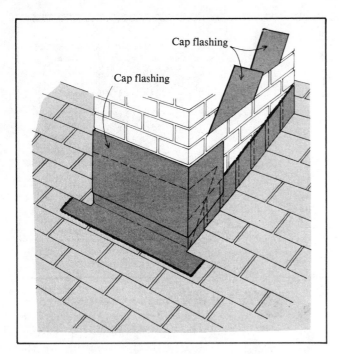

**Application of Cap Flashing at Front
and Side of Chimney
Figure 13-48**

Refill the joint with mortar. Finally, bend the flashing down so it covers the base flashing and lies snugly against the masonry.

Use one continuous piece of cap flashing on the front of the chimney as shown in Figure 13-48. On the sides and back of the chimney, use several pieces of similar-sized flashing, trimming each to fit the particular location of brick joint and roof pitch. (See Figure 13-49.) Start the side units at the

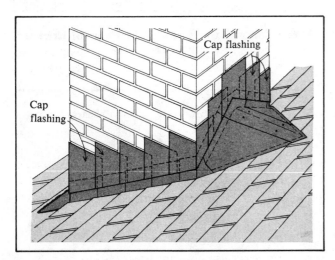

**Application of Cap Flashing at Side
and Rear of Chimney
Figure 13-49**

lowest point and overlap each at least 3 inches.

Hip and ridges: Apply shingles up to a hip or ridge from both sides of the roof before finishing the intersection. To make finishing easier, adjust the last few courses so that the ridge capping will cover the top courses of shingles equally on both sides of the ridge.

Some manufacturers supply special hip and ridge shingles and specify how they should be applied. Hip and ridge shingles also may be made from the 12-inch x 36-inch strip shingles used to cover the roof. Cut the strip shingles down to 12 x 12 inches on three-tab shingles or to a minimum of 9 x 12 inches on two-tab or no-cutout shingles. Taper the lap portion of each cap shingle slightly so that it is narrower than the exposed portion. This produces a neater job. (See Figure 13-50.)

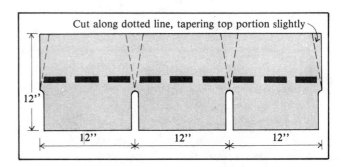

**Fabrication of Hip and Ridge Shingles From
Three-Tab Strip Shingles
Figure 13-50**

To apply the capping, bend each shingle along the centerline of the longer dimension so that it will extend an equal distance on each side of the hip or ridge. (A chalk line will ensure straight application.) In cold weather, warm the shingle until it is pliable before bending. Apply the shingles with a 5-inch exposure, beginning at the bottom of the hip or from the end of the ridge opposite the direction of the prevailing winds. (See Figure 13-51.)

Secure each shingle as illustrated in Figure 13-52 with one fastener on each side, 5½ inches back from the exposed end and 1 inch up from the edge.

Applying Strip Shingles on Low Slopes
Asphalt strip shingles may be used on slopes between 2 inches and 4 inches per foot if special procedures are followed. Never use shingles on slopes lower than 2 inches.

Low slopes can cause problems because water

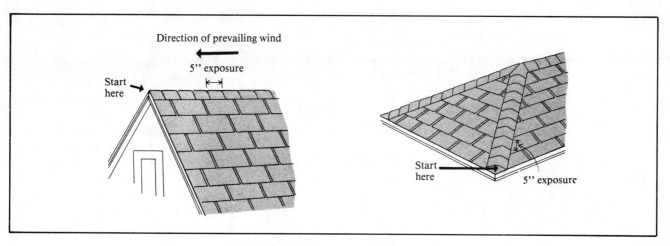

Application of Hip and Ridge Shingles
Figure 13-51

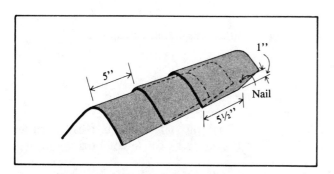

Nail Location for Hip and Ridge Shingles
Figure 13-52

drains slowly on a low slope, making back-up and damage from ice dams more likely. The procedure described below should produce a weathertight roof.

Underlayment: On low slopes, cover the deck with two layers of No. 15 non-perforated asphalt-saturated felt. Begin by nailing a 19-inch-wide strip of underlayment along the eaves, overhanging the drip edge by 1/4 to 3/8 inch. Place a full 36-inch-wide sheet over the starter, with the long edge placed along the eaves and completely overlapping the starter strip.

All following courses use 36-inch-wide sheets placed to cover the lower course by 19 inches. Use only enough fasteners to hold the course in place until the shingles are applied. End laps should be 12 inches wide and located at least 6 feet from end laps in the lower course.

Eave flashing: Wherever there is a possibility of icing along the eaves, cement all the laps in the underlayment courses from the eaves to a point at least 24 inches beyond the interior wall line of the building. The cemented double-ply underlayment serves as eave flashing. (See Figure 13-53.)

For eave flashing, cover the entire surface of the starter strip with a continuous layer of asphalt plastic cement. Use about 2 gallons per 100 square feet. Place the first course over the starter, pressing it firmly into the cement.

After the first course is in place, coat the upper 19 inches with cement. Place the second course and press it into the cement. Repeat the procedure for each course that lies within the eave flashing distance. It's important to apply the cement uniformly so the overlapping felt will float completely on the cement without touching the felt in the underlying course. But don't use too much cement.

After completing the eave flashing, secure each successive course with only enough fasteners to hold the underlayment in place until the shingles are down.

Shingle application: For increased wind resistance on low slope roofs, use either self-sealing shingles with a factory applied adhesive, or cement the tabs to the underlying course. Place a spot of asphalt plastic cement about the size of a quarter under each tab. Then press the tab into the cement. Be sure to cement all tabs throughout the roof. Any of the shingle application methods discussed previously may be used on low slopes.

Applying Strip Shingles on Steep Slopes
Asphalt shingles are popular on steep slopes. Mansard roofs are the best example of a steep slope.

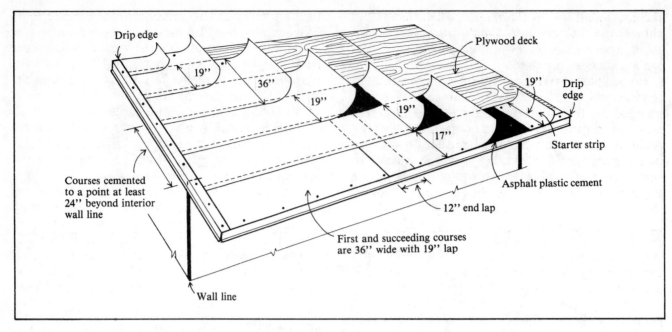

Application of Underlayment on Low Slopes Where Icing Along the Eaves is Anticipated
Figure 13-53

The problem with steep slopes is that the factory-applied self-sealing adhesive is less effective, especially on colder or shaded areas of the roof. Special methods are necessary.

The maximum slope for normal methods is 21 inches per foot. Over this limit you'll need to follow the special method described below for strip shingles with a nominal 5-inch exposure. Figure 13-54 illustrates this method.

Application procedures for underlayments, drip edges, eave flashing (if required) and other flashings are the same as those for normal slopes.

Shingle application: Apply the shingles with fasteners recommended by the roofing manufacturer. Also, follow the roofing manufacturer's

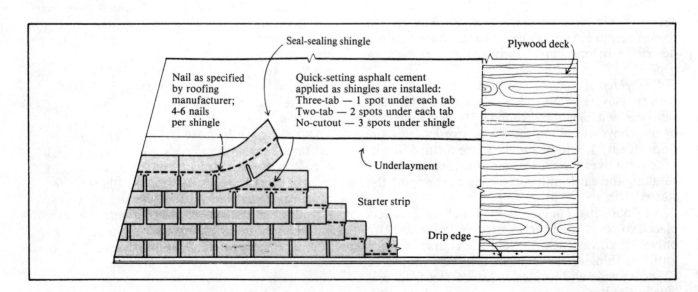

Application of Shingles on Steep Slopes
Figure 13-54

recommendations on the number of fasteners per shingle and their position. Don't drive fasteners into or above the factory-applied adhesive if self-sealing shingles are used.

Immediately after installation, cement each tab in place with a quick-setting asphalt cement recommended by the shingle manufacturer. Apply the adhesive in spots the size of a quarter. For shingles with three or more tabs, place one spot of cement under each tab. For two-tab shingles, place two spots under each tab. For no-cutout shingles, place three spots of cement under the exposed portion of the shingle.

It's also important that through-ventilation be provided in the space behind the roof deck to prevent moisture-laden air from being trapped behind the sheathing. If no ventilation is provided, the sheathing may buckle and the shingles may blister.

For strip shingles with more than a nominal 5-inch exposure, refer to the roofing manufacturer's application instructions.

Applying Strip Shingles for Reroofing

Many of the procedures for applying shingles on new roofs are also followed in reroofing. However, other procedures are used specifically in reroofing. The procedure you use will depend on the type of material on the existing roof, its condition, and whether the new roof can be placed over it.

Deck preparation: When starting a reroofing job, the first step is to decide whether the existing roofing should remain in place or be removed. Complete tear-off may not be necessary. In general, roofs covered with wood shingles, asphalt shingles, asphalt roll roofing or built-up asphalt roofing may be left in place. Unless your code says otherwise, a maximum of three roofs (original and two reroofs) can be installed before a tear-off is mandatory.

If a surface of old asphalt roofing or wood shingles is warped, curled or badly weathered, you won't be able to create a level surface with the new shingles. You may have to tear these off and start over. Asphalt shingle roofs that have lost most or all of their mineral surface may still provide a very good base for new asphalt shingles if the rest of the roof is sound.

Wood shingles can be left in place if you can level the surface enough so there is a relatively smooth base for the new asphalt shingles.

Tile, slate, or cedar shake roofs must be removed because of their irregular surface and the difficulty in nailing through them.

Your main worry will always be the condition of the deck and framing underneath the existing roof. New material installed on top of the old will add at least a ton to the roof on most jobs. Framing must be strong enough to support the additional load plus the weight of the roofers and their equipment. If the deck has deteriorated to the point where it won't hold a nail, it must be replaced.

If there are holes in the existing roof or if there are depressions or sagging that indicate structural problems, tear off the old roof and reinforce the deck and framing if necessary.

Regardless of whether the existing material remains or not, inspect adjacent parts of the building for defects before starting to reroof. Repair or replace rotting or deteriorated wood trim. If necessary, point up chimney joints and replace all worn flashing. Clean, rebuild or reline gutters. Make sure there is enough attic ventilation. After correcting or repairing any problems, clean all debris from the roof surface.

When all repairs are made and the deck is properly prepared, apply shingles following the same procedures as in new construction. Place underlayments, drip edges, eave flashing, valley flashing and other flashing where necessary. Roofing nails are the only recommended fasteners when reroofing over an existing roof.

Removing old roofing: If you have to do a tear-off, strip the materials down to the roof deck. If the deck under wood shingles or shakes consists of spaced sheathing, start removing old shingles at the ridge so that broken material will not fall through the open sheathing into the attic.

Flat-surfaced shovels are best for removing old asphalt shingles, wood shingles, built-up roofing and felt underlayments. Be sure that all old nails are removed from the surface. Use a pry bar to remove soil stack and vent flashing that has deteriorated. Don't damage old metal flashing that can be used as a pattern for new flashing. If metal cap flashings at the chimney and other vertical masonry wall intersections have not deteriorated, bend them up out of the way so that they can be used again. Be careful in these areas to avoid damaging reusable base flashing. For safety, keep the deck clear of waste material as work proceeds. Sweep the deck clean after all the old roofing has been removed.

At this point, inspect the deck to be sure it's sound. Make whatever repairs are necessary to roof framing to strengthen and level the deck. Replace rotted, damaged or warped sheathing or delaminated plywood. Cover all large cracks, knotholes, loose knots and resinous areas in the deck with sheet metal patches nailed to the sheathing. Remove loose or protruding nails or hammer them down.

If the deck consists of spaced sheathing, fill in all spaces with new boards of the same thickness as the old deck or cover the entire area with plywood sheathing.

Sweep the deck clean again. Then follow the shingle application procedures as for new construction. Flashing details generally follow those for new construction.

Old roofing remains in place: The procedure depends on the type of existing roofing. Four situations are common: asphalt shingles over old asphalt shingles, over wood shingles, over roll roofing and over built-up roofing.

Begin by taking measurements at both rakes. It's common to find a difference of several inches between rakes on the same roof. With these measurements in mind, adjust the courses to allow for any differences.

Asphalt shingles over old asphalt shingles— Inspect the existing roof for loose, curled, lifted or broken shingles. Remove them or nail them down. Replace all missing shingles with new shingles so you have an unbroken nailing base.

Buckled shingles are a sign of warped deck boards or protruding nails. Renail warped boards. Hammer in all loose or protruding nails or remove them and nail the shingle down in a new location. Remove all badly worn drip edges and replace them. Sweep the surface clean before applying new roofing.

If the existing asphalt shingles are the interlocking type, the new shingles may become uneven in time. If a smoother base surface is needed, remove the old shingles.

Asphalt shingles over old wood shingles— Remove all loose or protruding nails and renail the shingles in a new location. Nail down any loose shingles. Split badly curled or warped shingles and renail the two halves. Replace any missing shingles.

If shingles and trim at the eaves and rakes are badly weathered and subject to high winds, cut back the shingles at both locations far enough to install an edging strip. The edging should be a 1 x 4-inch to 1 x 6-inch board nailed firmly in place, with the outside edge projecting beyond the deck the same distance as the wood shingles. Install drip edges and eave flashing (if required).

Many wood shingles provide a smooth nailing surface and no further preparation is needed. But some codes require a 30-pound felt underlayment.

If an old wood roof does not give a good nailing surface but is still in fairly good condition, consider applying feathering strips or "horse feathers" along the butts of each course of old shingles. The 1 x 4-inch or 1 x 6-inch wood strips are beveled at the lower edge to meet the roof surface smoothly between each course of shingles.

*Asphalt shingles over old roofing—*Press the loose edges flat against the deck and fasten them down to provide a smooth surface. Remove loose or protruding nails and drive new nails nearby. Nail down any lap joints that have separated. Trim old roofing that has been torn or damaged. Inspect the underlying deck for knots or resinous areas which should be covered with sheet metal patches. Patch tears in the roofing with a new piece of roofing the same size as the trimmed area and nail the patch in place. Sweep the deck clean before applying new shingles.

Asphalt shingles over old built-up roofing— Built-up roofing on a slope between 2 inches and 4 inches per foot can be reroofed with asphalt shingles if there is no insulation between the deck and the felts. Remove old slag, gravel or other material that is in place, leaving the surface of the underlying felts smooth and clean. If there is no way to get a smooth surface, remove all the old roofing and start over. Apply the asphalt shingles directly over the felt the same way as in new construction on low slopes.

If there is rigid insulation under a built-up roof, install a plywood nail base over the old roofing and handle it like a shingle job after a tear-off. Check the condition of the insulation to make sure moisture is not trapped under the plywood. The nailing base should be the same as that for new construction.

Underlayment: If the old roofing has been removed, cover the deck with an asphalt-saturated felt underlayment as if it were new construction. If the old roofing is not removed and new shingles will be applied directly over the existing material, no additional underlayment is required because the old roofing serves the same purpose as the underlayment. However, some codes require 30-pound asphalt-saturated felt as an underlayment when reroofing over old wood shingles.

If reroofing over an existing roof and new eave flashing is required to protect against ice dams, remove the old roofing to a point at least 24 inches beyond the interior wall line and follow the procedure outlined previously.

How the shingles are applied depends on whether the existing roofing has been removed or

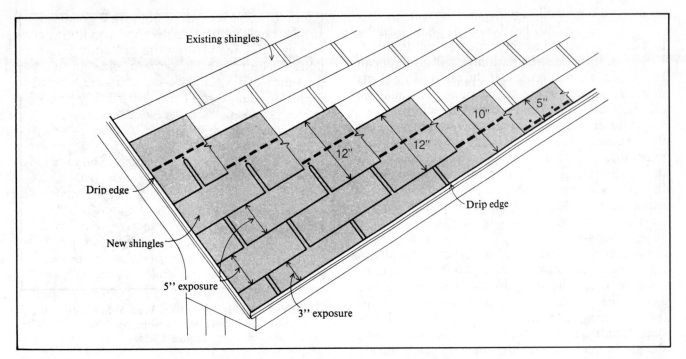

Application of New Asphalt Shingles Over Existing Asphalt Shingles
Figure 13-55

left in place and on the type of roofing material left in place.

If the old roofing has been removed or if you are reroofing over old roll roofing or built-up roofing, shingles go down the same as in new construction. Wood shingle roofs that are modified by feathering strips also are shingled like new construction.

The application that is different from new construction is when you are laying strip shingles over existing strip shingles. You want to minimize the unevenness that occurs when new shingles bridge over the butts of old shingles. What I recommend here ensures that the new horizontal nailing pattern is 2 inches below the old one.

The procedure illustrated in Figure 13-55 assumes that both the new roofing and old roofing have a 5-inch exposure and that the old shingles are laid straight. If new eave flashing has been added, snap chalk lines on it to guide installation of the new shingles until the courses butt against the existing courses.

Starter strips: Remove the tabs plus 2 inches or more from the top of the starter strip shingles so that the remaining portion is equal in width to the exposure of the old shingles (normally 5 inches). Apply the starter strip so that it is even with the existing roof at the eaves. If self-sealing shingles are used for the strip, locate the factory-applied adhesive along the eaves. Be sure the existing shingles overhang the eaves far enough to carry water into the gutter. If they do not, cut the starter strip to a width that will. Don't overlap the existing course above. Remove 3 inches from the rake end of the first starter strip shingle so that joints between adjacent starter strip shingles will be covered when the first course is applied.

For the first course, cut 2 inches or more from the butts of the first course of shingles so that the shingles fit between the butts of the existing third course and the eave edge of the new starter strip. Start at the rake with a full-length shingle. Use four nails per shingle, locating them in the same positions as in new construction—5/8 inch above the cutouts and 1 inch and 12 inches in from the sides. Don't nail into or above the factory-applied adhesive.

For the second and following courses use full-width shingles. Place the top edge of the new shingles against the butt edge of the old shingles in the course above. The full-width shingles used on the second course will reduce the exposure of the first course to 3 inches, but this area is usually concealed by gutters so appearance won't be a problem. For the remaining courses, the 5-inch exposure is automatic, the same as that of the existing shingles. As in new construction, apply the shingles across and diagonally up the roof.

If the old roofing has been removed, flashing details generally follow those for new construction. However, if the existing flashing is still good, it can be left in place and reused. If the old roofing is left in place, some flashing application details may be different from that in new construction:

• If the existing roof has an open valley, build up the exposed area of the valley with 90-pound mineral-surfaced roll roofing to a level flush with the existing roofing. Then install new open valley flashing as in new construction, overlapping the existing shingles. Better yet, build a woven or closed-cut valley, with the new shingles crossing over the valley filler strip.

• When flashing against the vertical sidewall using asphalt shingles over old asphalt shingles, continue to place the top edge of the new shingles against the butt edge of the existing shingles. Trim the new shingles to within 1/4 inch of the existing step flashing. Embed the last 3 inches of the end shingle in each course in asphalt plastic cement. Apply a bead of cement with a caulking gun at the joint between the ends of the new shingles and the sidewall.

• When flashing against a vertical sidewall using asphalt shingles over old wood shingles, place a 6- to 8-inch-wide strip of 50-pound smooth roll roofing over the wood shingles that butt the wall surface. Nail the strip along each side. Nails should be spaced 4 inches. Cover the flashing strip with a thin layer of asphalt plastic cement and embed the end shingle in each course firmly in it. Apply a bead of cement with a caulking gun at the joint between the ends of the new shingles and the sidewall. (See Figure 13-56.)

Applying Individual Shingles

There are three kinds of individual shingles: hexagonal, giant and interlocking. Your choice depends on the slope, wind expected, coverage, your customer's choice and what is available at a reasonable cost. Installation varies with the type of shingle used.

No matter what type of individual shingle you use on a new deck, preparation of the deck is the same as when applying strip shingles. Inspect the deck, framing and attic for enough ventilation. Cover the deck with an asphalt-saturated felt underlayment. Snap horizontal and vertical chalk lines. Install drip edges, eave flashing (if required) and valley flashing.

When individual shingles are used for reroofing, follow the same procedure as when reroofing with

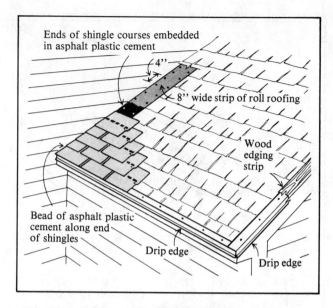

Flashing Against a Vertical Wall When Reroofing over Wood Shingles
Figure 13-56

strip shingles. Install flashing as needed.

Two types of hexagonal shingles are available: those that are locked together by a clip and those that have a built-in locking tab. (See Figure 13-57.) Both the clip-down and lock-down shingles are relatively lightweight and are intended primarily for reroofing. They are also used at times for new construction. For either application, the slope should be 4 inches per foot or more.

Giant shingles, shown in Figure 13-58, may be used for new construction or reroofing.

Dutch lap method—This method is intended primarily for reroofing over a smooth deck that will hold nails well. It can also be used to cover new decks where single coverage is enough. For either application, the slope should be 4 inches per foot or greater.

American method—This method of application is used for both new construction or reroofing. In either case, the slope should be 4 inches per foot or greater.

Interlocking shingles have a locking tab that gives extra wind resistance. The shingles can be used for reroofing over existing roofing on slopes recommended by the shingle manufacturer. They are also used in new construction, depending on whether single or double coverage is needed. In general, single coverage interlocking shingles are not recommended for new construction.

Nail or staple placement is important in interlocking shingles. For best results, follow the

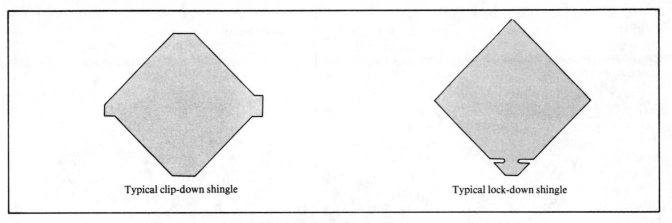

Individual "Hex" Shingles
Figure 13-57

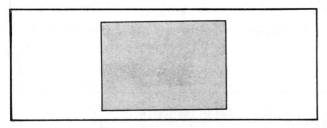

Giant Individual Shingle
Figure 13-58

manufacturer's directions for applying the starter, first and following courses.

Although interlocking shingles are self aligning, they are flexible enough to allow limited movement for adjustment. It's especially important to snap horizontal and vertical chalk lines to keep the work straight.

The integral locking tabs are manufactured to close tolerances so the shingles match up well. Be sure to lock them carefully and correctly. Figure 13-59 shows two common locking devices used in interlocking shingles.

During installation, locking tabs on shingles along the rakes and eaves may have to be removed in part or entirely. Where locking tabs are removed, apply cement or nail the cut edge in place according to the manufacturer's instructions.

Roll Roofing

Asphalt roll roofing is made in 36-inch-wide sheets in many weights, surfacings and colors. It is used both as a primary roof covering and flashing.

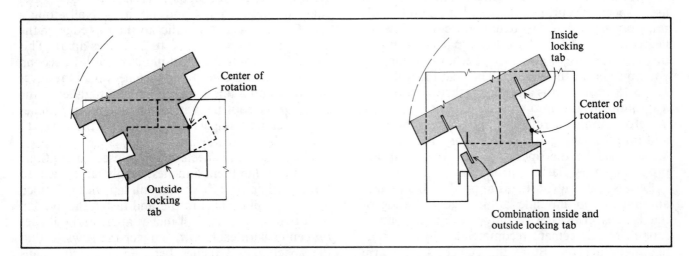

Common Interlocking Shingles and Their Methods of Locking
Figure 13-59

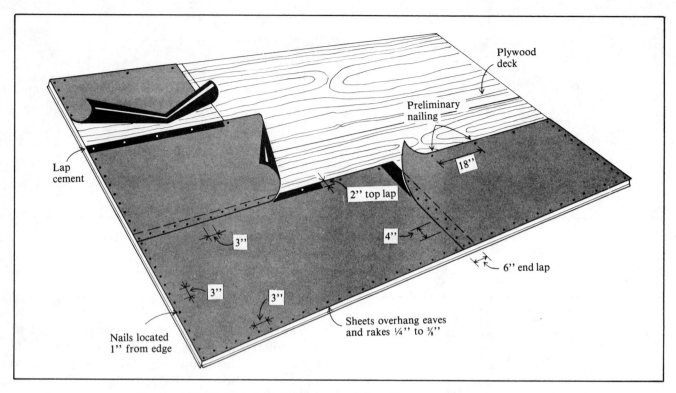

Exposed Nail Method of Applying Roll Roofing Parallel to the Eaves
Figure 13-60

As a primary roof covering, roll roofing may be used on slopes down to 2 inches per foot if applied by the exposed nail method or 1 inch per foot if applied by the concealed nail method. Where durability is important, the concealed nail method is best. With either method, the roofing can be applied parallel to eaves or parallel to the rakes.

Store the material in a warm place until ready for use, especially during cold weather. Don't apply roll roofing when the temperature is below 45 degrees. If cold rolls have to be put down below this temperature, warm them before unrolling so you don't crack the coating. Then cut the rolls into 12- to 18-foot lengths and spread them in a pile on a smooth surface until they flatten out.

Before applying roll roofing, prepare the deck and install flashing as described for strip shingles. Valleys will be the open type, so follow the appropriate valley flashing procedures.

Because all roll roofing is applied with a certain amount of top and side lapping, good sealing of the laps is critical. Use only the lap cement or asphalt plastic cement recommended by the roofing manufacturer. Store the cement in a warm place until ready for use. If it's necessary to warm the cement, place the unopened container in hot water. Never heat asphalt plastic cement directly over a flame, and don't try to thin it by diluting it with solvents.

Application parallel to the eaves with exposed nails is illustrated in Figure 13-60, including lapping, cementing and nailing.

First course: Position a full-width sheet so that the lower edge and ends overhang the eaves and rakes by between 1/4 and 3/8 inch. Nail along a line 1/2 to 3/4 inch parallel to the top edge of the sheet, spacing the nails 18 to 20 inches apart. This top nailing holds the sheet in place until the second course is placed over it and fastened. Nail the eaves and rakes on a line 1 inch from and parallel to the edges of the roofing, with the nails spaced 3 inches on center and staggered a bit along the eaves to avoid splitting the deck.

If two or more sheets must be used to continue the course, lap them 6 inches. Apply lap cement to the underlying edge over the full lap width. Embed the overlapping sheet into it and fasten the overlap with two rows of nails 4 inches apart and 4 inches on center with each row. Stagger the rows so that the spacing is 2 inches between successive nails from row to row.

Place the second course so it overlaps the first

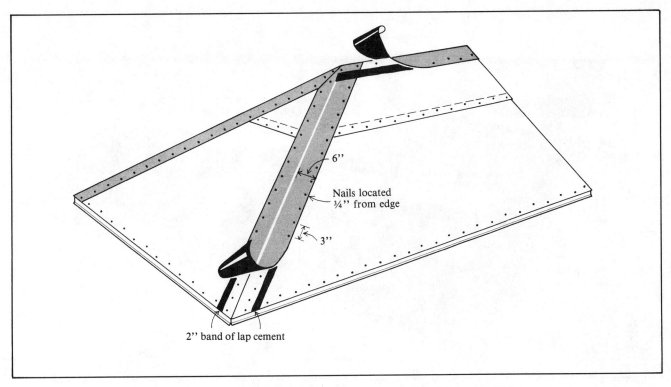

Exposed Nail Method of Applying Roll Roofing to Hips and Ridges
Figure 13-61

course by 2 inches. Fasten the second course along the top edge, following the same nailing procedure as the first course. Lift the lower edge of the overlapping sheet and apply lap cement evenly over the upper 2 inches of the first course. Then embed the overlapping sheet in the cement. Fasten the lap with nails spaced 3 inches on center and staggered slightly. Place the nails not less than 3/4 inch from the edge of the sheet. Nail the rake edges in the same manner as the first course.

Follow the same procedure for each successive course. End laps should be 6 inches wide and cemented and nailed the same way as the first course. Stagger end laps so that an end lap in one course is never positioned over the end lap in the joining course.

Hip and ridges: Trim, butt and nail the roofing as it meets at a hip or ridge. Snap a chalk line on each side of the hip or ridge and 5½ inches from the joint and parallel to it. Starting at the chalk lines and working toward the joint, spread a 2-inch-wide band of asphalt lap cement on each side of the hip or ridge. (See Figure 13-61.)

Cut strips of roll roofing 12 inches wide and bend them lengthwise along the centerline so that they lay 6 inches on each side of the hip or ridge. In

cold weather, warm the roofing before bending it. Lay the bent strip over the joint and embed it in the cement. Fasten the strip to the deck with two rows of nails, one on each side of the hip or ridge. The rows should be located 3/4 inch from the edges of the strip and the nails spaced 3 inches on center. Be sure the nails penetrate the cemented zone underneath. This seals the nail hole with asphalt. End laps should be 6 inches and cemented the full lap distance. Don't use too much cement.

Concealed nail method: When using this method, narrow edging strips are placed along the eaves and rakes before applying the roofing. Figure 13-62 shows the general installation procedure, including lapping, cementing and nailing.

Edge strips: Place 9-inch-wide strips of roll roofing along the eaves and rakes, positioning them to overhang the deck 1/4 to 3/8 inch. Fasten the strips with rows of nails located 1 inch and 8 inches from the roof edge and spaced 4 inches on center in each row.

First course: Position a full-width strip of roll roofing so that its lower edge and ends are flush with the edge strips at the eaves and rakes. Fasten the upper edge with nails 4 inches on center and slightly staggered. Locate the nails so that the next

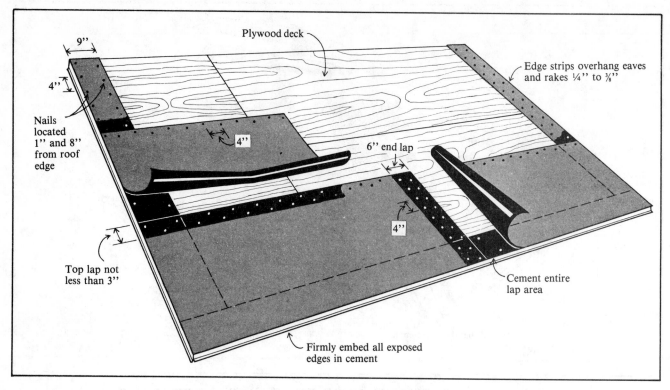

Concealed Nail Method of Applying Roll Roofing Parallel to The Eaves
Figure 13-62

course will overlap them a minimum of 1 inch. Lift the lower edge of the first course and cover the edge strips with cement according to the manufacturer's directions.

In cold weather, turn the course back carefully to avoid damaging the roofing material. Press the lower edge and rake ends of the first course firmly into the cement-covered edge strips. Work from one side of the sheet to the other to avoid wrinkling or bubbling.

End laps should be 6 inches wide and cemented over the full lap area with the recommended cement. Nail the underlying sheet in rows 1 inch and 5 inches from the end of the sheet, with the nails spaced 4 inches on center and slightly staggered. End laps in succeeding courses must not line up with one another.

Second and succeeding courses: Position the second course so that it overlaps the first course at least 3 inches. Fasten the upper edge to the deck, cement the laps and finish installing the sheet in the same manner as the first course. Follow the same procedure for each successive course. Don't apply nails within 18 inches of the rake until cement has been applied to the edge strip and the overlying strip has been pressed down.

Apply the cement in a continuous but fairly thin layer over the full width of the lap. Press the lower edge of the upper course firmly into the cement until a small bead appears along the edge of the sheet. Using a roller, apply pressure uniformly over the entire cemented area.

Hips and ridges: Trim, butt and nail the sheets as they meet at a hip or ridge. Next, cut 12 x 36-inch strips from the roll roofing and bend lengthwise to lay 6 inches on each side of the joint. Don't bend the strips in cold weather without first warming them. These will be used as "shingles" to cover the joint, each one overlapping the other by 6 inches as shown in Figure 13-63.

Start hips at the bottom and ridges at the end opposite the direction of the prevailing winds. Use a chalk line 5½ inches from and parallel to the joint on both sides. Apply asphalt plastic cement evenly over the entire area between the chalk lines from one side of the joint to the other. Fit the first folded strip over the joint and press it firmly into the cement, driving two nails 5½ inches from the edge of the end that will be lapped. Cover the 6-inch lap on this strip with lap cement. Then place the next strip over it. Nail and cement in the same way as the first strip. Cement uniformly so the overlap-

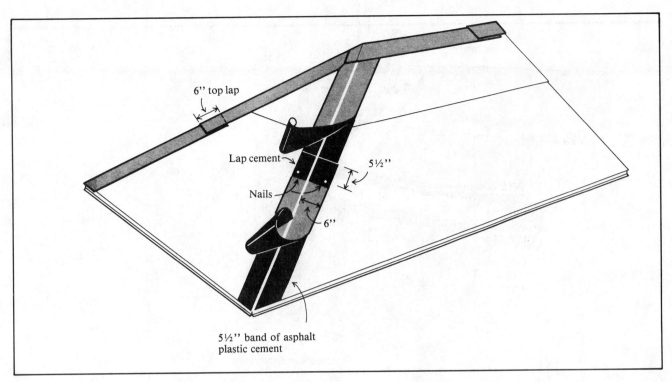

6" top lap

Lap cement

5½"

Nails

6"

5½" band of asphalt
plastic cement

Concealed Nail Method of Applying Roll Roofing to Hips and Ridges
Figure 13-63

ping felt will float completely on the cement without touching the felt in the underlying course.

After completing the eave flashing, lay each following course with only enough fasteners to hold the underlayment in place until the shingles are applied.

Figure 13-64 illustrates the exposed nail method of applying roll roofing parallel to the rake.

Applying double coverage roll roofing: Double coverage roll roofing is a 36-inch-wide sheet with 17 inches intended for exposure and 19 inches for underlay. It provides double coverage for the roof and may be used on slopes down to 1 inch per foot.

The 17-inch exposed portion is covered with granules. The 19-inch selvage part may be finished in several ways. Some manufacturers saturate the selvage portion with asphalt. Others saturate and coat it.

Be sure there is enough roof drainage so the water doesn't stand in puddles long after rain has passed. This is especially important on the low slopes where double coverage roofing is common.

Double coverage rolls can be put down parallel to the eaves or parallel to the rake. Although 19-inch selvage roll roofing is most common, any roll roofing can be applied the same way to get double coverage if the lapped portion of the sheet is 2 inches wider than the exposed portion. Figures 13-65 through 13-67 show applications of double coverage roll roofing.

Algae Discoloration

Roof discoloration caused by algae is usually called "fungus growth" and is a problem in high humidity areas. It's often mistaken for soot, dirt, moss or tree droppings.

The algae that cause this discoloration don't feed on the roofing and don't affect the life of the roofing. However, the algae can gradually turn a white or light-colored roof dark brown or black.

Algae discoloration is difficult to remove from roofing surfaces. But it can be lightened with a diluted solution of chlorine bleach. Sponge the solution on the roofing. Scrubbing will loosen the mineral granules. Apply the solution carefully so you don't damage other parts of the building or the shrubbery below. If possible, work from a ladder or walkboards to avoid walking directly on the roof surface. After sponging, rinse the solution off the roof with a hose.

Cleaning is only a temporary solution. The algae will grow back. Several types of algae-resistant roofing have been developed and are now available. These are specifically designed to control most algae growth for many years.

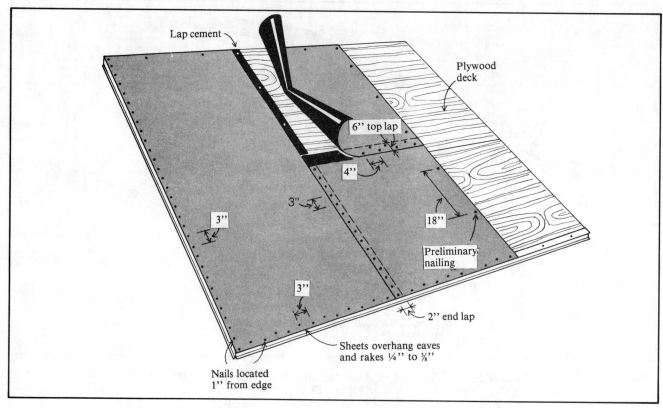

Exposed Nail Method of Applying Roll Roofing Parallel to The Rake
Figure 13-64

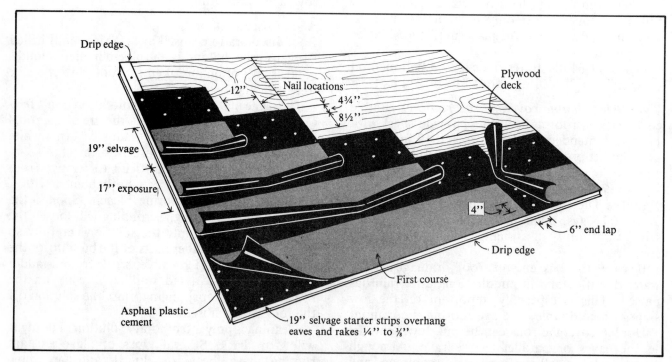

Application of Double Coverage Roll Roofing Parallel to The Eaves
Figure 13-65

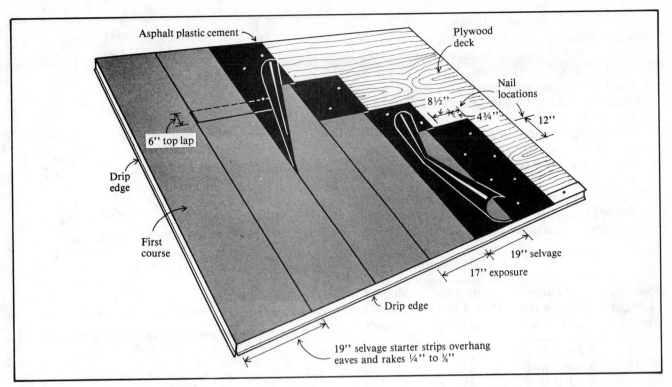

Application of Double Coverage Roll Roofing Parallel to The Rake
Figure 13-66

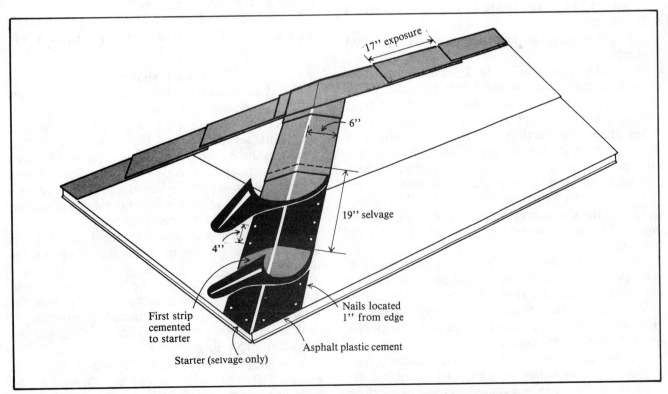

Application of Double Coverage Roll Roofing to Hips and Ridges
Figure 13-67

Wood Shingles

Wood shingles and shakes are still a good choice for many better-quality homes. Nothing sheds water like a good shingle roof. And wood shingles should last as long as asphalt shingles. The only major drawbacks are higher cost and increased risk of fire in fire-prone areas.

Wood shingles should be number 1 grade, which is all heartwood, all edge grain, and tapered. Cypress, redwood and western red cedar are the most common.

Widths of shingles in each bundle will vary, with the lower grades being narrower. The following are recommended exposures for common shingle sizes:

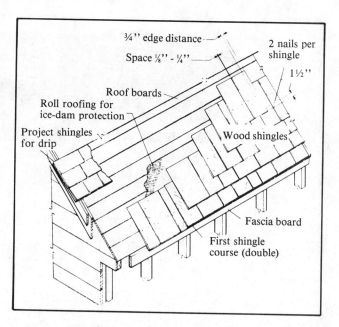

Application of Wood-Shingle Roofing Over Boards
Figure 13-68

		Maximum Exposure	
Shingle Length (Inches)	Shingle Thickness (Green)	Slope Less[1] than 4 in 12 inch	Slope 4 in 12 and over inch
16	5 butts in 2"	3¾	5
18	5 butts in 2¼"	4¼	5½
24	4 butts in 2"	5¾	7½

[1] Minimum slope for main roofs — 4 in 12. Minimum slope for porch roofs — 3 in 12.

Figure 13-68 shows the right way to apply wood shingles.

Here's a general guideline for laying wood shingles:

1. Extend shingles 1½ to 2 inches beyond the eave line and about 3/4 inch beyond the rake edge.

2. Nail each shingle with two rust resistant nails spaced about 3/4 inch from the edge and 1½ inches above the buttline of the next course. Use threepenny nails for 16- and 18-inch shingles and fourpenny nails for 24-inch shingles. Where shingles are applied over old wood shingles, use longer nails to penetrate through the old roofing and into the sheathing. A ring-shank nail is recommended where plywood roof sheathing is less than 1/2 inch thick.

3. Allow a 1/8- to 1/4-inch space between each shingle for expansion when wet. Lap vertical joints at least 1½ inches from shingles in the course above. Offset the joints in succeeding courses so that the joint in one course is not in line with the joint in the second course above it.

4. Shingle away from valleys, selecting and precutting wide valley shingles. The valley is 4 inches wide at the top and should increase in width at the rate of 1/8 inch per foot from the top. Use valley flashing with a standard seam. Do not nail

through the metal. Valley flashing should be a minimum of 24 inches wide for roof slopes under 4/12; 18 inches for roof slopes of 4/12 to 7/12; and 12 inches for roof slopes over 7/12.

5. Place a metal edging along the gable end of the roof to guide water away from the end walls.

Wood Shakes

Wood shakes go down the same way as shingles except that longer nails must be used because the shakes are thicker. Shakes have a greater exposure than shingles because of their length. Exposure distances are 8 inches for 18-inch shakes, 10 inches for 24-inch shakes and 13 inches for 32-inch shakes. Butts are often laid unevenly to create a rustic look. An 18-inch-wide underlay of 30-pound asphalt felt should be used between each course to prevent wind driven moisture from entering between the rough faces of the shakes. Position the underlay above the butt edge of the shakes a distance equal to double the weather exposure. Where exposure distance is less than one-third the total length, underlay is not usually required.

When reroofing over the old shingles, use the following procedures:

1. A 36-inch-wide strip of 15-pound minimum asphalt felt is applied over the old shingles at the eave line. 18-inch-wide strips of felt are then applied shingle style, at 10-inch intervals up the roof. Next, a starter course is applied at the eave line, ex-

The Look of Professional Work
Figure 13-69

tending approximately 1½ inches over the eave edge.

2. The first course of shakes is then put down directly over the starter course. The tip ends of each course of shakes tuck under the felt strips. With a 24-inch-long shake laid at 10-inch exposure, the top 4 inches of the shake will be covered by felt. Shakes should be spaced about 1/2 inch apart to allow for expansion. These joints or spaces between the shakes should be broken or offset at least 1½ inches in adjacent courses.

3. Felt covers the old shingles, and each felt strip overlaps the top 4 inches of the shake. Longer nails are needed in placing shakes over old roofing than in new roof construction.

4. For valleys, use at least 26-gauge galvanized and painted metal. It should be center and edge crimped. If new valley metal might come in contact with old metal, the two should be separated by a strip of lumber placed in each valley.

5. Shakes are cut parallel to valleys as courses are applied over the metal valley. Valley gutters should be approximately 6 inches wide. Metal valley sheets should be at least 20 inches wide with a 4- to 6-inch head lap.

6. Shakes are cut parallel to the hip line.

7. Factory assembled "hip and ridge" units are available for completing the hip and ridge. Apply them at the same exposure as the shingles.

Summary

Who knows which trend roofing will follow? In the past, wood shingles were the builder's choice. Then various styles of tin replaced the "unreliable" wood shingles. Asphalt shingles became popular on residential houses to get away from the "barn" look of tin. Now, wood shingles are back to give that "rustic" look that everyone was trying to get away from in the first place. Tomorrow the "barn" look of tin might be back again. Recognize that styles change—and it's in your interest to go along with the style.

As you can see, there's a lot involved in putting on a roof. Roofing has to be installed properly to be safe and weathertight. It also must look good. A sloppy roof can ruin the appearance of a house and can hurt your reputation as well.

When working on a roof, remember that no one should have to touch it for another 20 years. And you should be proud of the job you've done. Nothing builds a reputation for shoddy workmanship like a leaky roof. And nothing earns the respect of an owner better than a roof that lasts as long as he owns the house.

14

Fireplaces and Chimneys

Very few homes rely on a fireplace for heat. But builders are still putting fireplaces in new homes because buyers want them. A fire in the fireplace makes any house seem warm and cozy.

Old-fashioned fireplaces exhausted more warm air than they gave off. But new fireplaces use outside air for combustion and circulate room air through heating chambers in the fireplace. This makes modern fireplaces a popular remodeling project where cheap fuel is available.

If you're like most remodelers, you'll have an occasional fireplace and chimney project and will have to do repair work on chimneys fairly regularly.

Fireplace Repairs

Old fireplaces and chimneys sometimes need repairs. Cracks appear in the chimney, and the brick in the fireplace deteriorates and flakes off. In most older houses the chimneys do not have flue liners, so a chimney full of cracks can be a serious problem requiring repair or replacement. The fireplace may need the fine touch of a good mason to replace old brick with fire brick.

Some old fireplaces don't draw well and perhaps never did. A fireplace is not like a car that operates well when new and gradually wears out. An old fireplace should operate just as well as it did when it was first built. If it doesn't draw properly now, then it probably never drew right. It's best to replace old, defective and deteriorated fireplaces and chimneys with new ones that are properly designed.

Gas and oil heaters use chimneys. These, too, can become defective. Heaters can be vented properly with metal flues in most areas. Fabricated metal chimneys can usually be installed at a lower cost than the conventional masonry chimney.

Most codes require a flue liner in chimneys built today. Liners keep the flue tight even when cracks appear in the masonry chimney. Where flues are not lined, stainless steel flue lining can be installed. The lining is available in 2- or 2½-foot lengths and in sizes to fit most standard chimneys. The sections are connected and inserted from the top, with the installer joining an additional section as the liner is dropped into place. This only works, however, if the chimney is straight.

As we'll see later, a lot goes into building a good fireplace. If all the measurements and dimensions are not exactly right, the draft will be poor and the fireplace will leak smoke into the room.

To test a fireplace, light a rolled up newspaper and hold it near the hearth in front of the fireplace—not in it. Do the smoke and flame go into the fireplace? If not, something is wrong.

There are several things you can do. You can put a metal extension across the top of the fireplace opening, extend the height of the flue, or add a chimney cap to produce a venturi action which might boost the draft. But trying to make an improperly built fireplace work is usually hopeless. The only answer is to tear it down and build another one.

Fireplaces used to be built without dampers. Maybe a few still are. Since the fireplaces used today are seldom the sole source of heat, a damper is needed so the fireplace can be closed when not in use to keep heat from escaping up the chimney. If there isn't a damper, you can install glass doors on the fireplace to keep heat from escaping.

Chimneys often catch fire where creosote has accumulated. When this happens it sounds like the whole house is going up in flames. If the chimney is in pretty good condition, there's nothing to worry about. Folks used to set the soot on fire deliberately, just to burn it out. The more "lighter" knots (resinous wood) you burn in a fireplace, the more creosote you'll have. Green wood contributes to the creosote build-up, also.

Chemical soot removers are usually a waste of money. They cause soot to burn, which may be the very thing you don't want.

Proper Construction

Fireplaces that are built right don't smoke. They draw well, radiate heat freely, and burn wood evenly.

The illustrations of Plan, Elevation and Section (Figures 14-1, 14-2 and 14-3) show details of chimney and fireplace construction. The corresponding dimensions are listed in the Table of Dimensions, (Figure 14-4). These figures represent dimensions for typical sizes of conventional fireplaces.

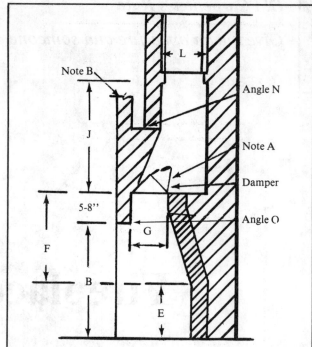

Notes

A — The back flange of the damper must be protected from intense heat by being fully supported and shielded by the back wall masonry. Also, the damper and lintel angles should not be built in solidly at the ends but given freedom to expand with heat, as shown in the front elevation view.

B — The drawing indicates thickness of brick fireplace front as 4''. No dimensions can be given for this because of various materials used, such as marble, stone, tile, etc., all of which have varying thicknesses.

Section X-X
Figure 14-2

First, select the opening of your fireplace from the left-hand column (A). The next column (B), to the right, gives you the appropriate height for that opening (width). The other columns give dimensions for the other parts and components of the fireplace.

Size of Flues

For the fireplace size listed in Figure 14-4, the flue size was determined by figuring the flue cross-sectional area as $\frac{1}{12}$ the area of the fireplace opening. If the fireplace is to vary from the size in Figure 14-4, the required flue size should be determined using the same fraction of $\frac{1}{12}$. The table in Figure 14-5 gives the cross-section area for standard flues.

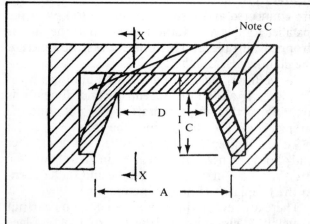

Note

C — The hollow, triangular spaces indicated in the plan, behind the splayed sides of the inner brickwork, should be filled to afford solid backing.

Plan
Figure 14-1

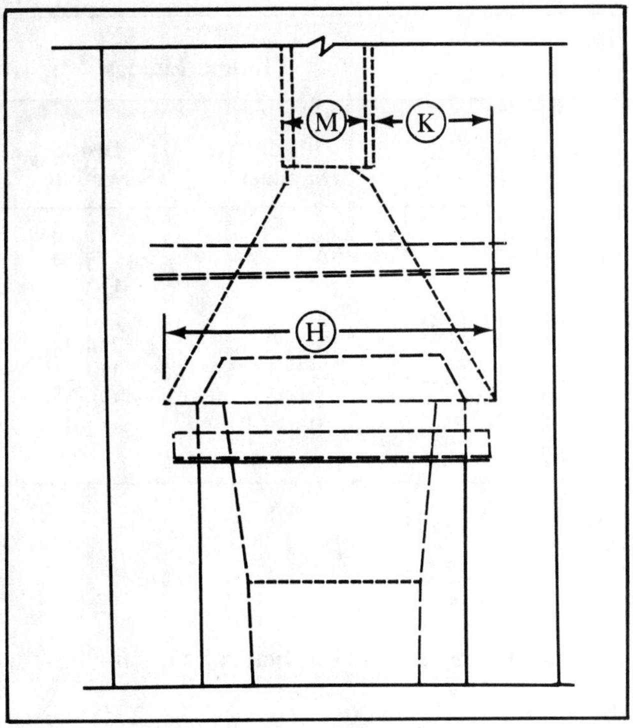

Elevation
Figure 14-3

When the fireplace is to be higher than it is wide, the flue should be 1/10 of the opening instead of 1/12. This is a safer proportion.

Note that the table in Figure 14-4 gives two columns of flue sizes, one designated "new" and the other "old." Don't substitute one style of flue for the other without thoroughly checking the size of the cross-section. The new or modular flue can't always be substituted for what seems to be the same size. Study Figure 14-4. Be slow to switch and quick to correct! The job has to be done right.

Three different measurements are used for the three types of flue lining: outside measurement of the old standard, "nominal" outside dimensions for the modular standard, and inside diameter for the round linings. Whatever type of lining is used, the sectional areas provide a guide for the desired flue capacity, which should be based on a sectional area not less than 1/12 that of the fireplace opening.

A Sloping Back

The back of the fireplace is sloped forward from a point 14 inches above the hearth to the level of the damper as indicated in Figure 14-2. This deflects flame and heat forward. Combustion gases pass through the damper (throat) but a maximum amount of heat is deflected into the room.

A Forward Damper

The sloped back contributes to other important design factors. It brings the damper (throat) forward of and not directly under the flue, and to a position just back of the breast wall of the fireplace. It also leaves room for an ample smoke shelf behind the damper.

When the damper is placed in the rear of the fireplace, no smoke shelf is possible and soot falling down the chimney comes through the damper opening and often into the room. There is no bar-

Finished Fireplace Opening										Rough Brick Work and Flue Size							
										New Flue Sizes			Old Flue Sizes			Steel Angles°	
A	B	C	D	E	F	G	H	I	J	K	L M	Round	K	L	M	N	O
24	24	16	11	14	18	8¾	32	20	19	10	8x12	8	11¾	8½ x	8½	A-36	A-36
26	24	16	13	14	18	8¾	34	20	21	11	8x12	8	12¾	8½ x	8½	A-36	A-36
28	24	16	15	14	18	8¾	36	20	21	12	8x12	10	11½	8½ x	13	A-36	A-36
30	29	16	17	14	23	8¾	38	20	24	13	12x12	10	12½	8½ x	13	A-42	A-36
32	29	16	19	14	23	8¾	40	20	24	14	12x12	10	13½	8½ x	13	A-42	A-42
36	29	16	23	14	23	8¾	44	20	27	16	12x12	12	15½	13 x	13	A-48	A-42
40	29	16	27	14	23	8¾	48	20	29	16	12x16	12	17½	13 x	13	A-48	A-48
42	32	16	29	14	26	8¾	50	20	32	17	16x16	12	18½	13 x	13	B-54	A-48
48	32	18	33	14	26	8¾	56	22	37	20	16x16	15	21½	13 x	13	B-60	B-54
54	37	20	37	16	29	13	68	24	45	26	16x16	15	25	13 x	18	B-72	B-60
60	37	22	42	16	29	13	72	26	45	26	16x20	15	27	13 x	18	B-72	B-66
60	40	22	42	16	31	13	72	26	45	26	16x20	18	27	18 x	18	B-72	B-66
72	40	22	54	16	31	13	84	26	56	32	20x20	18	33	18 x	18	C-84	C-84

Notes: °Angle Sizes: A — 3 x 3 x 3/16, B — 3½ x 3 x ¼, C — 5 x 3½ x 5/16.

Table of Fireplace Dimensions
Figure 14-4

Old Standard		Modular Standard		Round Linings	
Outside Dimensions	Inside Dimensions	Nominal Outside Dimensions	Inside Section	Inside Diameter	Inside Section
8½ x 8½	52.56	8 x 12	57	8"	50.26
8½ x 13	80.5	8 x 16	74	10"	78.54
8½ x 18	109.69	12 x 16	120	12"	113
13 x 13	126.56	16 x 16	162	15"	176.7
13 x 18	182.84	16 x 20	208	18"	254.4
18 x 18	248.06	20 x 20	262	20"	314.1
		20 x 24	320	22"	380.13
		24 x 24	385	24"	452.3
		12 x 12	87		

Size of Flues
Figure 14-5

rier to down drafts, thus the chimney will be smoky. And the rear position sacrifices a lot of heat. A sloping back, forward damper and roomy smoke shelf promote efficiency. The smoke shelf, with the upturned damper plate forming a wall at its front, forms a barrier to down-drafts which are trapped and rise upward along with the smoke.

The recommended vertical height of the damper is five to eight inches above the the breast wall of the fireplace (which is supported by a stiff steel angle). Sacrificing this breast wall margin to support the masonry with a damper flange will bring smoke into the room.

Smoke Chamber

Look at Figure 14-6. This is another critical item. At damper level, the enclosure narrows to form the smoke chamber. The slope of its two sides should be identical, the flue taking off from dead center. If the upright flue is not directly over the center, make the upward sloping connection with no reduction in area and not more than 7 inches to each 12 inches of rise. A form consisting of two boards with connecting braces helps give the brick a proper slope and provides a smooth surface which helps discharge the smoke.

Firebrick

Use firebrick to construct the fireplace interior back, sidewalls and bottom. Fire clay is used in the mortar mix. Follow manufacturer's mixing instructions.

Conventional Masonry Fireplace Footings

A chimney and fireplace put a lot of weight on a small area. Therefore, a sufficient footing is essential. Local codes usually cover this since the type of soil will have a lot to do with the depth and width of the footing.

I pour a concrete footing 1-inch thick for each foot of chimney height, up to 18 inches thick. Thus, a 10-foot chimney would sit on a 10-inch-thick footing. A 20-foot chimney would be on an 18-inch-thick footing. I extend the footing at least 12 inches beyond the chimney on all sides. And for each foot of height over 12 feet I add 1 inch of width. For a typical 30-foot chimney the footing would be 18 inches thick and extend beyond all chimney walls 30 inches.

Too much footing? Maybe. But as I said at the beginning of this book, I'm inclined toward "big feet" building. Frankly, I wouldn't give five cents for a dinky footing under anything. And I've never had a leaning chimney or a droopy corner.

Manufactured Fireplaces

According to separate surveys published in *Builder* and *Professional Builder* magazines, potential

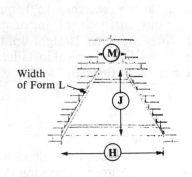

Smoke Chamber
At damper level, the enclosure narrows to form the smoke chamber. It is important that the slope of its two sides be identical, the flue taking off from the center. Where the upright flue is not directly over the center make the upward sloping connection with no reduction in area and not more than 7" sideways to each 12" of rise. A form consisting of two boards with connecting braces helps to give the brick a proper slope and assists in providing the important smooth surface which assists in discharge of smoke.

Smoke Chamber
Figure 14-6

home buyers consider fireplaces the most desired option in a new home. They even rate higher than carpeting. Supporting this claim are U.S. Census figures stating that construction of new single family homes with fireplaces has increased from 30 percent in the early '70s to nearly 75 percent in the late '70s.

Another 1981 survey, done by McGraw Hill's research department for *Housing Magazine*, states that 77 percent of today's home shoppers are willing to pay $2000 for a fireplace in their home.

The greatest jolt to traditional fireplace design has come from the current quest for energy efficient residences. Different types of dwellings—e.g., apartments, condominiums, and single-family homes—have different energy requirements.

The following fireplace research data was compiled by the John Morton Company, an independent research firm, for the Corporate Planning Department, American-Standard, Inc., on behalf of the Majestic Company:

1. New construction represents the major volume for the fireplace industry. However, remodeling is now the major growth area.

2. Factory-made, built-in fireplaces make up an estimated 75 percent of the fireplace remodeling market, and the majority of these built-in fireplaces are heat circulating models.

3. Homeowners don't see the energy savings potential of a fireplace, even though the vast majority have built-in fireplaces.

4. Only 50 percent think factory-made fireplaces are more heat efficient than masonry fireplaces.

5. But only 14 percent have never heard of factory-made fireplaces.

6. Homeowners ranked supplemental heating as the most important buying factor and appearance second.

7. After being informed of energy-saving, factory-made fireplaces, at least 70 percent of the homeowners would opt for a factory-made fireplace over a masonry fireplace.

8. Builders' preferences (excepting large planned community developers):

a. In new construction, masonry is the first choice in the East, but built-ins are the first choice in the West. Built-ins are the first choice nationwide for remodeling.

b. Builders rank appearance above heat efficiency in importance. Homeowners rank efficiency first and appearance second.

c. Over 70 percent of the time, the builder chooses or gives the consumer a limited number of fireplaces to select from.

d. In the remodeling market, the consumer chooses almost 50 percent of the time.

9. Buyers are willing to pay more for a premium product.

10. Higher heat efficiency and a more masonry-like appearance are important to both the small builder and the consumer.

That's information you can use the next time a potential customer asks you about fireplaces.

Now take a look at the factory-built fireplaces in Figures 14-7 and 14-8.

Installing the Built-in Fireplace

First, let's look at a few sketches to acquaint ourselves with various methods of installation. Figure 14-9 shows three methods: straight-up, two-elbow, and four-elbow. Other typical installations are shown in Figures 14-10 through 14-13.

Step-By-Step Procedures

Determine where selected accessories fit into this fireplace installation procedure by checking the installation instructions provided with the unit. Figure 14-14A shows the fireplace and accessories unpacked and waiting for you. Figure 14-14B

Factory-Built Fireplace
Figure 14-7

Built-In Heat Circulator
Figure 14-8

describes the parts. This is the unit we will be working with. The air flow of the unit is shown in Figure 14-14C.

Determining fireplace location: Some units can be mounted on a combustible (burnable) surface. The surface must be flat, hard, and have sufficient area to support the entire base of the fireplace uniformly. Do not support the fireplace at just a few points. The surface may be a raised wooden platform, as shown in Figure 14-15.

The fireplace may be placed against a combustible wall without providing any clearance between the fireplace and the wall. Nearby combustible floors and walls must be protected as we will discuss a little later. There are also some clearance requirements involved. (See Figure 14-16.)

Be sure that the area selected for the fireplace provides for clearance and framing restrictions. (See Figure 14-17.)

Installing metal safety strips: Thoroughly clean the intended fireplace location area. Move the fireplace to the exact location where it will be installed. Lift the fireplace front slightly to slide the furnished metal safety strips under the front bottom edge about 1½ inches, allowing the rest to extend in front of the fireplace. Overlap the strips at least ½ inch to provide a positive joint. (See Figure 14-18.)

Anchoring the fireplace: To prevent shifting of the fireplace and to maintain sealing (described later), anchor the fireplace to the floor. Turn the four fastening tabs located along the sides of the base pan and nail them to the floor. (See Figures 14-19A and 14-19B.)

Installing The Chimney

One of the first things you must do when installing a chimney is locate its centerpoint. (See Figure 14-20.) The procedure you use to find the centerpoint depends on whether you are installing a straight-up chimney or one that is offset.

Locating the Centerpoint

Straight-up installation: If the fireplace is positioned against the wall behind it, measure out 12½ in-

ches from the back wall and make a mark on the ceiling above the fireplace. Draw a line through this mark parallel to the back wall, defining the centerline of the chimney, as in Figure 14-21. Then, using a plumb bob (not a chalk line) positioned directly over the imaginary centerpoint of the fireplace flue collar, mark the ceiling to establish the chimney centerpoint.

Offset installation: To get past an obstruction it may be necessary to offset the chimney from the vertical plane. This is done by using chimney elbows of 15 and 30 degrees. Each offset requires a pair of elbows. You can't throw in 45- and 90-degree ells. Sharp turns don't work. Use the table in Figure 14-22 and the dimensions in Figure 14-23 to determine the offset that may be obtained using 15- and 30-degree elbows. The letters "T" and "S" in the table refer to the "T" and "S" dimensions in Figure 14-23.

The following safety rules apply to offset installation for our unit. The paragraph letters below match the letters in Figure 14-23:

A. Height of the chimney measured from the hearth to the chimney top:
Maximum: 90 feet
Minimum: 10 feet 6 inches without elbows; 12 feet 6 inches with 2 elbows; 28 feet with four elbows

B. Don't use more than four elbows per chimney. (Some units restrict elbows to two. Follow the manufacturer's instructions.) Attach the straps of all elbows, except the first elbow installed on the fireplace, to a structural framing member. If four elbows are used, the first pair of elbows must be separated from the second pair of elbows by a minimum of one vertical foot.

C. The chimney cannot be more than 30 degrees from the vertical plane.

D. The maximum length of the angled run of the total chimney system is 20 feet. (G plus H cannot exceed 20 feet.)

E. A chimney support is required every 6 feet of angled run of chimney. Chimney supports are required at 45 feet and 75 feet of chimney height above the hearth, also.

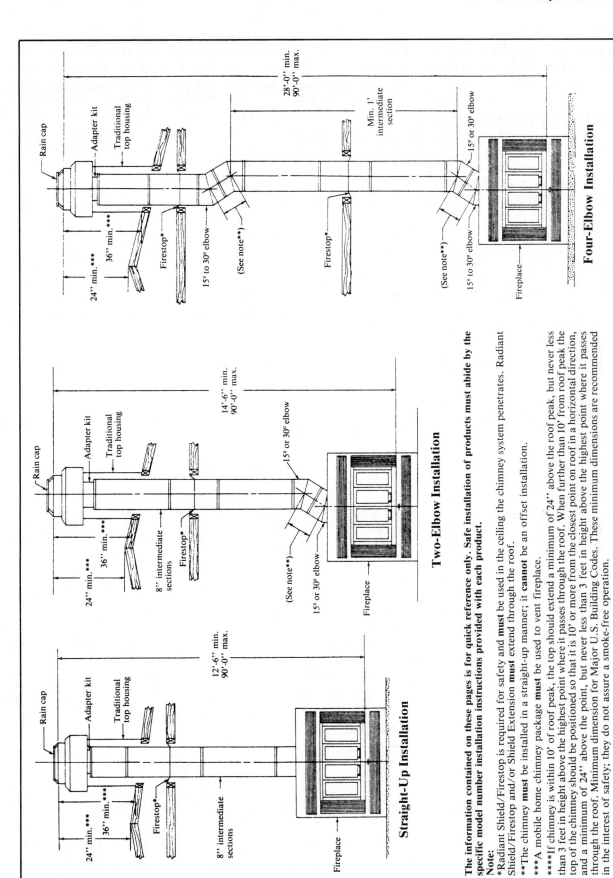

Straight-Up Installation

Two-Elbow Installation

Four-Elbow Installation

The information contained on these pages is **for quick reference only. Safe installation of products must abide by the specific model number installation instructions provided with each product.**

Note:

*Radiant Shield/Firestop is required for safety and **must** be used in the ceiling the chimney system penetrates. Radiant Shield/Firestop and/or Shield Extension **must** extend through the roof.

The chimney **must be installed in a straight-up manner; it **cannot** be an offset installation.

***A mobile home chimney package **must** be used to vent fireplace.

****If chimney is within 10' of roof peak, the top should extend a minimum of 24'' above the roof peak, but never less than 3 feet in height above the highest point where it passes through the roof. When further than 10' from roof peak the top of the chimney should be positioned so that it is 10' or more from the closest point on roof in a horizontal direction, and a minimum of 24'' above the point, but never less than 3 feet in height above the highest point where it passes through the roof. Minimum dimension for Major U.S. Building Codes. These minimum dimensions are recommended in the interest of safety; they do not assure a smoke-free operation.

Various Methods of Installation
Figure 14-9

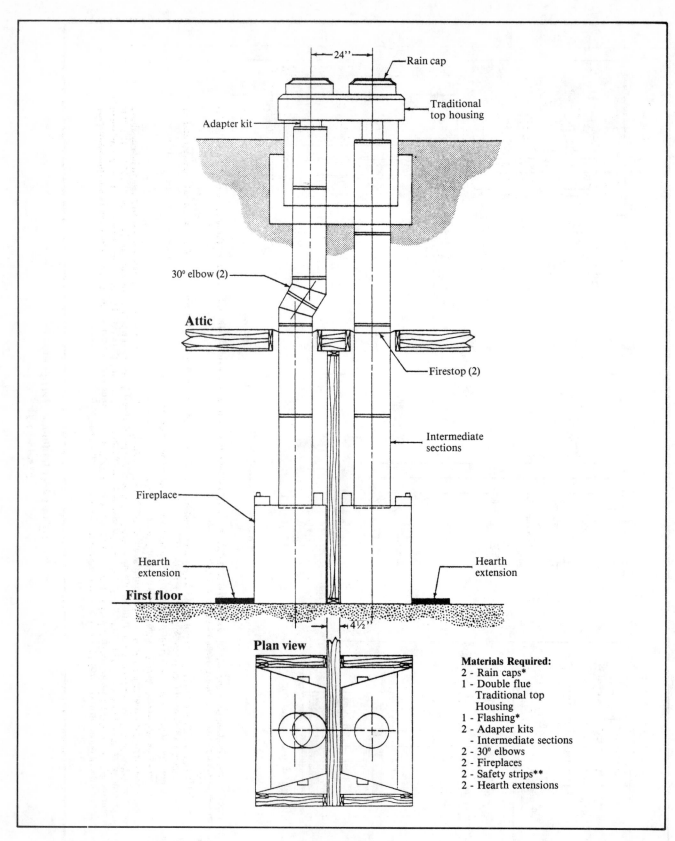

Two Series Fireplaces
Figure 14-10

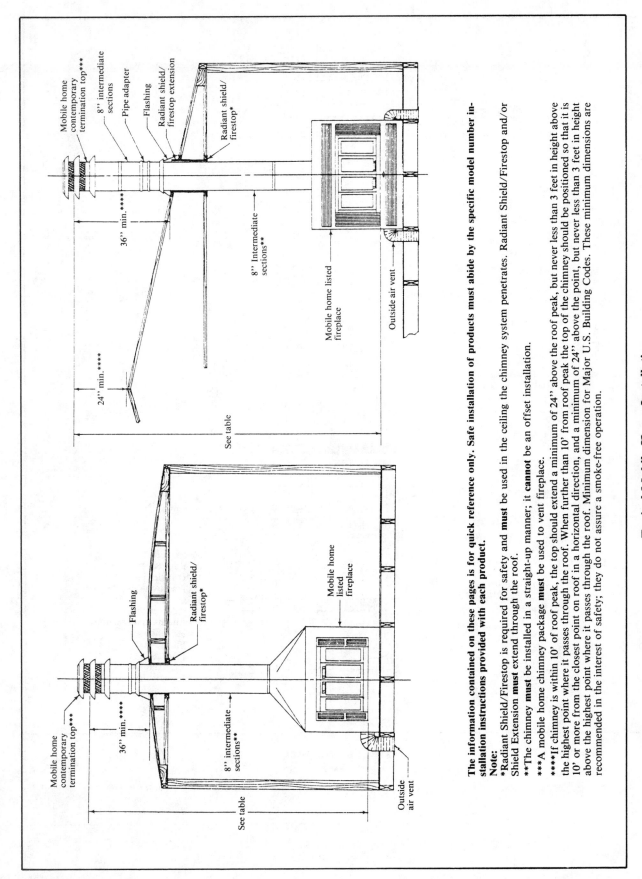

Typical Mobile Home Installation
Figure 14-11

The information contained on these pages is for quick reference only. **Safe installation of products must abide by the specific model number installation instructions provided with each product.**

Note:

*Radiant Shield/Firestop is required for safety and **must** be used in the ceiling the chimney system penetrates. Radiant Shield/Firestop and/or Shield Extension **must** extend through the roof.

The chimney **must be installed in a straight-up manner; it **cannot** be an offset installation.

***A mobile home chimney package **must** be used to vent fireplace.

****If chimney is within 10' of roof peak, the top should extend a minimum of 24" above the roof peak, but never less than 3 feet in height above the highest point where it passes through the roof. When further than 10' from roof peak the top of the chimney should be positioned so that it is 10' or more from the closest point on roof in a horizontal direction, and a minimum of 24" above the point, but never less than 3 feet in height above the highest point where it passes through the roof. Minimum dimension for Major U.S. Building Codes. These minimum dimensions are recommended in the interest of safety; they do not assure a smoke-free operation.

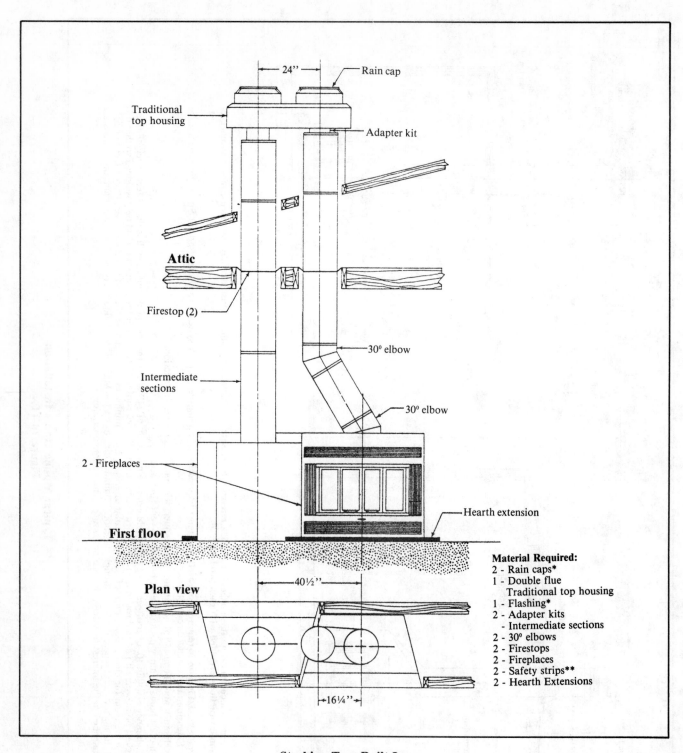

Stacking Two Built-Ins
Figure 14-12

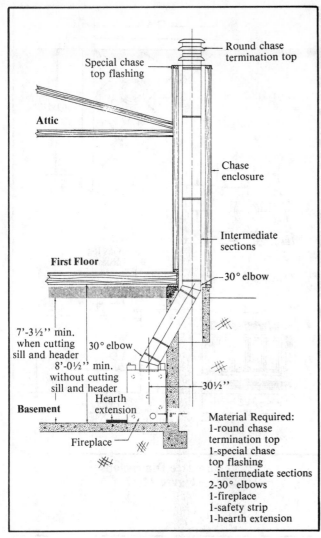

Round chase termination top

Special chase top flashing

Attic

Chase enclosure

Intermediate sections

First Floor

30° elbow

7'-3½" min. when cutting sill and header

30° elbow

8'-0½" min. without cutting sill and header

Basement

Hearth extension

30½"

8"

Fireplace

Material Required:
1-round chase termination top
1-special chase top flashing
-intermediate sections
2-30° elbows
1-fireplace
1-safety strip
1-hearth extension

Basement Installation
Figure 14-13

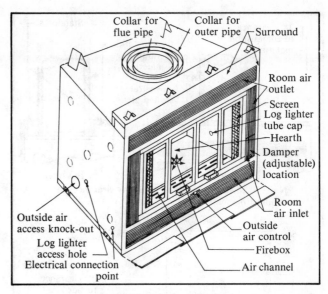

Collar for flue pipe — Collar for outer pipe — Surround

Room air outlet

Screen
Log lighter tube cap
Hearth

Damper (adjustable) location

Outside air access knock-out

Log lighter access hole

Electrical connection point

Room air inlet

Outside air control

Firebox

Air channel

Identification of Parts
Figure 14-14B

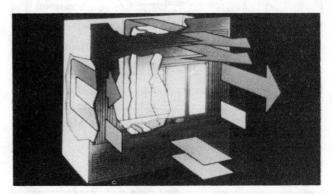

Air Flow
Figure 14-14C

The Unit and Accessories We Will Be Working With
Figure 14-14A

Installing Unit on Raised Wood Platform
Figure 14-15

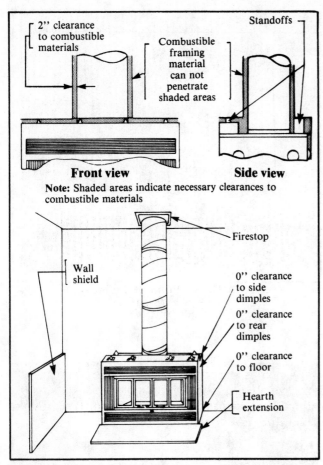

Minimum Clearances to Combustible Materials
Figure 14-16

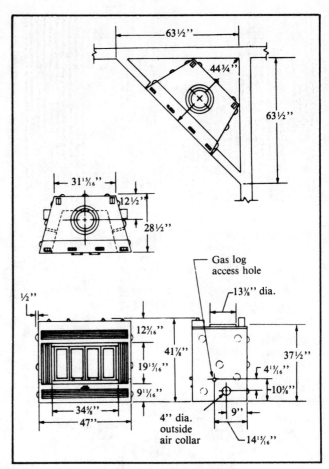

Fireplace Dimensions
Figure 14-17

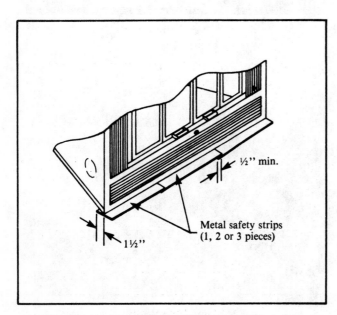

Install Metal Strip(s) Under Front of Fireplace
Figure 14-18

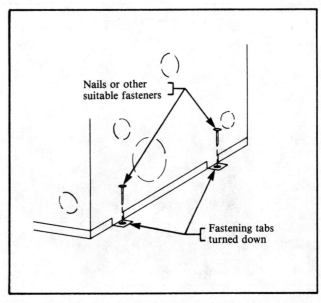

Fasten Fireplace in Position Using the Four Fastening Tabs
Figure 14-19A

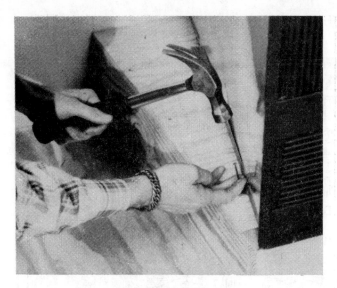

Nail Unit To Floor
Figure 14-19B

Locate Center Point of Chimney
Figure 14-20

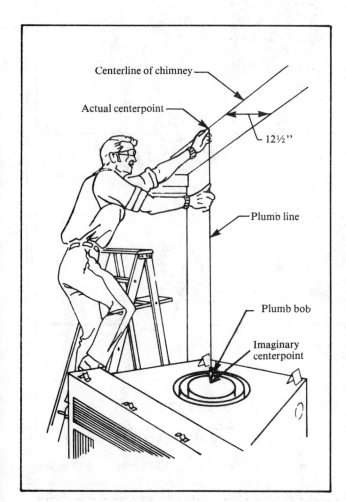

Locate Center Point of Chimney With Plumb Line
Figure 14-21

Determine the offset distance of your chimney arrangement from the centerline of the fireplace to the centerline of the chimney where it passes through the first ceiling. Note: This distance may not be your full offset distance (T).

	Dimensions			
	T		S	
Item	**15°**	**30°**	**15°**	**30°**
No intermediate section	0'- 2½''	0'- 5''	1'- 4¾''	1'- 7½''
One 1' section	0'- 5¼''	0'-10¾''	2'- 2¾''	2'- 4½''
One 1½' section	0'- 6¾''	1'- 1¼''	2'- 8¾''	2'- 9¾''
One 2' section	0'- 8½''	1- 4¼''	3'- 2½''	3'- 3''
One 1' and one 1½' section	0'- 9½''	1'- 6½''	3'- 6¾''	3'- 6¾''
One 3' section	0'-11½''	1'-10¼''	4'- 2''	4'- 1¼''
One 2' and one 1½' section	1'- 0¾''	2'- 0½''	4'- 6½''	4'- 5¼''
Two 2' sections	1'- 2½''	2'- 3½''	5'- 0¼''	4'-10½''
One 3' and one 1½' section	1'- 3¾''	2'- 6½''	5'- 6''	5'- 3½''
One 3' and one 2'' section	1'- 5¼''	2'- 9½''	5'-11¾''	5'- 8¾''
Two 2' and one 1½' section	1'- 6½''	2'-11¾''	6'- 4¼''	6'- 0¾''
Two 3' sections	1'- 8½''	3'- 3½''	6'-11½''	6'- 7¼''

Offset Dimensions
Figure 14-22

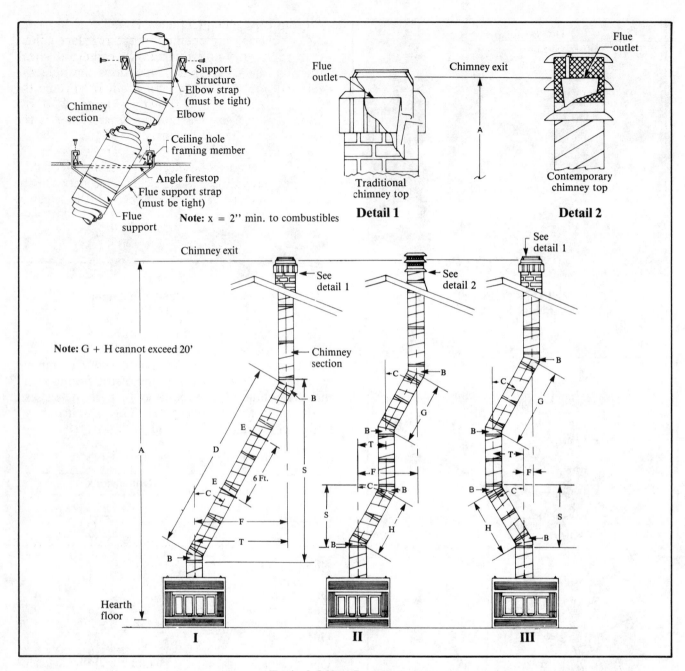

Typical Offset Installations
Figure 14-23

Locate on the ceiling the centerpoint of the chimney, as though a straight-up chimney arrangement were to be used. (See Figure 14-21.) Now measure your offset dimension from the straight-up chimney centerpoint on the ceiling. (See Figure 14-24.)

Marking the Area of Ceiling Chimney Hole
The size of the ceiling chimney hole will vary with the angle at which the chimney passes through the

ceiling. The following table will help here.

Size of Chimney	Angle of Chimney at Ceiling		
	Vertical	**15-Degree**	**30-Degree**
8" Flue	17½ x 17½	17⅞ x 22½	17⅞ x 29⅝

Drive a nail up through the ceiling at the marked chimney centerpoint. Go to the floor (or attic)

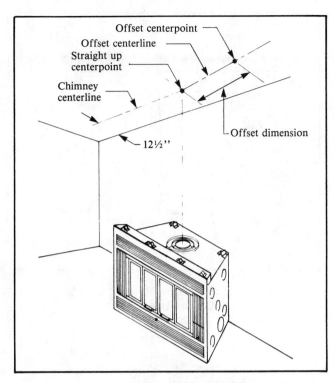

**Measure First Ceiling Offset Distance From
Straight Up Centerpoint
Figure 14-24**

Cutting the Ceiling Chimney Hole

After covering the opening of the fireplace collar, cut the chimney hole through the ceiling. Recheck the hole to be sure that it measures the hole size selected and that the chimney will fit. Frame the ceiling chimney hole as shown in Figure 14-26. It's good practice to use framing lumber that is the same size as the ceiling joists.

The inside dimension of the frame must be the same size as the hole selected from the preceeding table to provide the 2-inch clearance between the outside diameter of the chimney and the edges of the framed ceiling hole.

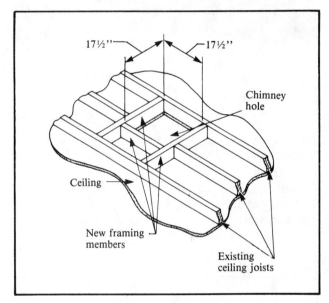

**Typical Frame for Ceiling Chimney Hole
Figure 14-26**

above and check where the hole will be cut relative to the ceiling joists and for any obstructions such as wiring or plumbing runs. If necessary, reposition the chimney and fireplace to better accommodate these joists or obstructions. (See Figure 14-25.)

Installing Firestop Spacers

Firestop spacers are required for safety. They are dished or angled to fit the angle of the chimney coming through the ceiling. Each provides the minimum clearance to the chimney pipe for ceiling thicknesses up to 8 inches. When the combined thickness of the ceiling material, ceiling joists and flooring materials exceeds 8 inches, adjustments must be made in the framing to ensure that the minimum clearances to the chimney are maintained. (See Figure 14-27.)

If the area above the ceiling is not an attic, position the firestop spacer with the flange on the ceiling side and the dished or angled portion extending up into the hole. If the area above the ceiling is in an attic, position the firestop spacer with the flange

**Locate Ceiling Cutout
Figure 14-25**

**Insure Proper Chimney Clearance
Figure 14-27**

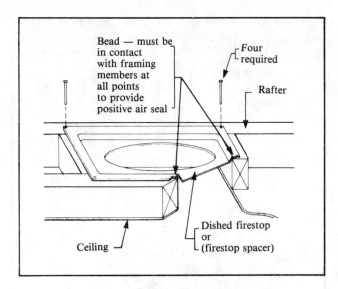

**Position of Firestop When Area Above
Ceiling Is An Attic
Figure 14-28B**

on the top of the framed hole and the dished or angled portion extending down into the hole. (See Figures 14-28A and 14-28B.)

Nail each corner of the firestop spacer to the framing members of the ceiling hole. *Note:* A firestop spacer is not required at the roof.

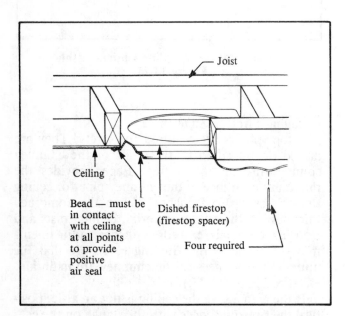

**Position of Firestop When Area Above
Ceiling Is Not An Attic
Figure 14-28A**

Attaching Chimney Sections
Attach the first straight chimney section to the fireplace collar, or elbow in offset installation. Mount the flue pipe first, using the built-in snap-lock fastener. (See Figure 14-29.) Then mount the inner pipe, and finally the outer pipe. Position each pipe section so the direction arrow is pointing up. Make sure each pipe is firmly snapped and locked together as it is mounted.

Continue installing chimney sections until one section (all three pipes) extends up through the ceiling hole. To extend through the ceiling, it may be necessary to assemble all three pipes, push them up through the ceiling hole, and then slide them down one at a time to connect them.

Attaching the First Elbow—Offset Installation
If the first elbow is not to be attached to the fireplace, install chimney sections as required.

Attach the first elbow where required. Note that only the outer pipe snap-locks. The flue and inner pipe telescope. (See Figure 14-30.)

Attach the straps of all elbows to a structural framing member. However, the straps for an elbow installed directly on the fireplace collar do not have to be attached.

How Many Chimney Supports Are Needed?
The chimney system is supported by the fireplace for chimney heights less than 45 feet above the hearth. Chimney supports are required if the

chimney height exceeds 45 feet. Additional supports must be provided again at 75 feet of vertical chimney height. Locate chimney supports at the ceiling holes or other structural framing at the 45-foot and 75-foot heights. Spacing between chimney supports must not exceed 30 feet. Special chimney supports are available. Support provided by elbow straps fulfills the support requirement only if the straps are spaced as described above.

Angled chimney runs require support every 6 feet in addition to the elbow straps. A chimney support is 3 inches long when installed. (Figure 14-31 shows a chimney support and a chimney elbow). This dimension must be considered when determining how many straight chimney sections are needed to provide the desired offset.

**Use Built-In Snap-Lock Fasteners To
Attach Chimney Pipe Sections
Figure 14-29**

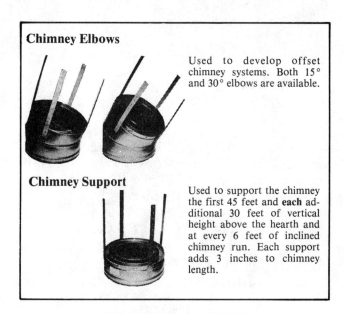

**Chimney Support and Ells
Figure 14-31**

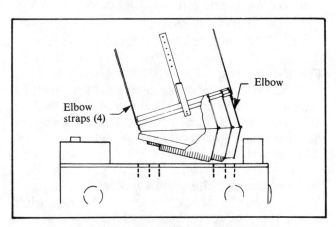

**Attach One-Piece Elbow to Fireplace
Collar or Chimney Section
Figure 14-30**

Installing Chimney Supports
Chimney supports are attached to chimney sections the same way elbows are attached. Nail the chimney support straps to the adjacent structural framing as shown in Figure 14-32. Bend the straps as necessary and make sure they are tight so they'll be able to support the weight of the chimney.

Locating the Centerpoint of Next Ceiling Hole
Locating and cutting the next ceiling hole is done the same way as the first ceiling hole. Install the firestop spacer and continue installing chimney sections and supports (as required) until the chimney passes through the second ceiling hole.

The chimney system must be vented out-of-doors and must be terminated in the proper top housing or termination.

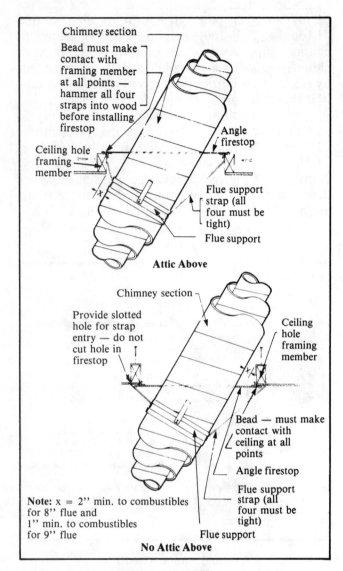

Note: x = 2" min. to combustibles for 8" flue and 1" min. to combustibles for 9" flue

No Attic Above

Mount Flue Supports at Ceiling Hole Frames or Other Structural Framing
Figure 14-32

Locate Chimney Center Point at Roof
Figure 14-33

Locating the Chimney Centerpoint at the Roof
Using your plumb bob, locate the centerpoint of the chimney (Figure 14-33), mark the centerpoint on the sheathing and drive a nail up through the roof at the point marked. The nail will mark the centerpoint on the outside of the roof.

Cutting and Framing the Roof Hole
The size of the roof hole varies with the type of chimney top that will be installed. Manufacturer's instructions for the chimney top housing or termination will give the size of the roof hole required. There must be at least a 2-inch clearance between the outermost portion of the chimney sec-

tion and any adjacent combustible surfaces. Combustible surfaces include such things as ceiling members, joists, flooring, combustible insulation and roof structures.

Mark the outline of the roof hole around the centerpoint nail. The hole dimensions given in the chimney top installation instructions are horizontal dimensions. The hole size must be marked on the roof accordingly.

Cover the opening of the installed chimney. You want to keep out the debris. Then cut the hole. (See Figure 14-34.)

Framing the Hole
Use framing lumber the same size as the rafters. Install the frame securely. Chimney top and flashing anchored to the frame must be able to withstand heavy winds. Figure 14-35 shows the chimney position through the roof.

Determining the Minimum Chimney Height
Most codes specify the minimum chimney height above the roof top. These specifications are summarized in the "Ten Foot Rule," Figure 14-36. Key points of the Rule are as follows:
- If the horizontal distance from the center of

Measure And Cut Thru The Roof
Figure 14-34

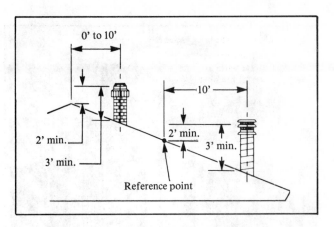

Ten Foot Rule For Chimney Height
Figure 14-36

the chimney to the peak of the roof is 10 feet or less, the top of the chimney must be at least 2 feet above the peak of the roof but never less than 3 feet above the highest point where it passes through the roof.

• If a horizontal distance from the center of the chimney to the peak of the roof is more than 10 feet, a chimney height reference point is established on the surface of the roof 10 feet from the center of the chimney in a horizontal plane. The top of the

chimney must be at least 2 feet above this reference point, but never less than 3 feet above the highest point where it passes through the roof.

These chimney heights are necessary in the interest of safety and do not guarantee a smoke-free operation.

Installing the Remaining Chimney Sections
Continue installing chimney sections up through the roof hole. Check your chimney top installation instructions for data on how high above the roof top the chimney sections (all three pipes) should go. Now you can install the chimney top housing. (See Figure 14-37.) The finished job is shown in Figure 14-38.

Chimney Position Through Roof
Figure 14-35

Install Remainder of Chimney Section
Figure 14-37

The Finished Job
Figure 14-38

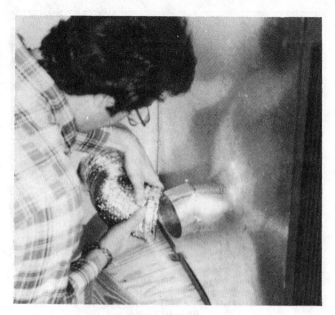

Hook-Up Outside Air System
Figure 14-39A

Installing a Chimney in a Chase

A chase is a vertical box-like structure built to enclose the fireplace or its chimney. Treatment of firestop spacers and construction of the chase may vary with the type of building. Check the code to determine the requirements for the chase.

When installing a fireplace in a chase, the outside wall of the chase should be insulated. This will reduce the volume of outside air that enters the room where the fireplace is located and will also prevent a cold fireplace. Upon completing the chase, install the chimney system, following procedures already discussed. The chimney top is installed following the steps used for through-the-roof installations.

Installing the Outside Air System

Outside air is drawn into the fireplace to keep the fire burning. This means that heated room air is not sucked into the fireplace—a major waste of energy in conventional fireplaces. However, this system is optional.

If the system is used, plan the location of the two duct terminations of the outside air system, and plan the path of the interconnecting duct. Two 16-inch sections of flexible duct are probably included with the fireplace. They make duct installation easy. (See Figures 14-39A and 14-39B.)

Outside Air System Duct
Figure 14-39B

If additional duct is required, 4-inch diameter straight duct and elbows may be added; or additional flexible duct may be purchased to complete the installation.

Duct length is limited to a *total* of twenty feet, with vertical height limited to three feet less than the flue termination height. (For typical duct installation, see Figure 14-40.)

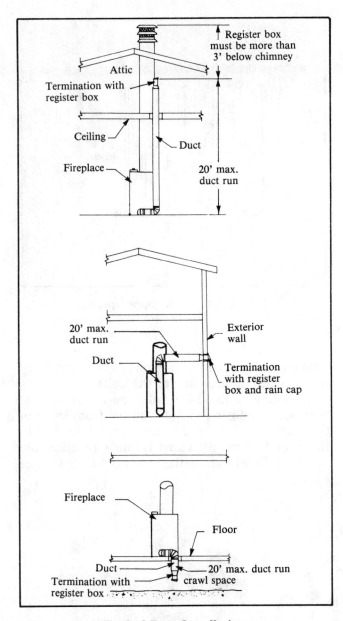

**Typical Duct Installations
Figure 14-40**

The duct termination may be installed in an outside wall, in an attic space or a crawl space. There must be a minimum of 100 square inches of unrestricted ventilation opening. The termination must be installed to avoid blockage from dirt or insulation. If installed on an outside wall, the termination should be located so it will not become blocked by shrubs or drifting snow. Don't install the termination in a garage or other area that could contain flammable liquids or fumes.

Installing the Terminations
The combustion air termination in an attic has to

have good support and must be placed where it won't be obstructed with dirt or insulation.

For an outside wall installation, locate the centerpoint of the penetration and draw a circle the size of each duct. Cut the marked hole. Put a bead of caulking around the hole, install the termination through the wall and attach it with small screws or nails.

Installing Duct Connection to Fireplace
Remove the two outside air access knock-outs and install the connecting collar with screws. (See Figure 14-41.) All units are not designed for outside air access and will not have knock-outs.

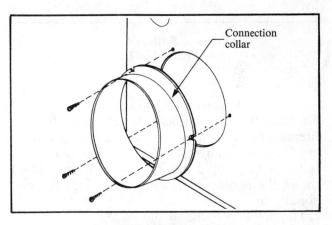

**Install Connection Collars
Figure 14-41**

If you're making this installation in New York State, better check the code. Only aluminum, non-combustible flexible or type "0" metal ducting is permissible.

Removing Room Air Panel
On some units the room air panel *must be removed* if the outside air system is *not* used.

Remove the bottom grille by removing the screws at each end. Then remove the room air panels from the left and right sides of the fireplace. (See Figure 14-42.) These panels may be discarded. Do not re-install the grille at this time.

Electrical Connection to the Fireplace
Follow the manufacturer's instructions. Most factory-built fireplaces have a fan which pulls in the room's cold air, circulates it through the fireplace to heat it and returns it to the room. After the electrical connections are made you can put the grille back on.

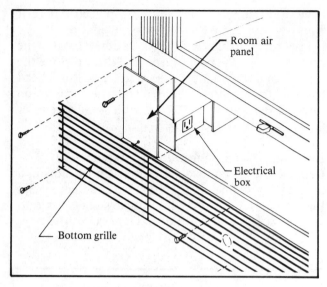

Remove Bottom Grille
Figure 14-42

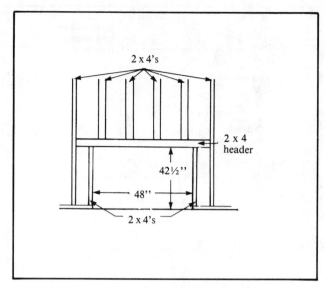

Dimensions of Fireplace Opening
Figure 14-43

Framing and Finishing

Fireplace framing can be built before or after the fireplace is set in place. Various models have varying framing requirements and the manufacturer's instructions should be followed. Figure 14-43 shows the minimum framing dimensions for one popular unit.

The fireplace frame should be constructed of 2 x 4 lumber or heavier. (See Figures 14-16 and 14-17 for fireplace dimensions for the unit).

On many units the header may rest on the fireplace standoffs (V-angles on the front ledge of the fireplace). The header must not be notched to fit around them. Standard framing practices are followed. Build right to the unit; it's a built-in. (See Figure 14-44.)

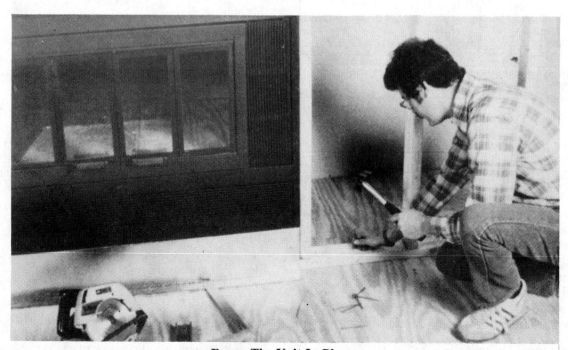

Frame The Unit In Place
Figure 14-44

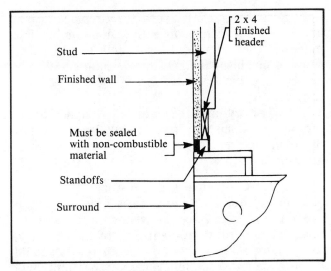

Fireplace Surround Flush With Finished Wall
Figure 14-45

Install Finish Wall
Figure 14-47

Framing should be positioned to accommodate wall covering and fireplace facing material. Figures 14-45 and 14-46 show typical arrangements.

Caution: All joints between the finished wall and the fireplace (top and sides) *must* be sealed with non-combustible material. Only non-combustible material may be applied as facing to the fireplace surround.

Finish Wall
Finish the wall with any material. (See Figure

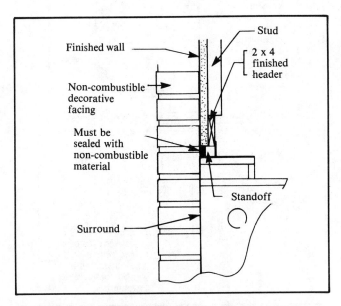

Fireplace Surround Flush With Finished Wall
Figure 14-46

14-47.) Don't install a combustible mantel or other combustible projection less than 12 inches above the outlet grille. If a combustible material is used, consult your local building code for the minimum clearance from the top of the fireplace opening to the bottom of the mantel.

All joints, (top, bottom and sides), where the wall or decorative facing material meets the fireplace must be completely sealed with a noncombustible material.

When finishing the fireplace, never obstruct or modify the air inlet and outlet grilles. Don't install decorative facing in a way that would prevent grille removal. The grilles must remain fastened to the fireplace surround as manufactured except when temporary removal is required for maintenance or installation of accessories.

Brick can be used as a finish material for a decorative wall. (See Figure 14-48.)

Adjacent combustible side walls within 24 inches of the fireplace opening must be protected with a non-combustible material. For fireplaces installed at 45 degrees to two sidewalls (corner installation), no sidewall protection is required.

Installing Hearth Extension
If the fireplace sits directly on the floor, a hearth extension may be required in front of the fireplace. Use ⅜-inch asbestos millboard, covered with a suitable non-combustible finish material. Secure the hearth extension to the floor to prevent shif-

Finish Off With Brick
Figure 14-48

ting. Seal the crack between the fireplace hearth and the hearth extension with a non-combustible material. (See Figures 14-49A and 14-49B.) The finished height of the hearth extension should not exceed $2^{11}/_{16}$ inches above the bottom of the fireplace.

Some factory-built fireplaces are designed to accept a ½-inch gas line for a gas log lighter. Follow the manufacturer's instructions for a gas log lighter hook-up.

Conventional Chimneys

A chimney should extend at least 3 feet above flat roofs and at least 2 feet above a roof ridge or the raised part of roof within 10 feet of the chimney. (See Figure 14-50.) A hood (Figure 14-51) should be provided if a chimney cannot be built high enough above a ridge to prevent trouble from eddies caused by wind being deflected from the roof. The open ends of the hood should be parallel to the ridge.

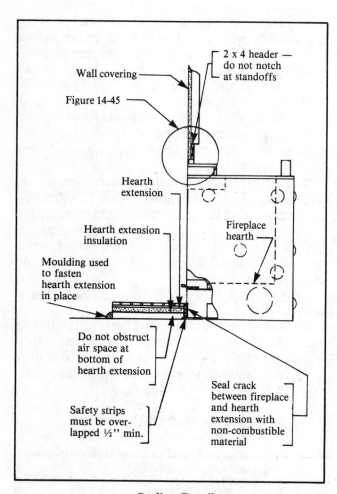

Sealing Detail
Figure 14-49A

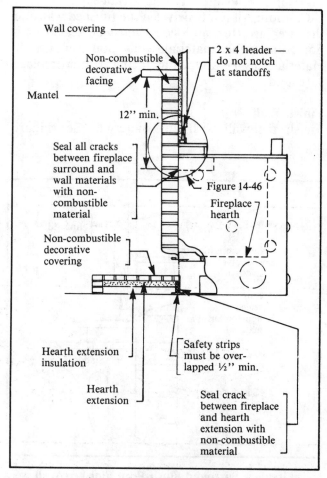

Sealing Detail
Figure 14-49B

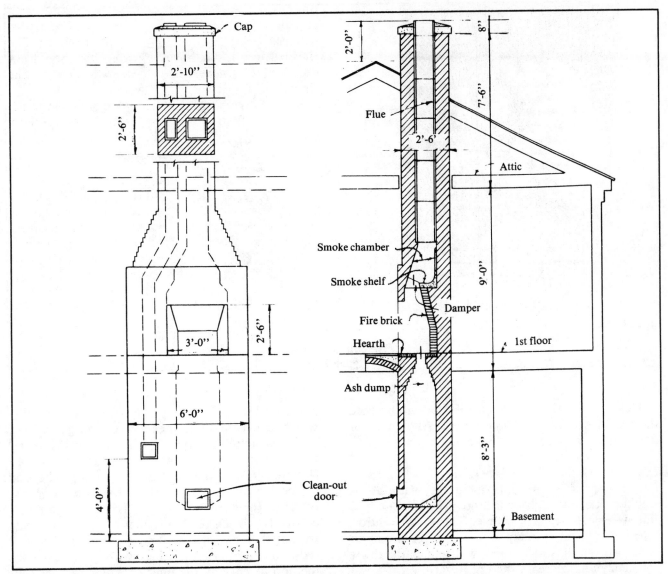

Diagram of an Entire Chimney Such as is Commonly Built to Serve the House-heating Unit and One Fireplace
Figure 14-50

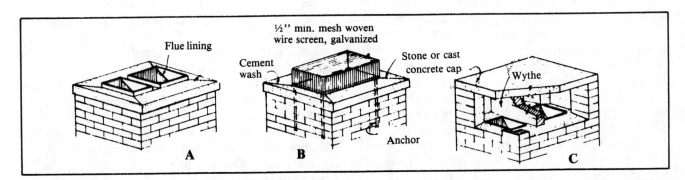

Top Construction of Chimneys: A, Good Method of Finishing Top of Chimney, Flue Lining Extends 4 Inches Above Cap; B, Spark Arrester or Bird Screen; C, Hood to Keep Out Rain
Figure 14-51

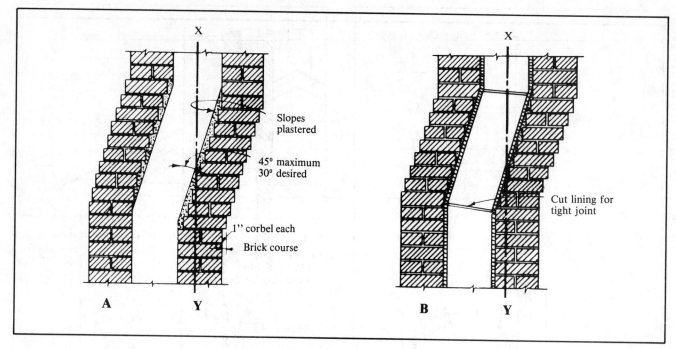

Offset in a Chimney
Figure 14-52

Metal pipe extensions are available that increase the height of snub-nosed chimneys. Anchor them down well when you install them.

Flue Liners

For best results, don't build a chimney without installing flue liners. Mortar and brick exposed to flue gases disintegrate in time. This disintegration plus that caused by temperature changes can open cracks in the masonry which will reduce the draft efficiency and increase the fire hazard.

The vitrified fire clay liners available at most supply houses are usually ⅝-inch thick and made to withstand rapid temperature changes and the action of flue gases. Rectangular-shaped lining is better adapted to brick construction; round is more efficient. Take the lining up as the work progresses. Each length of lining is placed in position—set in mortar with the inside joint struck smooth—and the brick laid around it. In masonry chimneys having walls less than 8 inches thick, there should be space between the liner and the chimney walls. Don't fill the space with mortar.

The first liner rests on brick on at least 3 sides. The brick projects to the inside surface of the lining. (See Figure 14-6.)

Flues should be as nearly vertical as possible. Any change in direction should never exceed 45 degrees. (See A and B in Figure 14-52.) A 30-degree

angle or less is better. Sharp turns set up eddies which affect the motion of smoke and gases.

For structural purposes, the amount of offset must be limited so that the center line of XY (Figure 14-52) of the upper flue will not fall beyond the center of the wall of the lower flue. Start the offset of the left wall (Figure 14-52 A) of an unlined flue two brick courses higher than the right wall so that the area of the sloping section will not be reduced after plastering. Figure 14-52 B shows the method of curving the liner to make a tight joint. The lining is cut before it is built into the chimney. Use a chisel.

Chimney Walls

Walls of chimneys not more than 30 feet high with flue liners should be at least 4 inches thick if made of brick or reinforced concrete and at least 12 inches if made of stone.

Figure 14-53 shows how a chimney can be built to change from 4-inch walls to 8-inch walls above the roof. The 8-inch wall, which is stronger and offers more wind-resistance, can prevent chimney damage at the roof or framing level.

Building codes usually require a separate flue for each fireplace, furnace or boiler. If a chimney contains three or more lined flues, each group of two flues must be separated from the outer single flue or group of two flues by brick divisions or wythes

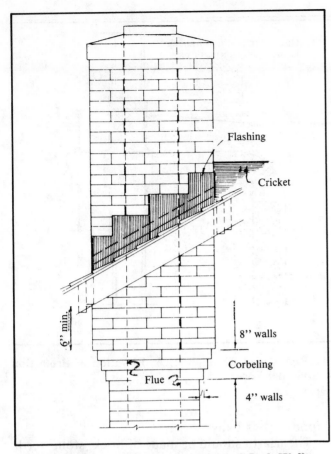

Corbelling of Chimney to Provide 8-Inch Walls For The Section Exposed to The Weather
Figure 14-53

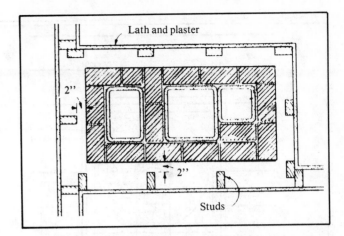

Plan of Chimney Showing Proper Arrangement of Three Flues. Bond Division Wall With Sidewalls by Staggering the Joints of Successive Courses. Wood Framing Should be at Least 2 Inches From Brickwork
Figure 14-54

soot. The pocket should be just below the flue where a cleanout door is installed. Fill or block the lower part of the chimney below the cleanout door. The cleanout door should be cast iron and should fit snugly.

A cleanout should serve only one flue. Where 2 or more flues are connected to the same cleanout, problems arise. Air drawn from one to another will affect the draft. Figure 14-55 illustrates a soot pocket and cleanout door.

at least 3¾ inches thick. (See Figure 14-54.) If two flues are grouped together without a dividing wall, the liner joints must be staggered so the joints of one flue will not be at the same level or position as the other flue. The joints must be sealed with mortar.

Where a chimney contains two or more unlined flues, the flues must be separated by a well-bonded wythe at least 8 inches thick.

Brickwork around chimney flues and fireplaces should be laid with cement mortar. Cement mortar is more resistant to heat and flue gas action than lime mortar. 1 part portland cement, 1 part hydrated lime and 6 parts sand, measured by volume, makes a good mortar to use in setting flue liners and chimney masonry. Some masons prefer a stronger mix—i.e., less sand—which is still good. Firebrick should be laid with a fire clay mixture.

Soot Pocket and Cleanout

Install a soot pocket and cleanout for each flue. Deep soot pockets permit the accumulation of

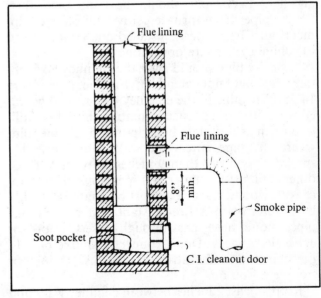

Soot Pocket and Cleanout For a Chimney Flue
Figure 14-55

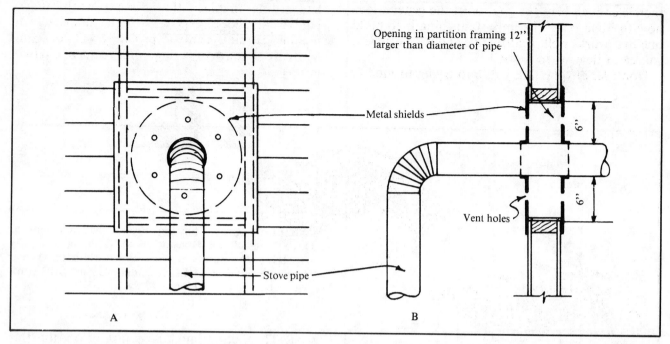

One Method of Protecting a Wood Partition When a Smoke Pipe Passes Through It. A, Elevation of Protection Around the Pipe; B, Sectional View
Figure 14-56

Figure 14-56 A and B illustrates methods of protecting a wood partition when a flue pipe passes through it. No range, fireplace, stove, or other equipment should be connected to the flue of the central heating unit. Any time more than one unit is connected to the same flue, fires may result from sparks passing into one flue opening and out another. Each unit should have its own flue. Smoke pipes from furnaces, stoves, or other equipment must be properly installed and connected to the chimney for safe operation.

A smoke pipe should enter the chimney horizontally. It should not extend into the flue. (See Figure 14-55.) The hole in the chimney wall should be lined with fire clay, or metal thimbles should be built into the masonry. Use boiler putty to seal the joint where the pipe enters the wall. Don't install a smoke pipe closer than 9 inches to woodwork or other combustible material. Pipes closer than 18 inches to woodwork or combustible material should be covered with a fire-resistant material. Smoke pipes should never pass through closets, floors, or concealed spaces. Don't connect a smoke pipe to the chimney in the attic. You're just asking for trouble if you do.

Keep at least 2 inches between chimney walls and all woodwork unless the chimney walls are 8 inches thick. If the chimney wall is at least 8 inches thick, a ½-inch clearance between the chimney and woodwork is okay.

The hearth should be made of brick, stone, terra cotta, or reinforced concrete at least 4 inches thick. It should project at least 20 inches from the fireplace and should be 12 inches wider on each side than the opening. The hearth can be flush with the floor or raised.

The back and sides of the fireplace should be at least 8 inches solid masonry. The fireplace should be lined with firebrick. Steel linings should be at least ¼-inch thick.

Factory-Built Inserts or Forms
Factory-built forms or inserts are constructed of metal and designed to be set in place and concealed by conventional brickwork or other suitable materials. The unit contains all the essential parts of a fireplace—firebox, damper, throat, and smoke shelf and chamber. Most are built with some duct and grilles to pass cold air from the room, through heating compartments, and back into the room. Some units use a fan to force the air flow. Where the unit is installed, only the grilles show. Figure 14-57 shows the various steps in installing such a unit.

Summary
You could write an entire book about fireplaces and chimneys and still not cover everything. But

there's enough here for you to get the general idea how to build one. The important thing is to build one that works well, is attractive and is an efficient source of heat when needed.

Don't let anyone talk you into trying to modify an existing fireplace that doesn't work—unless you've had a lot of experience with chimneys. Any number of things can and probably will go wrong when you start tinkering with someone else's mistake.

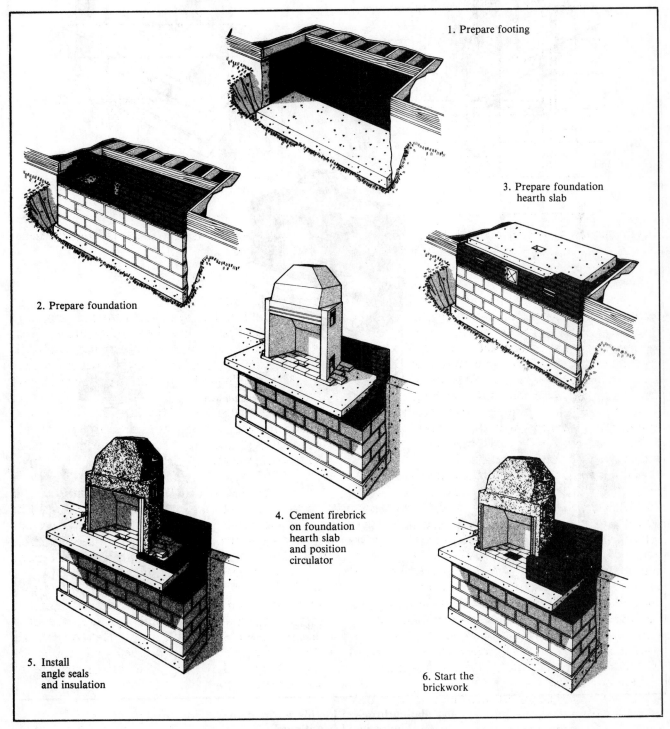

1. Prepare footing

3. Prepare foundation hearth slab

2. Prepare foundation

4. Cement firebrick on foundation hearth slab and position circulator

5. Install angle seals and insulation

6. Start the brickwork

Installation Process: Fireplace Circulator
Figure 14-57

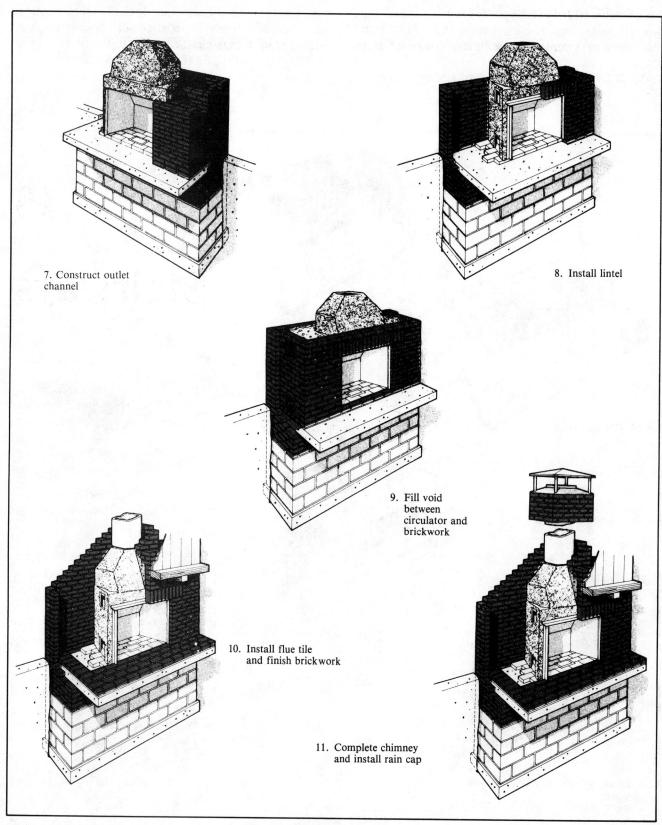

7. Construct outlet channel

8. Install lintel

9. Fill void between circulator and brickwork

10. Install flue tile and finish brickwork

11. Complete chimney and install rain cap

Installation Process: Fireplace Circulator
Figure 14-57 (Continued)

15

Stairways

A stairway has to allow people to go up and down in comfort and safety. It must have adequate headroom and should accommodate any type of furniture that will likely be needed on the higher floor.

Your building code prescribes stair dimensions. Since building codes differ, you should be aware of the requirements in the areas where you work. The dimensions given in this chapter will be acceptable in most communities, but the only safe way is to check the code before you start work.

Interior Stairs

Headroom
Stairs should have continuous clear headroom measured vertically from the front edge of the nosing to a line parallel with the stair pitch.
- Main stairs—minimum 6'8''.
- Basement and service stairs—minimum 6'4''.

Width, Clear of Handrail
- Main stairs—minimum 2'8''.
- Basement and service stairs—minimum 2'6''.

Run
- Main stairs, closed or open riser—minimum 9'' plus 1⅛-inch nosing.
- Basement and service stairs
1. Closed riser—minimum 9'' plus 1⅛-inch nosing.
2. Open riser—minimum 9'' plus ½-inch nosing.

Rise
- Main stairs—maximum 8¼''.
- Basement and service stairs—maximum 8¼''.
- All riser heights should be the same in any one flight. A change is a "flipper"—it'll flip you down the stairs.

Winders
Run at a point 18 inches from the converging end should not be less than the run of the same stairway if it were straight rather than winding.

Landings
- A landing should be provided at the top of any stair run having a door which swings toward the stairs.
- The landing should not be of a dimension less than 2 feet 6 inches.

Handrail
A continuous handrail should be installed on at least one side of stairs that exceed three risers. Stairs open on both sides, including basement stairs, should have a continuous handrail on one side and railing on the open portion on the other side.

Railings
A railing should be installed around all open sides of all stair wells, including those in the attic and

basement. (See Figures 15-1 through 15-3 for stair specifications and dimensions.)

Exterior Stairs

Entrance Stairs

• Stairs should be at least the width of the walk but not less than 3 feet.

• The run should be a minimum of 11 inches.

• The rise should be a maximum of 7½ inches. All riser heights should be the same in any one flight. (No flippers.)

• Stairs should have a handrail that is continuous on all open sides of stair flights to a platform more than 4 risers or 30 inches above the finish grade.

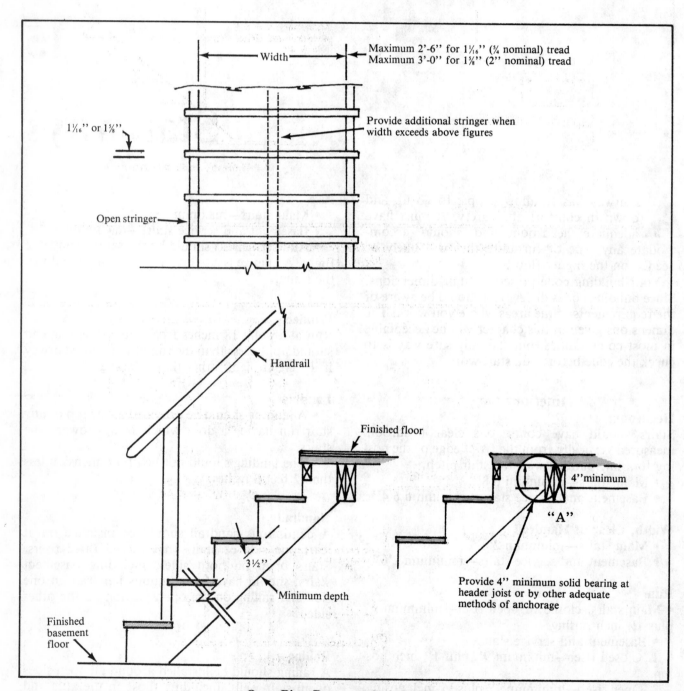

Width

Maximum 2'-6" for 1¹⁄₁₆" (¾ nominal) tread
Maximum 3'-0" for 1⅝" (2" nominal) tread

1¹⁄₁₆" or 1⅝"

Provide additional stringer when width exceeds above figures

Open stringer

Handrail

Finished floor

4" minimum

"A"

3½"

Minimum depth

Provide 4" minimum solid bearing at header joist or by other adequate methods of anchorage

Finished basement floor

Open-Riser Basement Stair-Interior
Figure 15-1

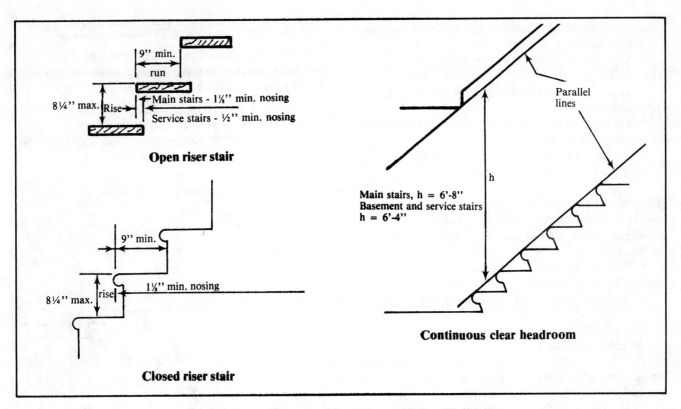

Interior Stairs Minimum Dimensions and Specifications
Figure 15-2

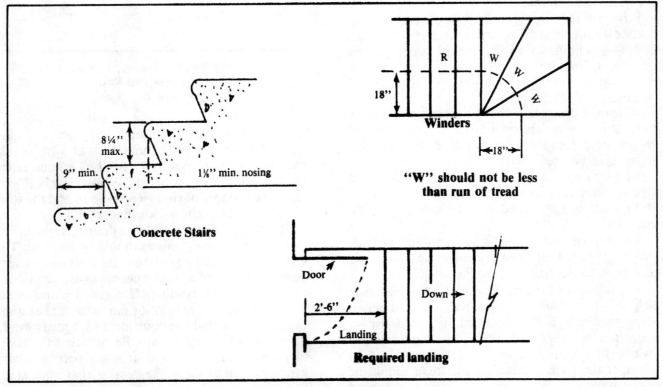

Windings and Landings
Figure 15-3

Stairs to Basement (Unprotected from the weather)
- Headroom of at least 6 feet 4 inches.
- A width of 2 feet 6 inches minimum (clear of the handrail).
- A run of a minimum 11 inches. The run may be 10 inches if a 1-inch nosing is provided.
- A rise of a maximum 7½ inches. Make all the risers the same height.
- The handrail should be installed on at least one side of stairs that exceed 4 risers.

Basement Stairs (Protected from the weather)
Same as for unprotected stairs except:
- Headroom should have a minimum of 6 feet 2 inches.
- The run should have a minimum of 8 inches plus a 1½ inch nosing.
- The rise should have a maximum of 8¼ inches.

Stair Construction

Treads should be hardwood, vertical grain softwood, or flat-grained softwood covered with other suitable finish flooring material such as carpeting or vinyl flooring. Stairs to basements or attics can be made of flat grain softwood.

Stringers are the strength of the stairs. The material used should be selected for strength and secured to the framing.

Stringers must provide solid bearing at the top and bottom of the stairs, and must have sufficient depth—a minimum of 3½ inches supported by other construction. Stringers used in open basement stairs should have a minimum thickness of 2 inches.

When the distance between the stringers is more than 2 feet 6 inches on stairways with 1¹⁄₁₆-inch treads, or 3 feet for 1⅛-inch treads, a center stringer must be used.

Factory-built stairs with wedged and glued treads and risers may be supported by two stringers if the width of the stairs does not exceed 3 feet 6 inches.

The top of open stair stringers should provide not less than a 4-inch end bearing or be adequately anchored to the header. (See Figure 15-1.)

The relation of the tread width (run) to the riser height is important in determining the number of steps required. For comfortable use, the rise of each step in inches times the width of the tread in inches should equal 72 to 75, as in Figure 15-4 A. Thus, if the riser is 8 inches, the tread should be about 9 inches. If the riser is 7½ inches, the tread should be about 10 inches.

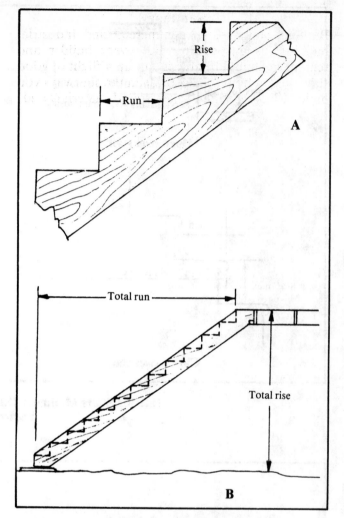

Riser-To-Tread Relationship. A, Individual Step; B, Total Rise and Run
Figure 15-4

The number of risers and treads can be found when the total height of the stairs is known. Divide the total rise in inches by 7½ (each riser) and select the nearest whole number. If the total rise is 107 inches, the number of risers would be 14 and the total run about 126 inches. (See Figure 15-4 B.)

Stairway landings should be double framed. The headers securing a stairway should be doubled. The stairway and landings should be designed to support the weight of as many people as can crowd onto the stairs and landing. If a grand piano needs support from a 5- or 6-joist span, what makes some builders think that the equivalent of 2 joists would hold it and the movers up? Be sure that the stairways you frame will hold as many people as you can crowd onto them. Remember that your stairway may be the only escape out of a burning building.

Summary

Building stairways in commercial and industrial buildings is a specialty. But every builder and remodeler should be able to put up a flight of good stairs in any home. Take pride in the stairways you build. Plan carefully, select sound materials and build according to plan. *Stair Builders Handbook*, published by Craftsman Book Company, P.O. Box 6500, Carlsbad, CA. 92008, will prove a valuable timesaver and guard against the expensive and embarrassing errors in calculations that are so easy to make in stairbuilding.

16

Skylights

For many years skylights have been an integral part of public buildings. Most large buildings constructed in the early part of this century had light wells and skylights to admit natural light to central portions of the structure. These skylights were made of steel and heavy wired glass.

Today most skylights are made of plastic and have only a light aluminum frame or no frame at all. Double and triple dome skylights are available so less heat and cold is transmitted through the translucent material. These improvements make skylights a popular and relatively inexpensive item for many remodeling projects.

The advantage of a skylight is obvious: free light any time the sun shines. In bathrooms and hallways a skylight may be all the lighting you need in daylight hours.

Rafter spacing will be your first consideration when planning a skylight. You don't want to reframe the entire roof just to get a skylight in. Skylights are available to fit roofs with rafters 16 and 24 inches on center. When the right width is used, little carpentry is needed.

To provide enough daylight under most conditions, you need about 1 square foot of skylight for each 20 square feet of floor. (See Figure 16-1.)

Skylight	Square Feet	Room Size
24" x 24"	4	80 S.F.

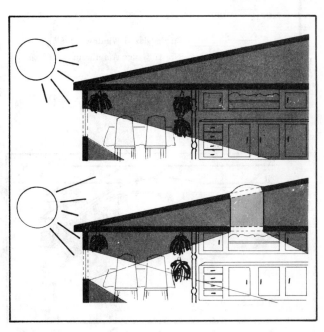

Light Distribution
Figure 16-1

Single or Double Dome

A single domed skylight can be used where insulation and condensation aren't a problem, such as a greenhouse or a patio.

If insulation and condensation are major concerns, a double domed skylight is preferable. The dead air space between the dome and diffuser creates a thermal insulation barrier, reducing the transfer of heat or cold through the skylight. (See Figure 16-2.) A double dome skylight transmits light just as well as a single dome skylight. Single and double dome skylights are shown in Figure 16-3.

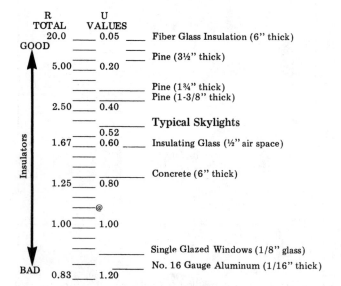

R TOTAL	U VALUES	
20.0	0.05	Fiber Glass Insulation (6" thick)
GOOD		
5.00	0.20	Pine (3½" thick)
		Pine (1¾" thick)
		Pine (1-3/8" thick)
2.50	0.40	
		Typical Skylights
	0.52	
1.67	0.60	Insulating Glass (½" air space)
		Concrete (6" thick)
1.25	0.80	
	@	
1.00	1.00	
		Single Glazed Windows (1/8" glass)
BAD		No. 16 Gauge Aluminum (1/16" thick)
0.83	1.20	

Insulating Values
Figure 16-2

(Insulators ← Good to Bad scale shown at left)

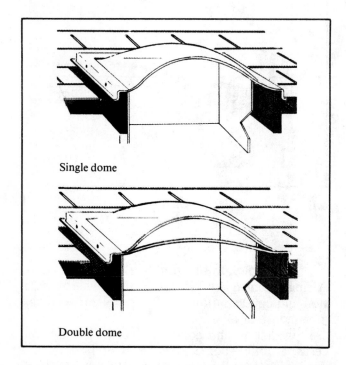

Single dome

Double dome

Single and Double Dome Skylights
Figure 16-3

Skylights are available with clear, frost and bronze domes and with combinations of these in multi-dome units. The clear dome with or without a clear inner dome gives maximum lighting and visibility when you want the most intense light.

The clear dome with a white diffusion panel is a good choice to light a larger area without harsh glare. This is the most popular model with homeowners.

The bronze skylight eliminates harsh glare and colors the area below with slightly tinted light. The bronze dome is usually used to make the dome exterior blend with an existing roof cover.

Curb Mounted Skylights

Curb mounted skylights have an integral flashing configuration which simply caps down over the supporting curb. Build a wood frame curb to the size recommended by the manufacturer and apply flashing. (See Figure 16-4.) A continuous flange rests on top of the curb. The skylight is secured to the curb from the side with corrosion-resistant screws and rubber washers placed over pre-drilled mounting holes.

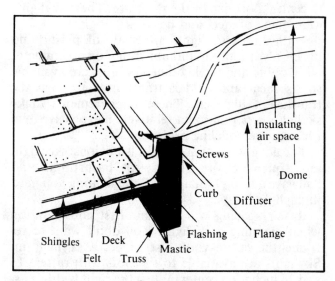

Curb Mount
Figure 16-4

Start the installation in an open beam ceiling by driving a 3-inch nail up through the roof at each corner of the proposed opening (Figure 16-5). If your skylight is to be placed in an attic, drive the nails up through the roof once you have located the skylight's position in relation to the ceiling. Make sure there are no electrical wires, water pipes, vent pipes, or ductwork in the way.

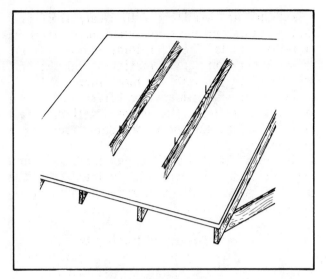

Locate Rafter Position
Figure 16-5

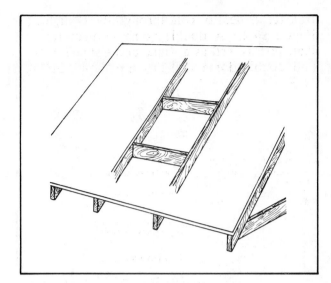

Frame The Opening
Figure 16-7

Now go up on the roof. Locate the nails you drove through the roof. Remove the roofing material back about 12 inches all around the skylight area. Use care if the roofing material is to be reused. Cut the hole through the sheathing. (See Figure 16-6.)

Apply asphalt plastic roofing cement around the opening, about 1/4 inch thick. Be sure to cover all exposed wood and felt. (See Figure 16-8.)

Build the curb out of 2 x 6-inch lumber. Make the inside dimensions the same as the opening in the roof. Use roofing cement to seal all joints. (See

Cut Thru Roof
Figure 16-6

Apply Asphalt Roofing Cement
Figure 16-8

Frame the inside of the roof opening at the top and bottom (Figure 16-7). If a rafter runs through the opening, cut it back and nail it to the skylight framing. Additional framing may be necessary if the skylight seams do not align with the rafters.

Figure 16-9.)

Apply roofing cement on the outside of the curb at the bottom on all four sides. Nail the cant strip in place as shown in Figure 16-10.

Then cover the entire outside of the curb with

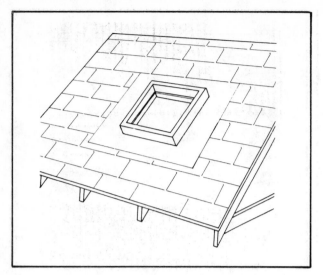

Construct The Curb
Figure 16-9

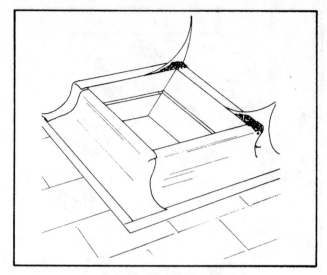

Apply Number 15 Pound Felt
Figure 16-11

roofing cement. Apply roofing felt in the still soft roofing cement. Use 15-pound building paper. The bottom strip at the low side of the skylight should go down first, then the sides and finally the strip at the top of the skylight.

Apply a bead of clear mastic weatherstripping around the tip edge of the curb. Press the skylight down into place. Drill small holes for the screws and secure the flange around the edge about every 3 inches. (See Figure 16-13.) Figure 16-14 is a cross

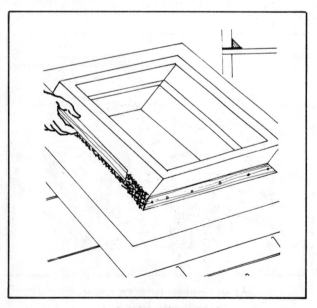

Seal Outside Curb
Figure 16-10

Replace Shingles
Figure 16-12

Apply roofing cement everywhere the felt overlaps. Cover exposed seams with roofing cement. (See Figure 16-11.) Then replace the shingles as shown in Figure 16-12.

section view of the finished installation.

Once installed as shown in Figure 16-15 the skylight should require no maintenance. Normal rainfall will keep the dome reasonably clean. If you

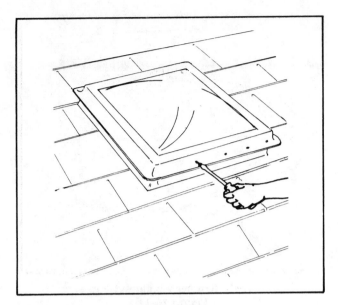

Install Skylight
Figure 16-13

The Completed Job
Figure 16-15

want to wash the unit so your customer gets a bright and shiny new skylight, use only soap and water. Petroleum-based cleaners will damage the dome surface.

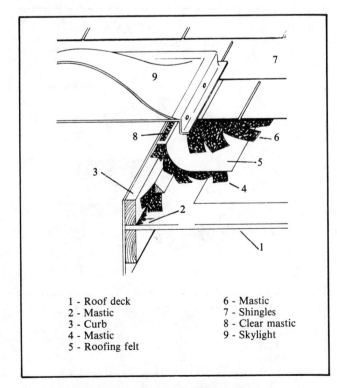

1 - Roof deck   6 - Mastic
2 - Mastic   7 - Shingles
3 - Curb   8 - Clear mastic
4 - Mastic   9 - Skylight
5 - Roofing felt

Cross-View of Installation
Figure 16-14

Self Flashing Skylights

Self flashing skylights install directly on a pitched roof. Installation without a curb gives the skylight a low profile and makes it less conspicuous. Attach it directly to the roof and flash it as described previously. Figure 16-16 shows this type of installation.

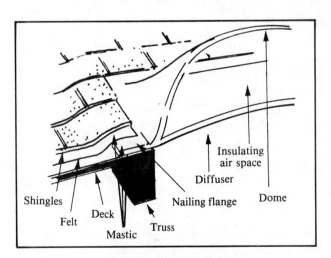

Self-Flashing Skylight
Figure 16-16

Locating, marking, and cutting for a self-flashing skylight are the same as for a curb-mounted skylight. Place the dome in a 1/4-inch-thick bead of roofing cement and nail in pre-

Install The Unit
Figure 16-17

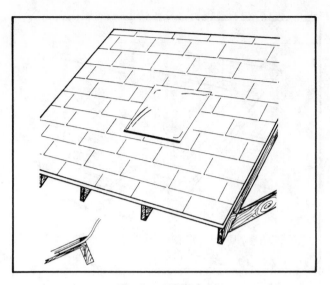

Replace Shingles
Figure 16-19

drilled holes. Most units come with special nails and rubber washers to secure the unit. (See Figure 16-17.)

Apply a thick coat of roofing cement over the edge of the skylight up to the bubble. Cut strips of roofing felt wide enough to go from the bubble to overlap the felt on the deck. Put the side pieces on first. Apply more cement over these strips at the top and apply the top strip of felt. Don't put a strip at the bottom. (See Figure 16-18.)

Apply mastic over the felt strips and replace the asphalt shingles. After the shingles are in place, apply roofing cement across the bottom of the skylight as shown in Figure 16-19. Figure 16-20 shows a cross-view of the finished installation.

Framing dimensions for various sizes of skylights are shown in Figure 16-21. Where more than one rafter is cut, the headers should be doubled so the roof is not weakened.

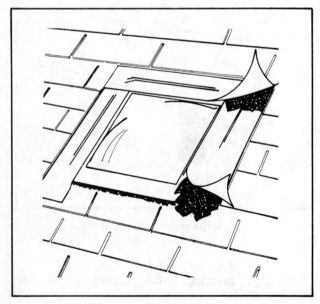

Apply Cement And Felt
Figure 16-18

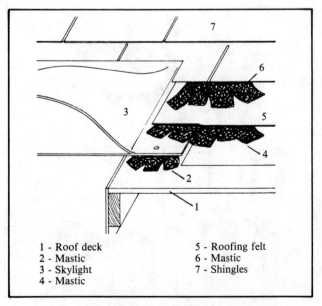

1 - Roof deck
2 - Mastic
3 - Skylight
4 - Mastic

5 - Roofing felt
6 - Mastic
7 - Shingles

Cross-View of Installation
Figure 16-20

Size	Max. Roof Opening	Size	Max. Roof Opening	Size	Max. Roof Opening	Size	Max. Roof Opening
16 x 16	14½ x 14½	24 x 24	22½ x 22½	32 x 32	30½ x 30½	36 x 48	34½ x 46½
16 x 24	14½ x 22½	24 x 32	22½ x 30½	32 x 48	30½ x 46½	48 x 48	46½ x 46½
16 x 32	14½ x 30½	24 x 48	22½ x 46½	36 x 36	34½ x 34½	48 x 72	46½ x 70½

Roof Opening Size Chart

Framing Dimension Chart

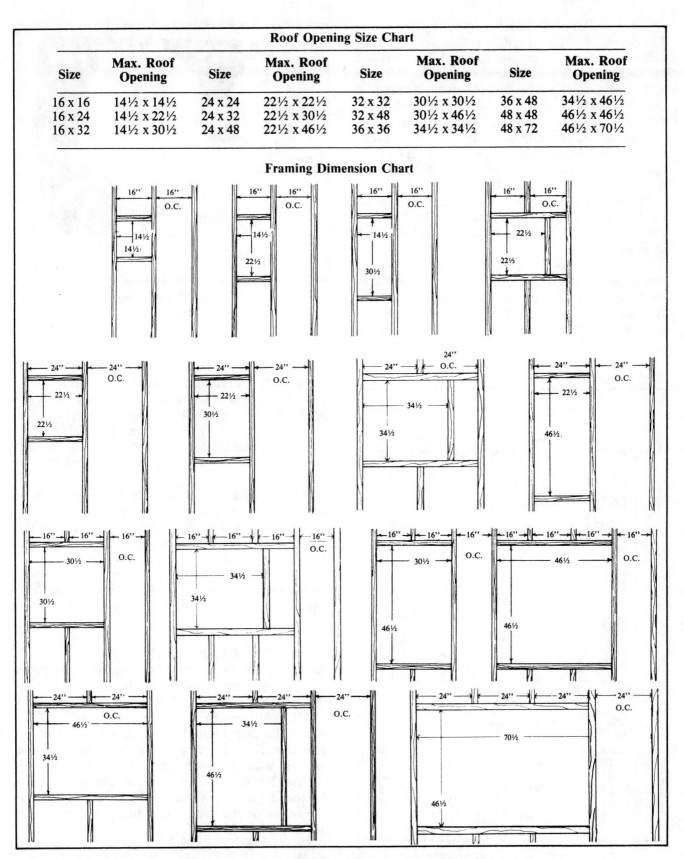

Roof Opening Size Chart and Framing Dimension Chart
Figure 16-21

The "Bubble" Skylight

The round "bubble" skylight is designed to be installed in the roof or in a vertical wall. It's installed the same way as a self-flanging skylight in the roof.

Begin by checking for any electrical, heating or plumbing runs through the point where the bubble will be placed. Locate the wall studs and determine the size of the opening necessary for the installation of a wood frame. (See Figure 16-22A.) Cut the opening in the outside and inside walls. Install the wood frame (Figure 16-22B).

Pre-drill nail or wood-screw holes in the flange of the bubble (the flange is nailed to the framing, not to the bender board casing), and tack the unit in place.

Using 1/4- to 1/2-inch-thick bender board, shape the circular casing. You'll need 40 linear feet to make three overlapping layers for a window 4 feet in diameter. Before installation, soak the boards (exterior plywood) in water to make them

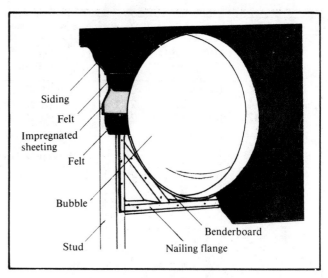

Install Wood Frame
Figure 16-22B

pliable. The boards should be soft enough so they bend easily. Bevel the ends to make smooth joints.

With the bender board casing installed, align the bubble in the opening. For weather protection, slip building paper under the bottom flange of the bubble. Fasten the bubble securely, and caulk all nail or screw heads. Overlap building paper on the top and sides.

When applying or reapplying exterior siding or interior gypsum board or paneling, first lay the material flat and scribe it with a compass to the diameter of the bubble. Then cut the material and fasten it in place.

Circular trim inside can be cut from particle board with a sabre saw. Put the trim on last. Round the edges by sanding. Pre-paint the piece before installing it.

Clustering Skylights

Skylights can be installed in clusters to achieve a very distinct and modern look. First, build individual curbs from 2 x 6-inch lumber. Then put 4 x 4-inch spacers between the curbs, and add 2 x 4-inch lumber around the outside to tie it all together. Each opening in the roof should be the skylight width (from the inside of the first 2 x 6 to the last 2 x 6) by the height (from the inside of the bottom 2 x 6 to the inside of the top 2 x 6). Be sure the inside of the roof is well braced. (See Figures 16-23A nd 16-23B.)

Clusters of skylights in room additions must conform to truss spacing. Using double (4 x 10) trusses spaced at more than 24 inches makes it

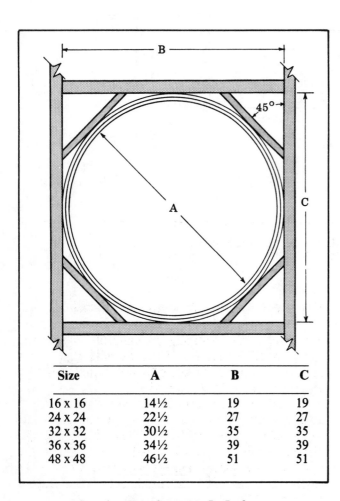

Size	A	B	C
16 x 16	14½	19	19
24 x 24	22½	27	27
32 x 32	30½	35	35
36 x 36	34½	39	39
48 x 48	46½	51	51

Opening Requirements In Inches
Figure 16-22A

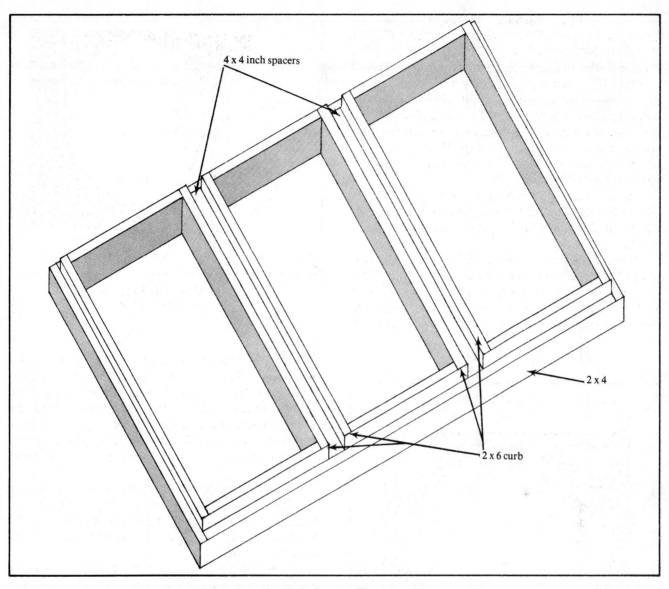

Framing For Cluster Installation on Existing Roofs
Figure 16-23A

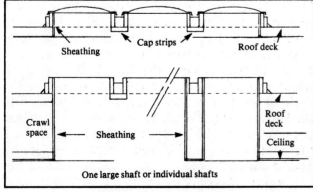

Optional Shaft Treatment
Figure 16-23B

easier to install the skylights because cutting and bracing aren't needed. After the trusses are in, build the curbs from 2 x 10 material. Fasten the curbs so that at least 6 inches extends above roof deck level. Frame complete clustered units with 2 x 4's. (See Figure 16-24 A and B.)

In a conventional attic home, you have to install a light shaft below the skylight. A straight shaft works best for a direct, overhead highlight. To distribute the light over a broader area, change the angle of the shaft. This works best when your customer wants soft diffused light. (See Figure 16-25.)

The shaft, from ceiling to roof, is just a frame

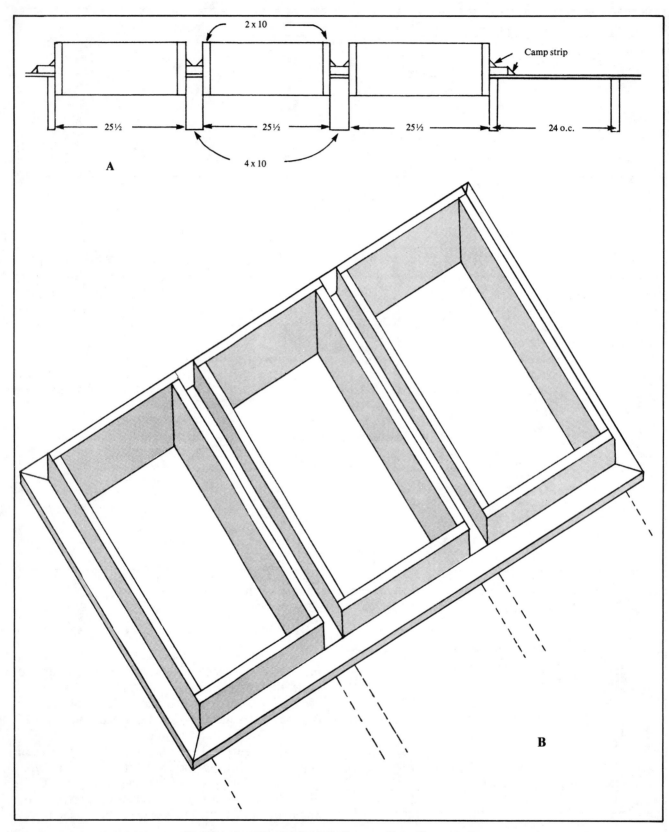

Framing for Cluster Installation on New Construction
Figure 16-24

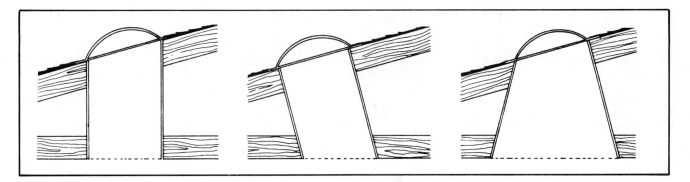

Shaft Control Light Direction
Figure 16-25

box. Be sure to locate its position before removing the finish ceiling in the shaft area. Follow the installation procedures illustrated in Figure 16-26. Insulate the shaft with glass fiber batts, stapling the batts between the framing in the same manner as in exterior walls, with the vapor barrier on the heated side (shaft side). The shaft may be finished with sheetrock or with the same material used to finish the ceiling. But be sure the box is painted white or a light color so it reflects light.

Hatch-type skylights are available where ventilation is needed. These are installed the same way as curb mount skylights. The hatch-type is usually equipped with a crank that opens and closes the unit. Automatic or remote controlled units are also available.

For industrial skylight needs, 4 x 6-foot and larger units are available with either a single or double dome. (See Figure 16-27.)

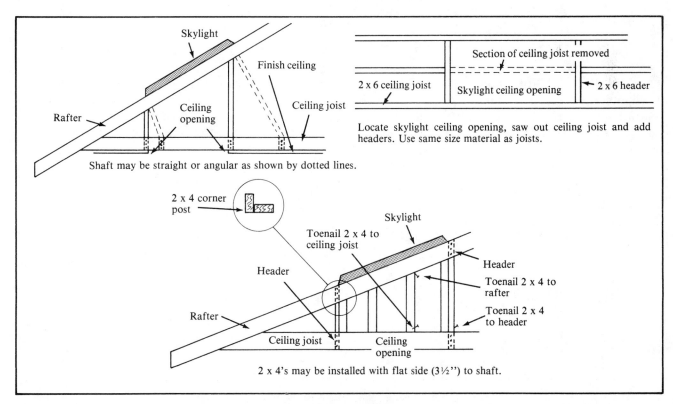

Framing the Shaft
Figure 16-26

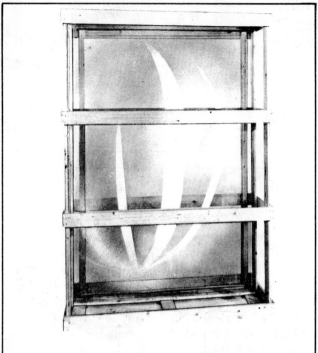

This new larger size curb unit will cover many of those big residential and industrial skylight needs. Each unit is shipped in a wood crate.

Maximum Roof Opening 46½ x 70½
Curb O.D. 49½ x 73½
Figure 16-27

Summary

Skylights are popular in studios or rooms where natural light is preferred over artificial light. Of course, skylights can be installed for other reasons, too. Whatever the reason, a skylight is a nice addition for most houses. Your customers will appreciate your professionalism when you suggest a skylight on a job that really needs one.

17

Paint and Painting

Nothing does more for a house in less time and at a lower cost than a fresh coat of paint. The right blend of colors can turn a neglected house into a thing of beauty. But, like everything else in construction, painting requires a certain know-how. Since most of your jobs require some painting and most small painting jobs aren't worth subcontracting, you better know some of the basics about paint and coatings.

Begin by understanding paint texture and color. Some colors and textures look good in certain situations and others do not. You don't need to be an interior decorator to know the difference. You're going to give advice on what paints and colors to use. Understanding a few key points will put you leagues ahead of your competition.

Using Color and Texture

Walls are usually the largest area to be painted. So the color used should be the dominant one in your plan. Carpeting, drapes and upholstery should complement or blend with each other and the walls.

Select colors that complement other colors and textures in the room. Look at the accessories in the room: porcelain, brass, bronze, gold, silver, copper, cork, wicker, rattan, and wood; and don't forget flowers and plants.

Furniture styles can also help you select colors, particularly when your client is tuned in to interior decorating. The designs of the Louis XIV and Louis XV era were complemented by soft greens, golds and other delicate hues. The furniture designs of the Queen Anne period and other 18th century styles appeared in rich, subtle color combinations of crimson reds, blues, greens and indigos.

You need to know a few things about texture. Texture adds dimension and interest to a room. Consider the paint finish as well as the color.

A flat finish is perfect for a formal setting because of its quiet elegance. Use a high quality latex, flat finish paint.

An eggshell finish works well in any setting, but is especially suitable for deep accent color and for high traffic areas.

A low lustre finish adds softness to casual, comfortable rooms. Use a satin enamel with a soft sheen.

A high gloss finish is the "wet look." High gloss is ideal for a contemporary room where a "loud" effect is what your client wants. Gloss finish has high washability which makes it a practical finish for trim, cabinets, doors and children's rooms. Enamel paint is usually used.

Textured paints are available in three finishes: sand, rough and stucco. This finish is ideal for a Mediterranean or Early American look.

Northern climates need warm colors to compensate for intermittent sunlight. Cool pastel tones are perfect for southern climates. They should, however, be bright enough so the color will not look weak in intense sunlight.

The house and its setting should blend. This

doesn't mean that a house on wooded property should be green. There are many subtle tones in wooded areas. The grays, browns, beiges and golds of bark, rock formations and undergrowth offer many color selection possibilities.

Notice that the immediate surroundings of a house affect its color. A green house, for example, surrounded by shrubs and grass will appear greener. Ideally, the contrast between color and surroundings should be just enough to retain house definition and color.

Combining light and dark shades gives interesting visual effects. Lighter colors command more attention and should be used to highlight the more attractive features. To highlight an appealing entry, for example, paint the area around the door a shade or two lighter than the rest of the entry. The lighter color will visually bring the entry forward. Then use a vivid, attention getting color on the door.

The same principle can be applied to larger sections of a house. A straight ranch style home can be given shape and proportion by highlighting a section of the long front. Paint the front entrance and window area, as defined by an eave, a lighter shade than the balance of the front. This brings the front visually forward and lifts it upward, framed by the darker subdued sections. If there are no natural stopping lines for the two shades, add vertical moldings.

Lighter shades give an impression of height, while darker shades minimize size. For a two- or three-level house which is too tall for its width, paint the top level darker than the bottom. When using several colors, keep the colors in the same color family.

An unattractive or poorly placed roof extension or dormer window can be made less obvious by painting it the same color as the roof.

Houses with large porches and overhangs need special attention. Try a dark trim color on the overhang and support columns. This gives a pleasing, shadow-box frame effect for the front entrance.

The Effects of Color

Color affects space, light and mood. Let's examine each of these.

Space

A small room will appear larger if walls, ceiling and woodwork are all the same color. A pale, bright color works best. The feeling of spaciousness is created by eliminating breaks in the flow of color at corners and ceiling lines.

A square room can be made more interesting by painting one wall a dramatic, contrasting color—usually the wall directly opposite the entrance to the room. Choose a color based on the dominant color in the furnishings—carpet, drapes or upholstery.

A long, narrow room can be squared up by painting the two end walls a deeper color. A room that is too large will seem better proportioned and more inviting if painted in darker shades.

Create a feeling of more space and continuity between rooms by using similar colors. This also applies to smaller houses—the total living space will appear bigger if all rooms are done in coordinated colors.

You can lower a ceiling with a color slightly darker than the adjacent walls.

Light

Northern light is cold and eastern light is harsh. Rooms facing either of these directions will look their best when warm colors are used—red, orange, yellow and brown.

Light from the south and west is warm and bright. Rooms facing these directions will need cool colors such as green, blue, violet and gray.

You can control the effect of a room with color. But color is affected by the light source in the room—daylight, fluorescent strips and incandescent bulbs. A color "pro" will look at the paint chips in the actual room under actual lighting conditions.

Mood

When you are helping a customer select colors, explain that colors create mood. For example, a kitchen done in gay, cheerful colors will promote a happy outlook for each new day. Here's a basic breakdown of moods supported by color, according to the experts:

Yellows and lime—bright and sunny.

Gold and bronze—rich and friendly.

Olive tones—somber and quiet.

Greens—cheerful and open.

Gray greens, aquas and jades—calm and serene.

Pinks and reds—lively and aggressive.

Purples and violet blues—sophisticated and daring.

Blues and turquoises—refreshing and tranquil.

Oranges and peaches—active and inviting.

Spice tones—advancing and bold.

Camels, beiges and taupes—subtle and elegant.

Earthtones—warm and cozy.

Grays, charcoals and neutrals—cool and receding.

Paint Materials

There's more to paint, of course, than just selecting color. You should know a little about the types of paint your dealer has on his shelf.

Most paints are based on a film former or binder which is either dissolved in a solvent or emulsified in water. When applied in a thin film, the paint will dry or cure to form a dry, tough coating. Solutions of such binders in a solvent have various names—clear finishes, varnishes (if they dry by oxidation), or lacquers (if they dry by evaporation). If opaque pigments or colors are dispersed in the binder, the product, which will produce an opaque white or colored film, is called a paint. Pigment concentration can also be varied to produce a high gloss, a semigloss or a lustreless (flat) finish. Special pigments such as red lead and zinc chromate can be used to provide corrosion resistance in primers. Metallic pigment can be added to varnishes to produce metallic coatings such as aluminum paints. The way a paint performs depends on the type of binder used.

Alkyd

Alkyd binders, for the most part, are oil modified phthalate resins which dry by reacting to oxygen in the surrounding air. Alkyd finishes are economical and available as clear or pigmented coatings. The finishes are available in flat, semigloss and high gloss, in a wide range of colors. These are easy to apply, and, with the exception of fresh (alkaline) concrete, masonry and plaster, may be used on most clean surfaces. Alkyd finishes have good color and gloss, and hold their color well in normal interior and exterior areas. Their durability is excellent in rural environments, but only fair in mildly corrosive areas.

Cement

Portland cement combined with several ingredients makes a good paint for some applications. It comes as a powder to which you add water. Cement paints are used on rough surfaces such as concrete, masonry and stucco. They dry to form a hard, flat, porous film that will allow water to pass through without damaging the surface. Since cement paints are powders, they can also be mixed with masonry sand and less water to form filler coats to smooth rough masonry before applying other paints.

You'll use cement paints most often on fresh masonry. The surface must be damp when they are applied, and must be kept damp for a few days for proper curing. When properly cured, good quality cement paints are durable. When improperly cured, they chalk heavily and soon need repainting.

Epoxy

Generally, epoxy binders come with two components, an epoxy resin and a polyamide hardening agent. You mix them before use. When mixed, the two ingredients react to form a hard final coating. Anything left at the end of the day must be thrown out. You can put epoxy paint on fairly thick to cover a textured surface. The best use of epoxy is to get a tile-like glaze coating for concrete and masonry. The cost per gallon is high, but you don't have to apply as many coats to get an adequate thickness.

Epoxy paints will chalk on exterior jobs. They lose their gloss and do fade. Otherwise, their durability is excellent

Latex

Latex paints dry by evaporation of the water. They have little odor, are easy to apply, and dry very rapidly. Interior latex paints can be used either as a primer or finish coat on interior walls and ceilings whether made of plaster or wall board. Exterior latex paints are used directly on exterior (including alkaline) masonry or on primed wood. These paints are non-inflammable, economical, and fade very little. Also, blistering caused by moisture vapor is less of a problem than with solvent-thinned paints. But they don't stick very well to chalked, dirty or glossy surfaces. Careful surface preparation is essential.

Oil

The major binder in oil house paint is linseed oil. Oil paints are used primarily on exterior wood and metal since they dry too slowly for most interior uses and can't be applied to masonry. They are easy to use and can be applied fairly thick. They also wet the surface very well so that surface preparation is less critical. They aren't particularly hard or resistant to abrasion, chemicals or strong solvents, but are durable in normal environments.

Rubber-Base

The so-called rubber-base paints are solvent thinned and are not the same as latex paints, which have rubber-base emulsions. There are four types: chlorinated rubber, styrene-butadiene, vinyl tolune-butadiene and styrene-acrylate. They are lacquer type products and dry rapidly to form

	Alkyd	Cement	Epoxy	Latex	Oil	Phenolic	Rubber	Moisture Curing Urethane	Vinyl
Ready for use	Yes	No	No[3]	Yes	Yes	Yes	Yes	Yes	Yes
Brushability	A	A	A	+	+	A	A	A	A
Odor	+[1]	+	—	+	A	A	A	—	—
Cure normal temp.	A	A	A	+	—	A	+	+	+
Cure low temp.	A	A	—	—	—	A	+	+	+
Film build/coat	A	+	+	A	+	A	A	+	—
Safety	A	+	—	+	A	A	A	—	—
Use on wood	A	—	A	A	A	A	—	A	—
Use on fresh conc.	—	+	+	+	—	—	+	A	+
Use on metal	+	—	+	—	+	+	A	A	+
Corrosive service	A	—	+	—	—	A	A	A	+
Gloss - choice	+	—	+	—	A	+	+	A	A
Gloss - retention	+	X	—	X	—	+	A	A	+
Color - initial	+	A	A	+	A	—	+	+	+
Color - retention	+	—	A	+	A	—	A	—	+
Hardness	A	+	+	A	—	+	+	+	A
Adhesion	A	—	+	A	+	A	A	+	—
Flexibility	A	—	+	+	+	A	A	+	+
Resistance to:									
Abrasion	A	A	+	A	—	+	A	+	+
Water	A	A	A	A	A	+	+	+	+
Acid	A	—	A	A	—	+	+	+	+
Alkali	A	+	+	A	—	A	+	+	+
Strong solvent	—	+	+	A	—	A	—	+	A
Heat	A	A	A	A	A	A	+[2]	A	—
Moisture permeability	Mod.	V.High	Low	High	Mod.	Low	Low	Low	Low

+ = Among the best for this property   A = Average
— = Among the poorest for this property   X = Not applicable

[1] Odorless type
[2] Special types
[3] Two component type

Comparison of Paint Binders Principal Properties
Figure 17-1

finishes which are highly resistant to water and mild chemicals. Rubber-based paints are available in a wide range of colors and levels of gloss. They are used for exterior masonry and for areas which are wet, humid or subject to frequent washing, such as swimming pools, kitchens and laundry rooms.

Urethane

Two types of urethane finishes are oil-modified and oil-free moisture-curing. Both are used as clears, but the oil free type is also available pigmented.

Oil-modified urethanes are more expensive and have better color than varnishes. They also dry quicker, are harder, and resist scuffing better. They can be used on all surfaces, as exterior varnishes or as tough floor finishes. But like all clear finishes, they are not very durable.

Moisture curing urethanes are the only organic products presently available which cure by reacting with moisture from the air. Be sure that all con-

tainers are kept full to exclude moisture. Otherwise the contents will turn to a gel.

Figure 17-1 shows the important properties of the major types of paint. Properties of binders are not included in this figure. The following are binders with properties similar to those of the paints listed in Figure 17-1.

Oil-alkyd: properties similar to oil and alkyd paints.

Cleoresinous: similar to alkyds but with less color retention.

Phenolic-alkyd: similar to phenolic and alkyd paints.

Oil-modified urethane: similar to phenolic and alkyd paints.

Vinyl-alkyd: similar to vinyl and alkyd paints.

Types of Paints

Up to this point we've been talking about paint bases. You can also describe paint by the type. You're probably more familiar with types of paints than types of bases.

Lacquer

All coatings which dry solely by evaporation of the solvents may be described as lacquers. Rubber-base coatings and vinyl coatings are lacquers. Lacquers dry rapidly even at low temperatures, making brushing difficult. The solids content usually is relatively low, so several coats may be needed. Recoating, especially by brush, must be done very carefully to avoid lifting the existing coat.

Varnish

Varnishes are alkyds or resins in solvent with driers added so that they dry by oxidation. The film is clear or nearly clear. Varnish is available in a variety of types such as spar varnish, aluminum mixing varnish, sealer and flat varnish.

Paint and Enamel

The difference between paint and enamel is mostly in the way it's used. When used on large areas such as walls, it's called paint. If it's fast drying, levels out to a smooth, hard finish and is used on relatively small areas or smooth materials such as woodwork, it's called enamel. Paints can be rolled on, but enamels rarely are.

The amount of pigment determines the gloss. Generally, gloss is reduced by adding non-opaque, lower cost pigments called extenders. Typical extenders are calcium carbonate (whiting), magnesium silicate (talc), aluminum silicate (clay), and silica.

Maximum gloss results from omitting all extenders. This produces maximum washability and durability, and is referred to as "high gloss."

Primer for Paint and Enamel

Generally two types of paints are required. The coat of paint next to the material painted is called the primer. The topcoat is used to produce the finish color and texture.

Primer-sealer is used to seal a porous or alkaline surface so that the topcoat is unaffected by loss of or damage to the binder. Some paints, such as interior latex wall paints, are self-priming and usually don't need a special primer.

Enamel undercoater is a coating which dries to a smooth, hard finish that can be sanded. When sanded, the smooth surface is ideal for a topcoat of smooth enamel.

Use primers with anti-corrosion pigments such as red lead, zinc chromate, lead silicone chromate and zinc dust when painting iron and steel. They slow corrosion of the metal but must be protected by other types of coatings.

What is "House Paint"?

"House paint" is a very broad term used to describe many paint formulations. But it is the most widely used type of paint. Some types are available for special needs such as chalk or mildew resistance. White is the most used color. House paint comes in both oil base and latex (water base) paint.

Aluminum Paint

Metallic paints, such as aluminum paints, are available in two forms: ready mix and ready-to-mix. Ready mix aluminum paint comes in one package and is ready for use after normal stirring. It is more convenient to use than the ready-to-mix form and eliminates errors in mixing.

Oil Stain

Oil stains use a drying oil, such as linseed oil, which is thinned to a very low consistency for maximum penetration when applied to wood. Interior stains should be applied only to a sanded, dust-free surface, and allowed to dry for a short time. Then the excess is wiped off so that only the stain which has penetrated the wood remains. Figure 17-2 will help you select the right stain for your job.

Estimating Paint Needs

Figure 17-3 shows how to calculate square-foot coverage of various shapes.

Exterior Surface Preparation
Wood Surface Preparation for Paint

New wood—All wood surfaces must be dry, free of grease, mildew, mortar and asphalt spatters. Knots and sappy spots must be sealed with a sealer. Rough surfaces should be sanded smooth, nailholes and cracks should be puttied, and window and door trim joints should be caulked after surfaces have been primed.

Repainting wood—Remove all blistered, peeling and scaling paint to the sound surface by scraping or sanding. Remove chalk deposits with high pressure spray or by hand scraping or wirebrushing. Loose or split caulking should be removed and replaced. Cracks, crevices and nailholes should be puttied. Use a commercial mildew remover if any mildew is present.

Figure 17-4 gives generally accepted commercial painting specifications.

Wood Surface Preparation for Stain

New wood—All surfaces must be dry and free of foreign matter. Roughen the surface with a stiff

	Aluminum	Cement Base Paint	Exterior Clear Finish	House Paint	Metal Roof Paint	Porch-and-Deck Paint	Primer or Undercoater	Rubber Base Paint	Spar Varnish	Transparent Sealer	Trim-and-Trellis Paint	Wood Stain	Metal Primer
Wood													
Natural finish	-	-	P	-	-	-	-	-	-	P	-	P	-
Porch floor	-	-	-	-	-	P	-	-	-	-	-	-	-
Shingle roof	-	-	-	-	-	-	-	-	-	-	-	-	-
Shutters and trim	-	-	-	P+	-	-	P	-	-	-	P+	-	-
Siding	P	-	-	P+	-	-	P	-	-	-	-	-	-
Windows	P	-	-	P+	-	-	P	-	-	-	P+	-	-
Masonry													
Asbestos cement	-	-	-	P+	-	-	P	P	-	-	-	-	-
Brick	P	P	-	P+	-	-	P	P	-	P	-	-	-
Cement & Cinder block	P	P	-	P+	-	-	P	P	-	P	-	-	-
Cement porch floor	-	-	-	-	-	P	-	P	-	-	-	-	-
Stucco	P	P	-	P+	-	-	P	P	-	P	-	-	-
Metal													
Copper	-	-	-	-	-	-	-	-	P	-	-	-	-
Galvanized	P+	-	-	P+	-	-	-	P	P	-	P+	-	P
Iron	P+	-	-	P+	-	-	-	-	-	-	P+	-	P
Roofing	-	-	-	-	P+	-	-	-	-	-	-	-	P
Siding	P+	-	-	P+	-	-	-	-	-	-	P+	-	P
Windows, alum.	P	-	-	P+	-	-	-	-	-	-	P+	-	P
Windows, steel	P+	-	-	P+	-	-	-	-	-	-	P+	-	P

"P" indicates preferred coating for this surface.

"P+" indicates that a primer or sealer may be necessary before the finishing coat or coats (unless the surface has been previously finished).

Exterior Paint Selection Chart
Table 17-2

brush to remove loose fibers and splinters.

Restain—It is possible to restain a wood surface. But you have to take the surface down to the base wood. All loose or split caulking should be removed and replaced. Mildew must be killed. Figure 17-5 is a staining specifications chart.

Metal

Iron and steel—Doors, trim, fire escapes, ornamental iron, handrails and other surfaces must be free of grease and oil. Use a solvent such as paint thinner. Surfaces with rust or scale should be cleaned by scraping and wire brushing or by sandblasting. Be sure to recoat with primer within 24 hours.

Figure 17-6 is a metal painting specifications chart.

Masonry

Concrete and masonry must cure for 60 to 90 days before painting. Brush the surface to remove loose sandy particles and mortar. Form release compound and salts that form on the surface of masonry will cause the paint to flake off. Clean form oil off with a solvent and remove residue with a wire brush. Fill cracks and voids by repointing and caulking.

Repainting—Surfaces that are peeling, scaling or have started to chalk must be cleaned by scraping, wirebrushing, or with a mechanical grinder or power brush. Fill all cracks with a patching compound. Cracked and loose caulking should be removed and replaced. Finally, kill and remove any mildew.

New masonry floors—Concrete slabs should cure 60 to 90 days before coating. Etch steel-troweled concrete with a 10 percent solution of muriatic acid, then rinse thoroughly. Let the surface dry completely before painting. Be sure to remove cement spatters, fill holes and crevices, then sweep out the dirt and dust.

Repainting masonry floors—Scaling and peeling paint must be removed by scraping and sanding. Then wash the floor with a strong detergent or solvent to remove grease and oily residue. Unpainted

Triangle

To find the number of square feet in any shape triangle or 3 sided surface, multiply the height by the width and divide the total by 2.

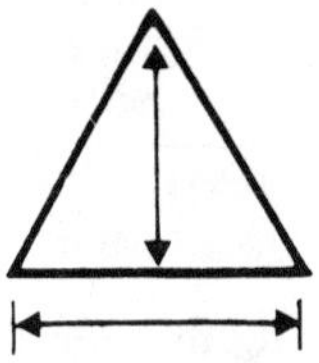

Square

Multiply the base measurement in feet times the height in feet.

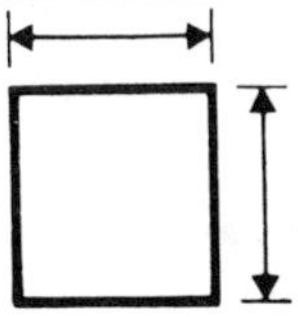

Rectangle

Multiply the base measurement in feet times the height in feet.

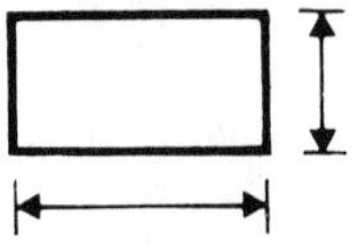

Cylinder

When circumference (distance around cylinder) is known, multiply height by circumference.

When diameter (distance across) is known, multiply diameter by 3.1416. This gives circumference. Then multiply by height.

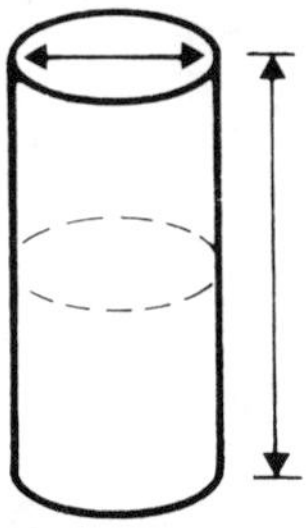

Circle

To find the number of square feet in a circle multiply the diameter (distance across) by itself and then multiply this total by .7854.

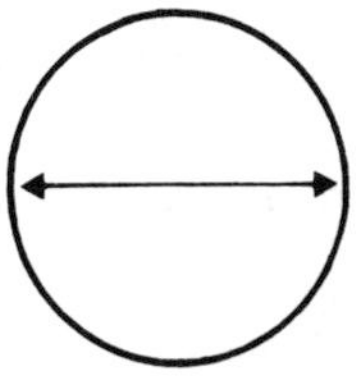

Arch Roof

Multiply length (B) by width (A) and add one-half the total.

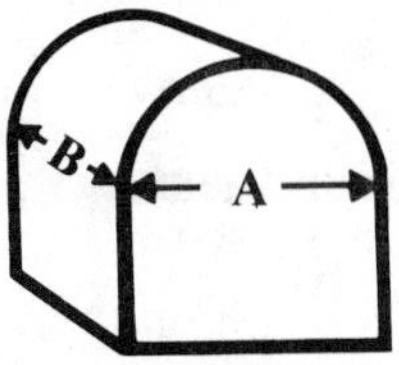

Gambrel Roof

Multiply length (B) by width (A) and add one-third of the total.

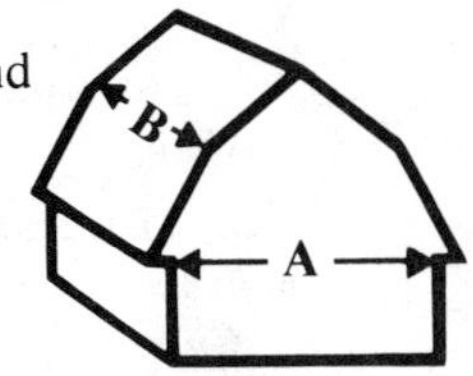

Cone

Determine area of base by multiplying 3.1416 times radius (A) in feet.

Determine the surface area of a cone by multiplying circumference of base (in feet) times one-half of the slant height (B) in feet.

Add the square foot area of the base to the square foot area of the cone side for total square foot area.

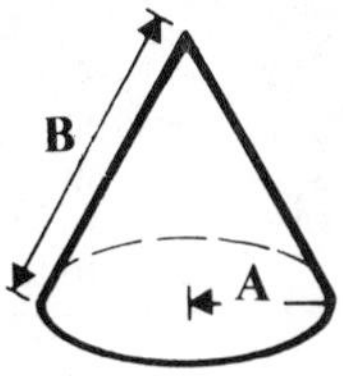

Calculating Square Foot Coverage
Figure 17-3

floors can be treated like new floors. But any floor with oil and grease spots must be cleaned with a grease-dissolving compound. Figure 17-7 shows painting requirements for masonry.

Interior Surface Preparation
Wood Surface Preparation for Paint
New wood—The surface should be dry, sanded smooth, and free of dust, dirt or grit. Putty nailholes, cracks and blemishes after the undercoat has been applied. Knots should be coated with a primer or sealer before the primer goes on. Be sure each coat is completely dry before the next coat is applied. Sand lightly between coats. Tops and bottoms of doors need a coat of primer.

Repainting—The surface should be clean and free of wax, grease and grime. Glossy surfaces should be dulled by sanding. Remove all loose paint by scraping and sanding. Repair holes, crevices and cracks with a patching compound. Wash surfaces that have been defaced with marking pens, lipstick, etc., with a solvent. Then apply a primer over those spots to control bleeding. Scarred or chipped spots should be sanded to feather flush with the surface. Figure 17-8 shows requirements for wood painting.

Wood Surface Preparation for Stain
New wood—All surfaces must be dry, sanded smooth, and free of dirt, dust or grit. Fill nailholes, cracks and blemishes with filler tinted to match the color of the stain or color of the wood if left natural. Open-grained wood such as oak or walnut can be filled with a suitable wood filler. Tops and

Surfaces	Primer Selection	Type	Finish Coatings Selection	Finish	Type
Painted: Clapboard siding, hardboard siding, shingles, shakes, framing, trim, doors, fascia	Primer	Long-oil alkyd	Latex house paint	Low lustre	Latex
			Eggshell finish house paint	Eggshell	Long-oil alkyd
			Latex house & trim paint	Soft gloss	Latex
			House paint	High gloss	Long-oil alkyd
Enameled: Window/door framing, trim, sash, doors	Primer	Long-oil alkyd	High gloss enamel	High gloss	Alkyd
			Q. D. industrial enamel		
Porches, steps, platforms, decking, railings	Porch & floor enamel, thinned	Alkyd	Porch & floor enamel, as packaged	High gloss	Alkyd

Wood Painting Specifications - Exterior
Figure 17-4

Surfaces	Primer Selection	Type	Finish Coatings Selection	Finish	Type
Stained: Clapboard siding, shakes, shingles, textured or rough sawn siding, trim, fencing, decking	None. If necessary to suppress bleeding, use latex exterior primer	Latex	Exterior stain Solid colors	Flat	Alkyd
			Exterior stain semi-transparent	Penetrating	
			Vinyl acrylic latex stains	Flat	Latex
Clear Finish: Doors, trim and misc.	Spar varnish, thinned	Phenolic-modified tung oil	Spar varnish, as packaged	High gloss	Phenolic-modified tung oil
Clear Finish: Clapboard siding, shakes, shingles, textured or rough sawn siding, fencing, decking, trim	None		Penetrating clear wood finish	Penetrating	Phenolic-tung oil modified linseed oil

Wood Staining Specifications - Exterior
Figure 17-5

Surfaces	Primer Selection	Type	Finish Coatings Selection	Finish	Type
Ferrous: Structural steel, storage tanks, doors, trim, sash, fire escapes, ornamental iron, catwalks, railings	Metal primer	Alkyd	Optional	Stain	Alkyd
			Optional	Soft gloss	Latex
			Enamel		Alkyd
	Zinc Chromate primer		Q.D. industrial enamel	High gloss	
			Enamel		Epoxy ester
Galvanized Iron: Gutters, leaders, vents, doors, ducts, framing, siding	Galvanized metal primer	Latex	Latex house paint	Low lustre	Latex
			Latex house & trim paint	Soft gloss	Long-oil Alkyd
		Alkyd	House paint		
			Enamel		Epoxy ester
			Q.D. industrial enamel		
	Zinc Chrome primer	Alkyd	Enamel		Linseed-coumarone indene
			Aluminum	Bright	
Aluminum metal	Zinc chromate primer	Alkyd	Aluminum	Bright	Linseed-coumarone indene
Factory finished aluminum siding	If required, base primer	Long oil alkyd	Latex house paint	Low lustre	Latex
			Latex house & trim paint	Soft gloss	
			Eggshell house paint	Eggshell	Alkyd
			House paint	High gloss	
Factory finished steel buildings	If required, metal primer	Alkyd	Enamel	High gloss	Alkyd
			Q.D. industrial enamel		

Metal Painting Specifications
Figure 17-6

Surfaces	Primer Selection	Type	Finish Coatings Selection	Finish	Type
Poured and precast concrete, cement and cinder block	Block filler, except under satin. Where required for repaint, use primer	Latex	Latex house paint	Low lustre	Latex
			Latex house & trim paint	Soft gloss	
			Satin	Satin	
Stucco, brick, unglazed asbestos-cement siding/ shingles, flexboard, concrete	None. Where required for re-paint, use primer	None	Latex house paint (2 coats)	Low lustre	Latex
			Latex house & trim paint (2 coats)	Soft gloss	
			Satin	Satin	
Concrete floors, platforms	Porch and floor enamel, thinned	Alkyd	Porch & floor enamel, as packaged	High gloss	Alkyd

Masonry Painting Specifications - Exterior
Figure 17-7

Surfaces	Primer Selection	Type	Finish Coatings Selection	Finish	Type
Painted/ Enameled: Doors, trim cabinets, ceilings, paneling, framing, sash	Alkyd enamel underbody or Latex enamel underbody	Alkyd Latex	Wall satin Flat Optional	Flat Eggshell	Latex Alkyd Latex
			Optional	Satin	
			Optional		
			Optional	Semi-gloss	Alkyd
			Enamel		
			Q.D. industrial enamel	High gloss	
Fire Retardant	Alkyd enamel underbody	Alkyd	Latex fire retardant paint	Flat	Latex

Wood Painting Specifications - Interior
Figure 17-8

bottoms of doors should be sealed with a clear finish coating.

Restaining—Surfaces that have been treated with wax or oily furniture polish should be cleaned with a solvent. Change wiping cloths often so you don't wipe the oil or wax back onto the surface. Scaling paint can be removed by scraping and sanding with fine grit paper. Glossy surfaces should be dulled with fine grade steel wool or sandpaper. Surfaces that need complete restoration must have the finish removed by power sanding or by treating with a paint and varnish remover. Figure 17-9 shows requirements for staining.

Plaster and Drywall
New surfaces—All plaster surfaces must be dry and

Surfaces	Primer Selection	Type	Finish Coatings Selection	Finish	Type
Stain/ Clear Finish: Doors, trim, cabinets, open roof decking, trusses, paneling framing, floors	See Product Descriptions		Stain Penetrating stain†	Flat	Alkyd
			Coating finishes Urethane finishes	Gloss/low Lustre/ flat*	Urethane
			Satin finish varnish	Satin	Alkyd
			One hour clear finishes	Gloss/low lustre*	Vinyl-toluene alkyd
			Latex finishes*	Gloss/satin	Latex
			Sealer finishes Q.D. sanding sealer●		Vinyl-toluene
			Oil finish	Satin	Castor-rosin ester

† Fill open-grained wood before staining, if desired.
* Not recommended for floors.
● Do not use as a finish on floors.

Wood Staining Specifications - Interior
Figure 17-9

Surfaces	Primer Selection	Type	Finish Coatings Selection	Finish	Type
Keene's cement plaster, textured/ sand finish plaster, drywall, composition/ wood pulp board	Alkyd primer Sealer*	Alkyd	Wall satin Flat Optional	Flat Eggshell	Latex Alkyd
			Optional	Satin	Latex
	or latex Q.D. prime seal	Latex	Optional		
			Optional	Semi-gloss	Alkyd
			Enamel		
			Q.D. enamel		
	Latex Q.D. prime seal or latex enamel underbody	Latex	Enamel	High gloss	Epoxy-ester
			Tile like enamel		Epoxy-polyester
Fire retardant	Latex enamel underbody	Latex	Latex fire retardant paint	Flat	Latex

* Not for new drywall/composition board

Plaster/Drywall Painting Specifications
Figure 17-10

clean before you begin painting. Let plaster cure for 30 days before painting. Fill cracks and voids with plaster filler and finish the filler to match the wall texture. The repaired surface should be spot-primed with a latex quick-dry primer seal before applying the overall coat of primer-sealer.

Gypsum drywall should be free of sanding dust before you start painting. Make sure the joint treatment cement is thoroughly dry. Repair damaged or defective joints with joint treatment compound. Coat steel corner beading with sealer before applying water-thinned coatings.

Repainting plaster and drywall—Remove any peeling or scaling paint. Sand these areas to feather the edges. Cracks, holes and blemishes should be filled and sanded flush, then spot-primed to control residual "bleeding." Wash greasy walls and ceilings with a strong detergent solution. Dull glossy areas by sanding. Ceilings with water stains should be sealed with primer. Figure 17-10 shows requirements for painting plaster and drywall.

Masonry
New masonry—Interior masonry surface preparation is the same as for exterior masonry. Follow the procedures previously discussed under that heading.

Repainting—Follow the same procedures outlined for surface preparation of exterior masonry. Figure 17-11 shows requirements for interior masonry painting.

Surface Problems
Under-Eave Peeling (Scaling)
Under-eave peeling is scaling that begins between paint layers. Top coats of paint peel away in paper-thin layers from previously painted coats underneath. Under-eave peeling generally occurs on surfaces protected from normal exposure to weather: under eaves, overhangs and on porches. (See Figure 17-12.)

This scaling results from a build-up of chemical substances on protected surfaces where rain cannot remove the accumulation. On vertical, unprotected surfaces these deposits are normally washed away by rain. But protected surfaces never benefit from a good shower, allowing salt deposits to accumulate. These deposits attract moisture even after new coats of paint are added to the surface. In freezing temperatures, moisture lingering in the salt layer freezes and expands, forcing the top coat of paint outward. The bond between layers breaks and the top coat peels away from the rest of the paint. (See Figure 17-13.)

Surfaces	Primer Selection	Type	Finish Coatings Selection	Finish	Type
Poured/ precast concrete, cement, and cinder block walls and ceilings	Block filler	Latex	Wall satin	Flat	Latex
			Alkyd flat		Alkyd
			Optional	Eggshell	
			Optional Satin Enamel	Satin	Latex
					Alkyd
	Latex, Q.D. prime seal*	Latex	Optional	Semi-gloss	
			Enamel	High gloss	Alkyd
			Q.D. enamel		
	Waterproofing masonry paint†	Vinyl tolune-butadiene	Enamel		Epoxy ester
			Tile-like enamel		Epoxy polyester
			Ceilings: Sweep-up spray finishes	Flat or eggshell	Alkyd
			Texture: Texture paint	Sand/ rough/ Spanish	Latex
Fire retardant (new unpainted block). New unpainted smooth surfaces or repaint	Block filler	Latex	Latex fire retardant paint	Flat	Latex
	Latex enamel underbody				
Concrete floors, platforms, stairs	Same as finish coat. (1st coat thinned)		Porch & floor enamel	High gloss	Alkyd
			Enamel		Epoxy ester
			Latex floor & patio finish	Satin	Latex-epoxy modified

* For smooth surfaces or repaint priming.

† For new/unpainted wall surfaces subject to excessive moisture. Particularly effective under solvent thinned coatings and primer for the tile like enamel.

Masonry Painting Specifications - Interior
Figure 17-11

Scaling may also occur in protected areas where gloss paints have been applied. Lack of weathering leaves these surfaces hard and glossy. A new coat of paint won't adhere to the slick surface. This condition is sometimes called "intercoat" peeling.

Correcting the problem: Under-eave surfaces and weather-protected areas should be washed with detergent and rinsed with a strong stream from a garden hose. When dry, remove all loose paint with a scraper or wirebrush. Sand glossy surfaces so the new paint will bond well to the old paint.

Usually under-eave peeling doesn't affect the underlying surface, so priming is not required. However, if you plan a drastic change in color, one or two coats of primer tinted toward the finish color may be needed.

Under Eave Peeling
Figure 17-12

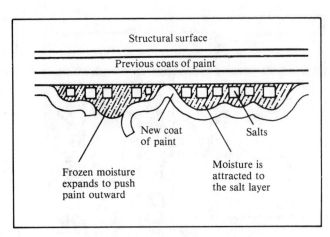

Beginning of Scaling Conditions
Figure 17-13

Paint Blistering
Figure 17-14

Paint Peeling
Figure 17-15

Blistering and Peeling

Blistering and peeling is common when gloss house paint is put on a wood surface that has not dried out. The moisture rises to the surface as the temperature changes, causing the paint to blister and eventually peel. (See Figures 17-14 and 17-15.)

If there is a blistering and peeling problem, excess moisture lingering in the wood surface is almost certainly the culprit. There are various ways that moisture can accumulate in the wood:

• Uncaulked cracks and joints allow moisture to seep into adjoining wood surfaces. (See Figure 17-16.)

• Worn out caulking that crumbles and falls out, no longer protecting the wood surfaces beneath from moisture.

• Leaking roofs, damp basements, loose siding and other defects that provide access for moisture.

• Ice or trash-choked gutters that force water up

339

Moisture Enters Thru Uncaulked Cracks and Joints
Figure 17-16

under shingles and into nearby walls.

• Moisture buildup inside the house, especially during the winter months, that condenses on the inside of cold exterior sidewalls.

Correcting the problem: The first step is to eliminate the source of moisture. Carefully inspect the outside of the house to find what caused the moisture or dampness. Next, repair the gutter, roof, or siding.

Check window and trim areas and other joints for caulking that might be in poor condition. Remove any loose or cracked caulking, prime the exposed wood, and allow it to dry. Without a sealing coat of primer, the wood will absorb the caulking and cause it to dry out. Apply caulk and smooth it out. (See Figure 17-17.)

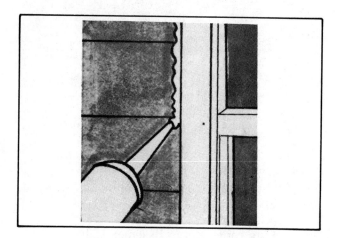

Seal Cracks With a Good Caulk
Figure 17-17

If you notice blistering and peeling paint near the masonry foundation, moisture is probably migrating from the ground through the masonry and into the wood. You'll need to waterproof the foundation area to keep this problem from happening again.

Excess humidity can be a problem in kitchen, bath, and laundry areas, causing blistering and peeling on nearby exterior windows and walls. A vent will stop the moisture buildup.

To prepare for painting, remove all loose paint with a scraper or wire brush. Sand rough surfaces smooth with medium-grade sandpaper. Countersink protruding nailheads and spot prime to seal against rusting. Fill nail holes, cracks and large imperfections with filler, then sand smooth.

Seal all bare wood with primer tinted to the finish color. Let the primer dry one or two days before applying the finish coat. If boards are damp, wait until they are dry before priming.

Non-Moisture Blistering
Some surfaces will blister even when there has been no accumulation of moisture. This is most common when deep color solvent-thinned house paints and enamels are applied in the direct rays of a hot summer sun. Dark colors retain heat which causes the paint to surface-dry quite rapidly, trapping the solvents before they have a chance to evaporate. These solvents eventually vaporize, raising the outer layer in a pattern of blisters.

Correcting the problem: These "dry" blisters are usually brittle and paper-thin and are easily removed by scraping and sanding. Generally the prime coat is not affected, so it isn't necessary to prime before repainting. You can prevent non-moisture blistering by planning the progress of your paint job. Be sure direct sunlight does not fall on a freshly painted surface. The best method is to paint a side after the sun has moved around the house.

Peeling Gutters and Downspouts
New galvanized gutters and downspouts are coated with a fabricating oil that must be removed prior to painting. Otherwise the oil will interfere with paint adhesion, and peeling will develop. (See Figure 17-18.)

Eliminating the surface oil is your objective. If you plan to paint immediately, remove all traces of surface oil. Moisten a rag with mineral spirits and thoroughly wipe all surfaces to be painted. Change to fresh rags frequently to thoroughly remove all oily residue.

Aluminum gutters are factory coated and should

Peeling Due to Improper Surface Preparation
Figure 17-18

Checking — Paint Has Lost Elasticity
Figure 17-19

stay in good condition for several years. When it's time to paint the aluminum, use the same procedure as for previously painted galvanized gutters. Here are the steps to follow:

Remove all loose paint with a scraper or wirebrush. Sand edges to smooth the transition from sound paint to bare spots. If rust is present, sand away loose particles and prime exposed surfaces with a rust inhibitive paint. When the gutter is dry, prime the entire surface with a metal primer. Then apply the finish coat.

Checking and Cracking

Checking is a pattern of short narrow breaks confined to the top layer of paint. Cracking occurs in later stages as the breaks in the paint deepen to form a split that reaches the underlying wood surface. The split film allows moisture to enter and eventually loosen the paint from the surface. (See Figures 17-19 and 17-20.)

Checking and cracking usually develop as paint begins to lose its elasticity. Newly applied paint is remarkably flexible, capable of expanding and contracting with the wood surface in response to changes in temperature and humidity. On older homes which have been painted many times, the paint layers become brittle with age and no longer expand and contract with the wood surface. As the wood swells, stress breaks the bond between layers

to form checks.

Correcting the problem: Remove as much loose paint as possible with a scraper and wire brush. Smooth rough surfaces and bare wood with

Checking Leads to Cracking
Figure 17-20

Excessive Chalking Indicates Time to Repaint
Figure 17-21

medium-grade sandpaper. In some cases the best solution may be to remove all paint down to the bare wood. This avoids the unevenness that can result from removing many layers of paint in some areas and none in others.

Inspect the house for loose boards to be renailed. You may need to replace an occasional board that shows signs of deterioration.

Countersink new and exposed nailheads and prime. When dry, fill the depression with filler. Spot prime if your finish is a latex product. This will minimize nailhead staining. Where necessary, scrape out old caulking, prime the area, then recaulk.

You'll need to prime all bare wood with one or two coats of primer tinted to the color of the finish coat.

Excessive Chalking
Chalking is a powdery residue on the surface. It is the normal result of weathering and occurs in time with all paint. On most surfaces moderate chalking is helpful, particularly with white. When it rains, dirt and dust are washed away with the chalky residue, keeping the painted surface relatively clean. Chalking also prepares the surface for repainting by gradually reducing the thickness of the outer paint layers. This minimizes paint buildup

and the possibility of future checking and cracking. Normal chalking will usually leave traces of paint dust on your fingertips. (See Figure 17-21.)

To test for excessive chalking, wipe a gloved hand over the surface. If paint dust covers your entire hand, it's time to think about repainting. Paint applied over a heavily chalked surface without removing the chalk residue cannot bond firmly to the undersurface and will soon start to peel.

Correcting the problem: Use a stiff fiber or wire brush to remove the powdery deposits. Then spray with a strong stream from a garden hose to wash away the loosened dust. (See Figure 17-22.)

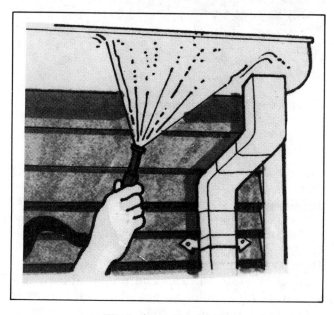

Wash Away Loose Dirt
Figure 17-22

When the surface is thoroughly dry, apply primer to uniformly seal the porous surface. If you don't prime, the finish coat may lose gloss or show a noticeable color difference in spots that are more porous than others.

Wrinkling
Solvent-thinned paints can develop a random pattern of bumps and depressions in the painted surface. This wrinkling usually happens when something interrupted the normal drying of the paint. To form a smooth, protective film, a coat of paint must dry at a regular rate all the way through. If paint is put on too heavy and not brushed out thoroughly, the surface will dry while

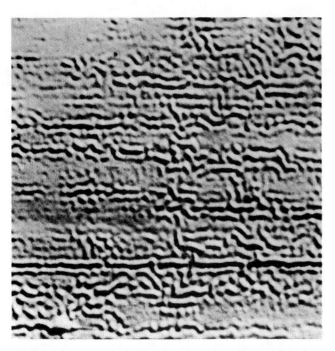

**Wrinkling Needs To Be Removed Before Repainting
Figure 17-23**

the layers below remain wet. This lets the surface film move and form wrinkles on the still soft undercoat. (See Figure 17-23.)

It's tempting to put on a thick coat, especially in cool weather when the paint thickens by itself. But putting on a heavy coat is always a mistake. Avoid thick application in cool weather. And don't thin the paint; just make it more spreadable by warming the can in a pail of warm water.

Wrinkling also occurs when the temperature is very high. Excess heat dries the surface much faster than normal, sealing a wet layer of paint underneath.

Correcting the problem: Start by removing the wrinkled layers of paint. Scrape and sand away paint until you get down to a smooth surface.

For difficult-to-remove paint accumulations (such as in tight corners) use paint remover to soften the paint and make scraping easier.

If wrinkling has occurred in isolated areas, feather sand edges of spots where you have removed the paint to smooth the transition to the sound paint surfaces.

Mildew

Mildew is a fungus growth that can disfigure painted surfaces in almost any climate. It's common where humidity is high and temperatures are warm. Houses in southern and coastal regions are particularly vulnerable to mildew. Often thick shrubbery and trees close to the house will block out the sun, providing the dampness mildew needs.

Sometimes it's hard to distinguish mildew from dirt accumulation. One way to be certain is to soak a piece of cloth with household bleach and dab it on the surface without wiping. If it's mildew, the spots will bleach out in a few minutes. Ordinary dirt won't be affected by the bleach.

Correcting the problem: Mildew must be removed completely or the spores will continue to grow right through the new paint. Prepare a solution of 1 cup trisodium phosphate or other powdered detergent (don't use ammonia-containing detergents), and 1 quart of household bleach in 3 quarts of warm water.

Scrub the solution into the surface and allow it to remain for five minutes. Then thoroughly hose the surface with clean water. Where mildew is a persistent problem, clean the surface with high-pressure steam equipment.

A coat of primer is generally recommended after surfaces have been treated for mildew and have dried thoroughly. A prime coat will seal the surface and provide a firm foundation for the finish coat.

Surface Discoloration

Brownish stains sometimes appear on redwood or cedar siding. These woods have natural water-soluble staining agents. Moisture dissolves the compounds and moves them to the surface. There the moisture evaporates, leaving a stained, discolored surface. (See Figure 17-24.)

If a prime coat on wood siding or shingles is put on carelessly, some under edges will not be covered completely. Eventually rainwater will draw out the water-soluble compounds, causing stains to appear under the edges of shingles or under the overlap of siding.

Correcting the problem: Here's how to stop further discoloration. Coat the surface with one or two coats of primer tinted to the color of the finish coat, and allow the primer to dry for several days. If some spots still have a stained appearance, reprime them. Pay special attention to the areas between rows of shingles and siding. Spread enough paint under the laps, and be sure shingle edges are completely coated.

Paint over stained surfaces with one or two coats of solid color stain. If the present color is a light shade, consider using a darker tone that will make future staining less obvious.

Surface Discoloration
Figure 17-24

Weathered Unpainted Shingles

Some homeowners leave new shingles unpainted and untreated to get a naturally weathered look. Eventually the shingles will darken. This discoloration will be uneven and darker on the shady side of the house and under the eaves. When this weathered look begins to annoy your customer, the shingles can be stained with a solid color vinyl acrylic latex stain. Wire brush the surface to remove any loose wood particles that would cause the stain you apply to flake off. Generally, one coat is all that's needed unless it's a very light color.

Surface Preparation

Careful preparation of the surface is essential for every painting job. Skipping over this stage to save time and money is always a mistake.

To get the best performance from paint, start with a smooth, clean surface that is caulked and filled where needed. Take time to identify existing surface problems. Let's run through the preparation procedure:

1. Put drop cloths in place under areas where you are working. Cover sidewalks as well as shrubs. This will save cleanup time later.

2. Be sure all surfaces are smooth and free of cracks and crevices. Countersink exposed

nailheads and spot prime. Fill nailhead depressions, cracks and surface imperfections with putty.

3. Lightly sand glossy surfaces with medium-grade sandpaper, particularly trim areas on the north side of the house and protected surfaces under eaves and overhangs, to provide better anchorage for subsequent coats of paint.

4. Clean off dust and dirt with a cloth or soft brush. Loosen any chalky residue with a stiff brush and remove it with a spray of water from a garden hose.

5. Remove any mildew with a scrubbing solution. Hose down the surface and allow it to dry thoroughly before priming.

6. Hose down the surfaces under eaves and overhangs to remove salt accumulations.

7. Scrape out old caulking with a putty knife and prime exposed wood. Replace old oil-base caulking with quality grade caulking for better flexibility and longer life.

8. Prime all exposed areas and surfaces where paint has worn very thin with a primer tinted to the approximate color of the finish coat. Be sure to allow the primer enough time to dry; one or two days may be necessary.

Choosing the Primer and Finish Coat

The primer and finish coat work hand-in-hand on exterior painted surfaces. Together they are known as the paint system.

Primer

The primer's job is to seal the structural material to prevent absorption from the finish coat of paint, and to protect the finish coat from damage caused by chemicals in the material painted. It also provides an ideal surface to which the finish coat can adhere.

There are two types of primers: general purpose sealing types for wood, metal and masonry, and anti-corrosive primers to protect metal against rust and corrosion.

Primers usually come in white. But they should be tinted to the color of the finish coat. This avoids a color contrast that may be difficult to cover later.

All new, unpainted surfaces of wood, metal and masonry will need a coat of either a general primer or a rust-inhibitive primer. You'll need to prime previously painted surfaces where the bare wood or metal is exposed after you have removed deteriorating paint or loose rust.

Finish Coat

The finish coat of paint is designed to protect the

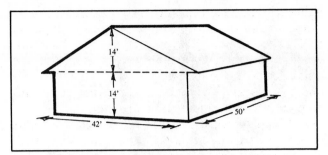

Before you can bid, you must know how much paint is required for the job.
Figure 17-25

surface from weathering. It also provides the color and texture that make a new paint job so attractive.

Flat and low-lustre paints can be used on most surfaces of wood, metal or masonry. These finishes will not highlight rough, uneven or textured surfaces.

Soft-gloss and gloss house paints are best for smooth surfaces of wood or metal. They are not recommended for rough-surfaced shingles and other textured surfaces because of the unattractive highlights that develop.

Painting the Outside of a House

Figure 17-25 shows a house you want to paint. How much paint will it take? Figure 17-26 shows how to figure the job.

To prevent drips and spatters from spoiling previously painted areas, work from the top down, doing gutters and eaves first if they are to match the siding color. (See Figure 17-27.)

Work from side to side. It makes no difference whether you work from left to right but to minimize lapping, coat 4 or 5 boards across the house until completed (instead of from top to bottom) before moving the ladder. (See Figure 17-28.)

When using a roller, pour a small amount of paint into the well of the roller tray. Work paint into the roller by rolling it in the paint in the tray. Be sure the entire roller surface is covered. Remove excess paint by rolling it over the ribbed portion of the tray. Apply paint in light even strokes, rolling first in one direction, then over-rolling in the other,

---

To calculate the amount of paint needed, you must first determine the number of square feet of surface to be painted. The following steps provide a simple formula for determining square footage:

1. Figure the general area to be painted.
A simple formula for an entire house would be:
A. (Length of house + width of house) x 2 = total distance around house.
B. Distance around house x height of house = total sidewall area.
C. (½ gable height x width of gable base) x 2 = total gable area.
D. Total sidewall area + total gable area = total surface to be painted.
Note: You don't need to subtract window and door areas from the total surface figure if each opening is less than 100 sq. ft.

Example:
A. (50' + 42') x 2 = 184'
B. 184' x 14' = 2,576 sq. ft.

C. ½ (14') x 42' x 2 = 588 sq. ft.
D. 2,576 sq. ft. + 588 sq. ft. = 3,164 sq. ft.

2. Figure special areas:
These simple formulas will help you estimate paint requirements for the following special areas. Add the figures you get for each area together for the total square footage.

Balustrades
Measure the front area and multiply by 4.

Stairs
Count the risers and multiply by 8.

Lattice Work
Measure the front area and multiply by 2.

Porches
Multiply the length by the width

Cornices
Measure the front area and multiply by 2.

Eaves
Measure the areas and multiply by 2.

Gutters & Downspouts
Measure the front area and multiply by 2.

Eaves with Rafters
Measure the area and multiply by 3.

3. Total surface figures to determine gallons of paint required.

To do this, divide the total surface area by the spreading rate (coverage rate) of a gallon of the paint of your choice. Spreading rates are usually included on the direction panel. Figure primer and finish coat needs separately.

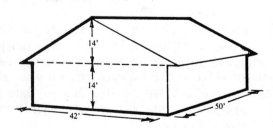

How Much Paint Will Be Required?
Figure 17-26

Work From Top Down
Figure 17-27

Work From Side to Side
Figure 17-28

Roll in Even Strokes
Figure 17-29

as shown in Figure 17-29.

When using a brush, take care to coat under the edges of the clapboard (Figure 17-30). "Feather" the ends of your brush strokes to assure smoothness where one painted area joins another. Don't bear down too hard on the brush. (See Figure 17-31.)

Trim and shutters are usually painted last, often in contrasting colors and with a different product. If shutters can be removed, paint them separately and replace them when the rest of the job is completed. Paint window sash and recessed parts of the window frame first, then the frame and the sill. A good painter only uses trim brushes for the closest work. For most work he turns a large brush sideways and paints a "trim" line with room to spare. (See Figure 17-32.)

Clean off dust and dirt with a sturdy cloth or stiff brush. (See Figure 17-33.) If there's a nail protruding, drive it below the surface of the wood with a nail set (Figure 17-34). Spot prime these areas; when dry, fill them with putty. Seal any knots or pitch spots with an appropriate knot sealer to avoid brown spots later on.

Look for scaling, flaking paint. Where old paint is damaged, scrape to a sound surface with a broad knife (Figure 17-35) and wire brush. Smooth all rough areas with sandpaper.

Coat Under Edges of Each Board
Figure 17-30

"Feather" Ends of Brush Strokes
Figure 17-31

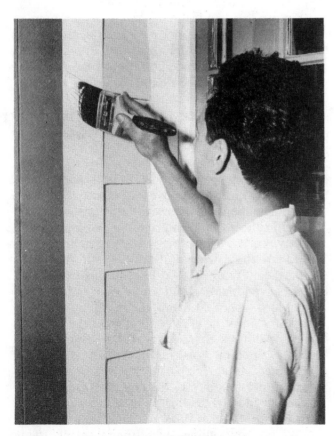

Paint Trim and Shutters Last
Figure 17-32

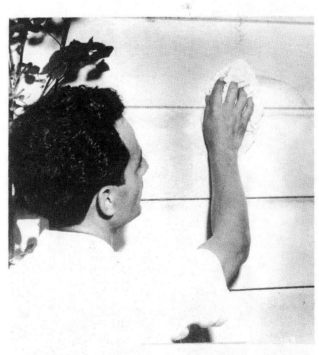

Clean Off The Dirt Before Starting To Paint
Figure 17-33

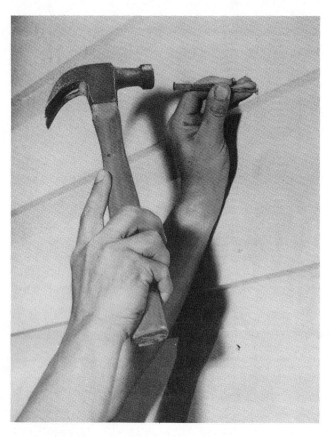

Set Nails and Putty
Figure 17-34

Scrape Off Damaged Paint
Figure 17-35

Remove Rust From Gutter and Prime
Figure 17-36

Remove rust and peeling paint by scraping or wire brushing. Wash protected areas—under eaves for example—with water. A strong stream from a garden hose does a good job. Apply primer when the surface is dry. (See Figure 17-36.)

Painting Walls, Ceilings and Trim

Carefully examine surfaces for nailholes, cracks and gouges. These have to be corrected before the paint goes on.

Rake out large plaster cracks in walls and ceilings with a putty knife to remove loose particles (Figure 17-37). Dampen those cracks and any large wall and ceiling holes with water before patching. Using a putty knife, firmly press latex patching compound into the crevices and smooth until flush with the surface (Figure 17-38). Fill mitered trim joints which have opened, and door or window trim which has separated from the wall, by pressing compound into the crevices and smoothing with a finger. Allow them to dry, then sand lightly. Because patching compound shrinks as it dries, large holes and cracks usually require a second application of compound after the first has dried.

Remove loose or scaling paint with a putty knife. (See Figure 17-39.) Where it has been removed from walls or ceiling, sand paint edges for a smooth surface. If it has been removed from sash, trim or doors, sand the entire surface with fine

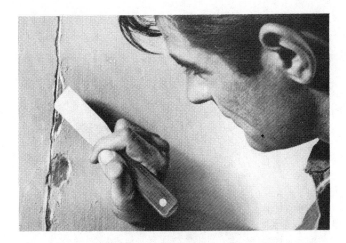

Rake Out Plaster Cracks
Figure 17-37

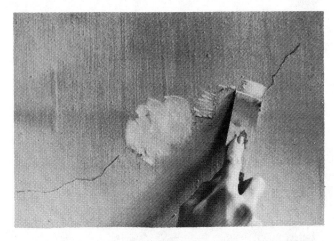

Use Patching Compound To Seal Cracks And Holes
Figure 17-38

Remove Loose And Scaling Paint With a Putty Knife
Figure 17-39

emery cloth or sand paper. Spot-prime patched wall surfaces and wood surfaces where sealing paint has been removed to expose raw wood.

Start where the ceiling meets the wall. Apply the coating along the joint of the ceiling and wall, beginning in a corner of the room. Coat a strip about 4 inches wide along the perimeter of the ceiling and the walls beneath it. (See Figure 17-40.)

Now, with a roller, begin near a corner of the ceiling, blending the coating into the ceiling line painted previously. (See Figure 17-41.) Paint across the width rather than the length of the ceiling. Don't stop until the ceiling is completely covered.

When the ceiling is dry, return to the spot where you began and, using a trim brush, "cut in" the wall/ceiling line. Then coat along the baseboard and around the door and window trim. Paint one wall at a time. Return to your starting place and, using the roller, coat the wall with slow, even strokes to minimize spattering. Paint the remaining walls, following the same procedures. (See Figure 17-42.)

Painting Double Hung Sash
Remove the sash lock. Lower the upper part of the sill and raise the bottom sash out of the way. Coat

Begin Painting In A Corner
Figure 17-40

Roll Across Width of Room
Figure 17-41

Use Slow, Even Strokes, Over Rolling The Paint Out
Figure 17-42

the window sash, then the rails. Don't paint the sash tracks. Return the upper sash to a near-closed position, lower the bottom section and paint it the same way. To complete the window, coat the check rails, frame and sill. The best tool for this work is a 2- or 2½-inch angular sash brush (Figure 17-43). It's suitable for flat trim work as well. Leave upper and lower sash slightly open until paint has dried completely. When dry, remove any paint smears or spatters from the glass with a razor blade.

Painting Doors

Remove all the hardware. Open the door and place a block of wood on the floor between the door and its casing to brace it open. If the door is paneled, first paint the panels, then the horizontal sections and finally, the vertical sections. (See Figure 17-44.) If the door is the flush type, begin at the top and coat about 1/3 of the door. Continue until completely coated. Keep the brush fully loaded and work rapidly, always brushing into the wet areas. Coat the door edgings. Keep a rag handy to stop paint that might run onto the other side of the door.

Painting Door and Window Trim

Using the edge of the sash brush, carefully coat the

Use The Right Brush For The Job
Figure 17-43

350

There's A Right And Wrong Way
Figure 17-44

Save The Trim For Last
Figure 17-45

edge of the window trim nearest the wall. Then paint the facing of the trim. (See Figure 17-45.) Coat door trim the same way. Extend the color over the door casing to include the door stop.

Baseboard and Molding

These are the last trim surfaces painted. Protect the floor or carpet with a metal trim guard or rigid piece of cardboard. (See Figure 17-46.) Before coating the shoe molding, brush the baseboard surface a few times to remove any excess paint which might run between the trim guard and the molding. Wipe paint from the trim guard before moving it to an adjacent location. Immediately remove any paint that falls on the rug or floor with a rag dampened with the appropriate solvent—either water or mineral spirits.

Summary

Most remodelers do their own painting. You're not likely to build an addition or rework an old house from foundation to ridge and then let someone else take over the "light" work.

But painting requires know-how just as the other

Protect The Finish Floor
Figure 17-46

trades do. It is the "finish" of a lot of hard work and planning. It must be a professional job, with the proper attention given to the many small details involved. Not everyone who walks around with a paint brush stuck in the hip pocket of his white overalls is a qualified painter. And qualified painters are not minimum wage workers, contrary to what some homeowners may think. Painting is a highly skilled trade, and painters get paid accordingly. So, if you do your own painting, charge the going rate plus your regular overhead and profit.

From here we go to additions. But hold onto your paint brush because we'll need it over there, too.

 Old Carpenter's Rule

 Above ground is sound. Below ground is waterbound.

18

Basement Conversions

The addition market is wide open. The small builder who concentrates on additions is usually well paid for his efforts. The competition is stiff in some areas, but careful planning and evaluation will keep you hip deep in good, profitable jobs. If you do quality work, you don't have to worry too much about competition.

Basically there are three types of additions: downward, outward and upward. There are also three basic types of customers: downward, outward and upward. The type of addition is not necessarily the same as the customer. There is one customer trait that might be found in any one of the three types: the "I got" syndrome. It works like this: "I got the carpet thrown in at no extra charge." Or, "I got him to add that hand-carved mantel from Spain at half price."

I don't know whether it's strictly an American trait or if it applies world wide. But a builder can only tolerate so many "I gots." I don't mind a couple. Most people are happy if they can get you on a few items. I usually allow for at least one "I got" in every job, but one has to draw the line somewhere. I got to live, too!

Some people can be downright nasty about their "I gots." They don't care what the contract or agreement says, it's just a piece of paper. "I've got to have everything I can get" is the name of the game.

So, the smart thing to do is let them slip an "I got" or two by you and everybody will be happy. If you really want to please your customer, drop a few hints toward the end of the project that you'll be lucky if you break even on the job. That way they'll recommend you with the greatest enthusiasm to their friends: if you only broke even on their job, you'll surely go bankrupt on the next one.

Basement Conversions

Basement conversion is an example of a "downward" addition. Of course, you're not really "adding to" the structure. If you dig a basement under an existing house adjacent to the foundation, the entire structure would probably collapse. With basement conversions the idea is to improve or increase the habitable space of the building.

We've already discussed paneling a basement, covering a basement floor with tile and applying ceiling material. But we haven't discussed basement construction in detail. Let's take a look at that.

Basement Construction

Excavation of the area around the basement should allow enough room for laying and sealing the exterior wall.

The basement should be designed and constructed to adequately protect against moisture penetration. In locations with a high water table or where surface or ground water drainage is a problem, extra precautions must be taken. Remember that the basement is a bathtub in reverse. It will have to be tight as a ship.

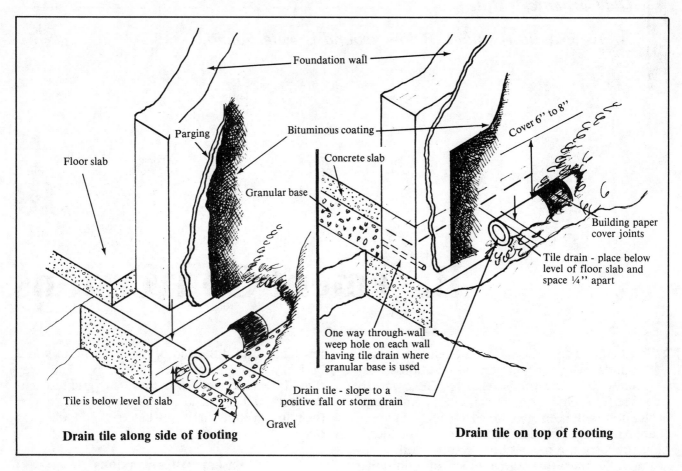

Drain Tile along side of footing

Drain tile on top of footing

Drain Tile
Figure 18-1

Foundation Drains

Foundation and footing drains should be provided around foundations enclosing the basement or below grade of habitable living space. Drains should be installed at or below the area to be protected and should discharge by gravity or by mechanical means to a suitable discharge area, such as a ditch or storm drainage system. Bituminized-fiber pipe, clay or concrete drain tile work well for this purpose.

The top joints of the drain pipe should be protected with strips of building paper. After the building paper is applied, cover it with 6 or 8 inches of coarse gravel or crushed rock. A minimum 2-inch layer of gravel or crushed stone should be under the pipe. Figure 18-1 illustrates drain tile installation and dampproofing.

Dampproofing

The exterior foundation walls of a basement masonry wall should have at least 1 coat of portland cement parging from footing to finish grade, with a 3/8-inch minimum thickness. Apply at least 1 coat of bituminous dampproofing material over the parging. Follow the manufacturer's instructions.

Where habitable rooms below grade must be enclosed with waterproofed walls, install a waterproof membrane extending from the edge of the footing to the finish grade. The membrane should be 2-ply hot-mopped felts, 6-mil polyvinyl chloride, 55-pound roll roofing or similar material. All laps should be sealed. The membrane must be firmly affixed to the wall.

Be sure there is a continuous membrane below the slab. The membrane should be turned up at the edges, to the top of the slab. Place a 4-inch-thick base course under the slab below the dampproofing membrane. This base course should be clean graded gravel, crushed stone or crushed blast furnace slag. All vapor barriers should comply with prescribed codes, and should have at least a 6-inch

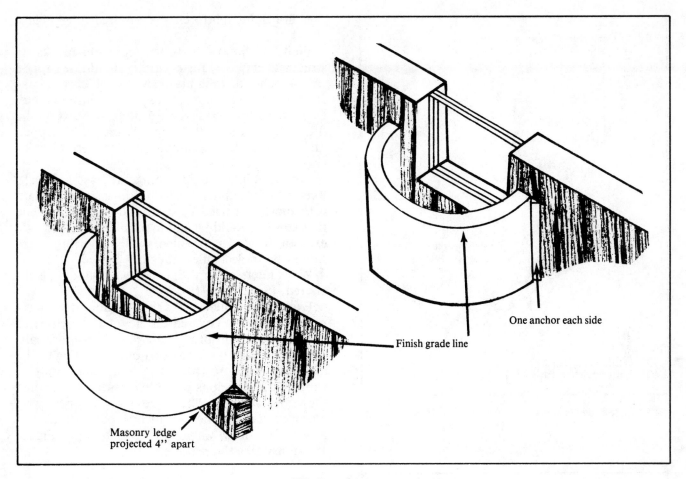

One anchor each side

Finish grade line

Masonry ledge
projected 4'' apart

Window Areaways
Figure 18-2

sealed overlap.

A membrane consisting of two layers of 15-pound asphalt-saturated felt mopped together and mopped over with mopping asphalt usually makes a tight seal. Windows should have areaways. Figure 18-2 illustrates one method of installing areaways at windows.

Masonry foundation walls supporting wood framing should extend at least 8 inches above the finish grade. Generally, the thickness of the foundation wall should not be less than that of the wall supported. The minimum thickness should be enough to resist lateral pressure from adjacent earth and to support design loads. Figure 18-3 offers some figures for a conventional wall where soil conditions are average.

Exterior walls of masonry veneer over frame construction may be corbelled 1 inch maximum. (See Figure 18-4.)

The upper portion of foundation walls may be reduced in thickness to 4 inches to permit veneer application. When the 4-inch portion exceeds 4 in-

Foundation Wall Construction	Maximum Height of Unbalanced Fill (Feet)[1]	Minimum Thickness (Inches)	
		Frame	Masonry or Masonry Veneer
Hollow Masonry	3	8	8
	5	8	8
	7	12	10
Solid Masonry	3	6	8
	5	8	8
	7	10	8
Plain Concrete	3	6[2]	8
	5	6[2]	8
	7	8	8

[1]Height of finish grade above basement floor or inside grade.
[2]Provided forms are used both sides full height.

Foundation Walls
Figure 18-3

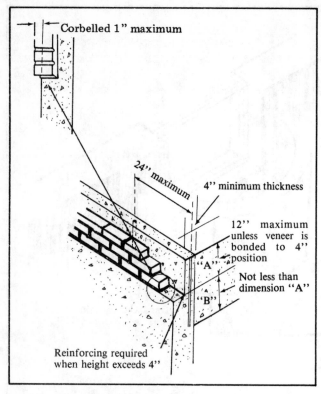

Reduced Thickness - Foundation Walls
Figure 18-4

ches in height, reinforcing is required. (See Figure 18-4.)

Walls of hollow masonry (concrete blocks or structural clay tile, for example) should be capped with 4 inches of solid masonry (cap blocks) or concrete or the top course cells filled with concrete. Capping isn't needed where a sill plate bears on the full width of the wall. The walls should be capped under girders with a minimum thickness of 6 inches.

Mortar joints should average at least 1/2 inch. Keep all joints under 3/4 inch. Block should be laid with mortar applied to head and bed joints. The first course should be laid in a full mortar bed. All exterior joints in basement walls of living areas should be tooled unless they are parged.

Walls intersecting with other walls should be anchored. (See Figure 18-5.)

Existing basement walls that need a little dampproofing may be covered on the interior by one of the many waterproofing materials, such as asphalt. They are brushed or troweled on, according to the manufacturer's instructions. You can apply a vapor barrier such as polyethylene directly against the asphalt paint. Then add furring strips with masonry nails, concrete anchor bolts or a stud driver. The finish wall can then be applied. Figure 18-6 illustrates the process.

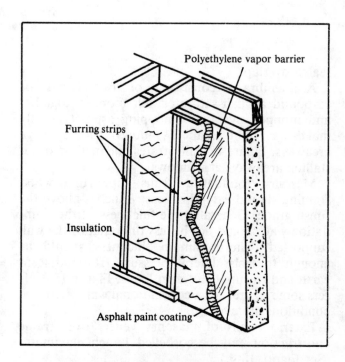

Basement Wall Dampproofing with Asphalt Paint
Figure 18-6

Method of Anchoring Masonry Walls
Figure 18-5

Summary

If the basement leaks water through the walls or floor, the problem is probably faulty construction, and little can be done about it from the inside. If the wall leaks, it must be repaired from the outside by removing the dirt and locating the problem. There may not be a drain pipe, or the walls may need sealing by mopping and installing a water barrier as previously discussed.

If the floor leaks, it may require mopping with asphalt, installing a vapor barrier, and pouring a thin slab over the floor. But it may still require drains at the footing, or water will eventually work up through the joint where the wall and floor meet.

Basements can be a constant problem. Solving the problem will often require removing the earth from around the basement walls and laying drain tiles with runoff to a suitable storm drain or ditch. If this is done, re-mop the walls and waterproof as previously discussed.

If the house is in an area with a high water level, the problem may not have a solution. If this is the case, consider passing up the job. Go to work where problems can be solved.

19

Attic Conversions

Some houses have enough attic space for additional rooms. Usually there are two deficiencies: not enough window space and not enough floor joist support. But these aren't difficult to remedy. Dormers can be built for windows, and floor joists added for required strength.

Usually, building codes require a floor design live load of 30 pounds per square inch for attics used as sleeping quarters. Some ceiling joists are designed for a 10- or 20-pound per square inch live load. Usually, depending on the joist span, you can double the existing joists and be in the ballpark.

You need a minimum height of 7 feet 6 inches from floor to ceiling, at least over one half of the required room width. Knee walls must be at least 5 feet high for general use. (See Figure 19-1.)

A good method of gaining more room in the attic is with a shed dormer. (See Figure 19-2.)

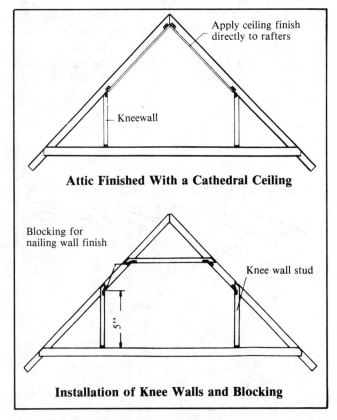

Attic Finished With a Cathedral Ceiling

Apply ceiling finish directly to rafters

Kneewall

Blocking for nailing wall finish

Knee wall stud

5"

Installation of Knee Walls and Blocking

Two methods of Converting an Attic
Figure 19-1

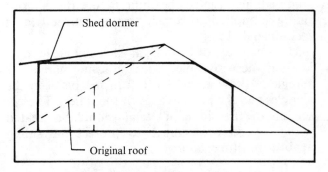

Shed dormer

Original roof

Shed Dormer for Additional Attic Space
Figure 19-2

Gable Dormers

The gable dormer, Figure 19-3, is another way to gain more head room and window space. The roof of the gable dormer usually is given the same pitch as the main roof. When locating the dormer, let both sides fall over an existing rafter, if possible. If not, add an extra rafter at that point. Then double the rafters to provide support for the side studs and short valley rafters. Tie the valley rafters to the existing roof framing with a header as illustrated.

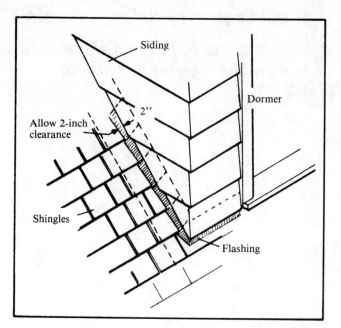

Flashing at Dormer Walls
Figure 19-4

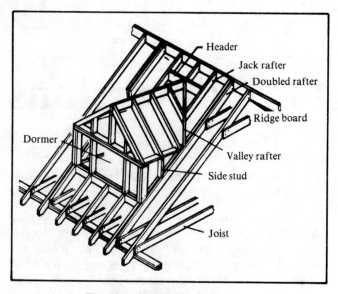

Framing for Gable Dormer
Figure 19-3

Figure 19-4 shows the proper flashing of dormers. Do it right or you'll have problems that can plague you forever. Use felt and metal flashing (felt first, metal on top). Install felt and flashing under the shingles on the existing roof and seal with asphalt roofing cement as was discussed in the section on roofing. Turn the flashing up on the stud wall against the sheathing before the exterior siding is applied. A 6-inch turn-up on the wall is recommended.

Use one piece of flashing at each shingle course, lapping successive pieces in the same manner as shingles. Allow about 1½ to 2 inches between the shingles and the bottom edge of the siding. The cut section of the siding material should be treated with a wood preservative to protect the siding against moisture damage.

Shed Dormers

Shed dormers can be built to extend all the way across the width of the house. They can be on the front or back of the house, depending on the particular style of the building. A lot of people consider the shed dormer a type of "lean-to," and prefer it to be on the back of the house. Sides of the shed dormer are framed the same way as gable dormers. (See Figure 19-5.)

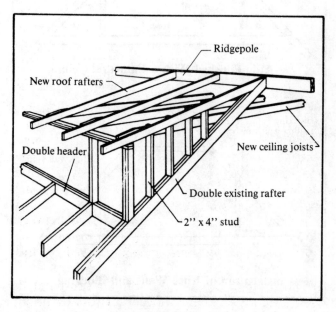

Framing for Shed Dormer
Figure 19-5

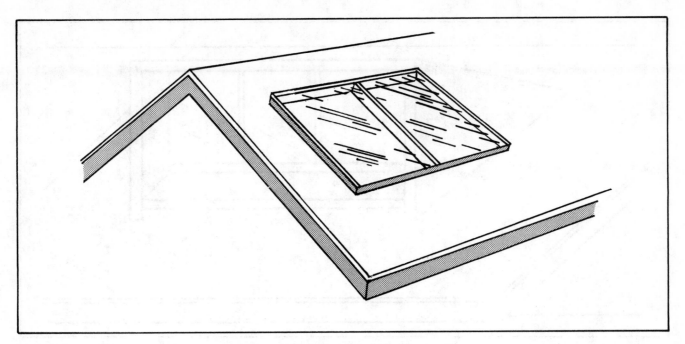

Skylight Installation for Attic Conversion
Figure 19-6

Low-slope roof rafters are framed directly into the ridge plate. The ceiling joists rest on the outer wall of the dormer, with the opposite ends of the joists securely nailed to the main roof rafters on the other side of the ridge plate. The low-pitch roof should be covered as discussed for low-slope roofs in the section on roofing.

Follow standard construction procedures in framing attic rooms except as noted above. Attic walls must be no lower than 5 feet (knee walls). Studs are nailed to a sole plate in the same way as other partitions. Blocking is nailed between the rafters and studs at the top of knee walls to provide a nailing surface for the wall and ceiling finish.

Collar beams are nailed between opposite rafters to serve as ceiling framing. (See Figure 19-1.) Again, these should be at least 7 feet 6 inches above the finish floor. Nail blocking between the rafters and collar beams for a ceiling nailing surface. For a cathedral ceiling, the collar beams may be left off and the ceiling finish applied to the rafters. In this case ventilation space must be provided between the rafters, from the knee wall closet area to the ridge plate. A ridge vent is recommended for this arrangement.

The strength of the roof framing must not be sacrificed when adding dormers. Cross-partitions or some other suitable bracing is required.

A chimney passing through the attic can be enclosed at a closet area or otherwise worked into the layout to be as inobtrusive as possible. Keep framing and combustible materials at least 2 inches from the chimney.

Attic conversions are ideal for skylight illumination. See the section on skylights for installation instruction. (See Figure 19-6.)

Windows in a gable end offer a pleasing solution to window requirements. A light scoop is another alternative. (See Figure 19-7.) Light scoops are shed-dormer type construction and do not offer much more headroom.

The important thing to remember in attic conversion is to reinforce the floor with joist members to guard against squeaks and "give."

Adding a Second Story

There are few one-story houses that won't support a second story. If the house has 2 x 4 studs on 24-inch centers or less, it will support a second story. The important thing is for the existing footing to support the additional load. Code requirements and the soil type are the key considerations here. You'll probably need the help of an engineer. Generally, a footing 6 inches thick with a 3-inch projection on each side is adequate for a two-story house without a basement. A 6-inch-thick footing with a 4-inch projection on each side is adequate for a masonry veneer two-story house

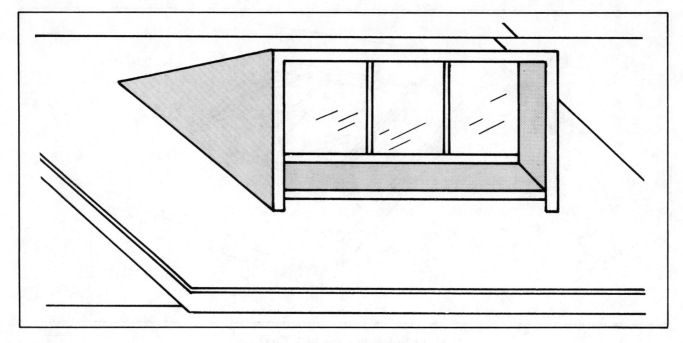

Light Scoop Attic Conversion
Figure 19-7

with no basement.

In adding a second story, the roof and roof framing of the existing house are removed down to the ceiling joists. The ceiling joists are doubled or larger joists such as 2 x 10's are added and the second floor built from that point with standard construction procedures. The big problem in adding a second story is the weather. The job calls for a crew large enough to get the second story sealed before it rains. That might not be easy in an area of sudden cloudbursts.

Adding a second floor to a one-story garage is another matter. Again, the footing must be checked to determine if it is heavy enough. The roofing material and roof framing must be removed and the joists reinforced or new and larger members installed. This type of addition is ideal where the one-story garage is joined to an existing two-story house. (See Figure 19-8.)

Summary

Your building code probably has a section on additions. That's the place to start. Of course, most areas require inspection of plans and specs before issuing a building permit, and the code will often limit what can be done on a particular house in a particular area. Be aware of what the code requires. Nothing impresses your customer more than your technical knowledge of what the building department will require on the job he wants done.

When you bid on an attic conversion job, keep in mind the labor and time involved in getting the materials up there. Considerable time is also required in providing extra support for the floor. Labor and time are important factors in attic conversion.

Remember, too, that some jobs don't have room for cutting lumber where it's needed. If you have to cut lumber outside and take it inside, carpentry manhours will include several hours of walking back and forth. Keep that in mind when preparing your bid. Every step in building takes time and effort. That's what you're charging for.

What formula do you use to find the total manhours required to finish the job? In remodeling a house, about all you can rely on is experience. There are so many variables involved, and each remodeling job is different. Each has its own problems. Building a house is like fitting the pieces of a puzzle together. Remodeling a house is like separating two puzzles and then piecing them together. Only experience can teach you the number of manhours required.

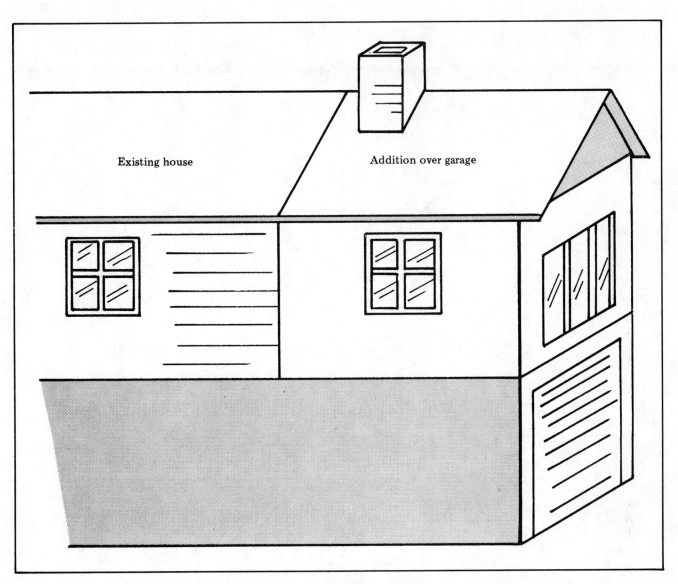

Adding A Room Over A Garage
Figure 19-8

20

Adding Outward

Where space on the lot is available and codes permit, the outward addition is by far the most popular choice. It keeps the work out of the house until the last phase of the construction, when access is provided between the house and the addition and the tie-in finish is made. There are many ways to add on to a house at ground level.

It is important that the addition "fit in" with the existing house—that the addition not have that "add-on" look—and that the general appearance of the house will be improved with the new wing or extension. On many lots you can only add to the back of the house because the width of the lot won't permit extending on either side without violating the code. Most houses on a street are already "aligned" to the street, with the code prohibiting any addition on the front of the house. In the country there are usually no such restrictions.

Sometimes the "satellite" addition is the easiest of several approaches available to the builder because it offers fewer tie-in problems. (See Figure 20-1.)

Let's start with the footing and work our way up.

Footings

An adequate footing is essential for every house or addition. You don't want your work to become an example of what not to do.

Footing depth is probably specified in your code. It has to be on undisturbed soil and below the frost line. In Bangor, Maine it needs to be down 48 in-

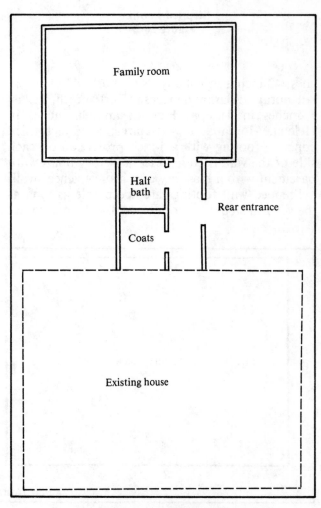

Addition Using the Satellite Concept
Figure 20-1

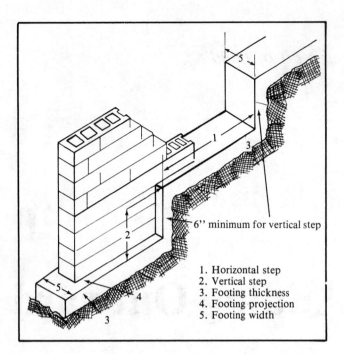

Stepped Footing
Figure 20-2

6" minimum for vertical step

1. Horizontal step
2. Vertical step
3. Footing thickness
4. Footing projection
5. Footing width

distance between steps. The horizontal distance between steps should not be less than 2 feet. The horizontal and vertical steps should be placed at the same time, with no break in the masonry. The vertical connection should be the same width as the footing and not less than 6 inches thick. Figure 20-2 shows the right way to do it.

If you dig a footing too deep, don't try to fill it back to the right level by replacing the dirt. That might be the area where a bathroom will be located, and the ceramic tile walls will crack when the house settles. Instead, fill the over-dug footing with concrete.

To find out how many yards of ready-mix you'll need for a footing, multiply the length in feet, times its width in feet, times its thickness in inches, then divide by 314.

Cement blocks can be used in most areas for constructing foundation walls and pillars. To estimate how many 8 x 8 x 16-inch concrete blocks are needed in a wall:

Take the height of wall (A) in feet, times 1½, equals the number of courses. The length of wall (B) in feet, times 3/4, equals the number of blocks in each course. Multiply A times B to find the number of block required.

To lay 100 blocks, you need a 70-pound bag of masonry cement and 3 cubic feet of sand.

A word about squaring the addition. Some houses aren't square. You must establish your key line with the front of the house if you're adding to either end, and to a common wall if adding flush with an outside wall (see Figure 2-5 in Chapter 2). The key line on the end addition is shown in Figure 20-3.

ches; 42 inches in Albany, New York; 42 inches in Anchorage, Alaska; 6 inches in Fort Worth, Texas; 6 inches in Tampa, Florida; and 36 inches in Helena, Montana. In most areas an 8-inch-thick concrete footing with a 5-inch projection on each side of the wall is enough for a 2-story house with basement with a masonry or masonry veneer wall.

Stepped wall footings are acceptable as long as the vertical step does not exceed 3/4 the horizontal

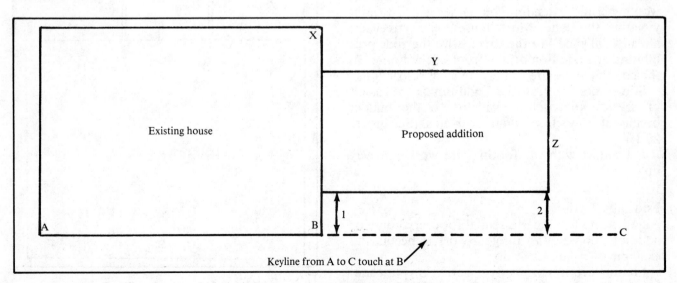

Existing house

Proposed addition

Keyline from A to C touch at B

Squaring End Addition with House
Figure 20-3

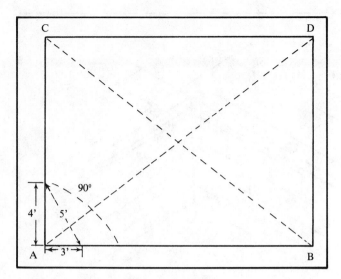

Working the Square
Figure 20-4

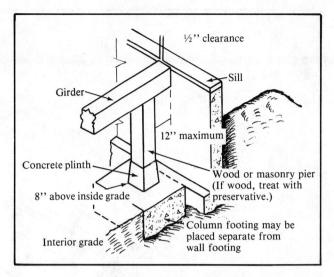

Column Supported Girder
Figure 20-5

Run a string from the corner of A, touch the corner at B and pull the string to C. The distances at "1" and "2" must be the same. Your key is "C." You square from this key line and not from wall XB which may or may not be square. Wall Z is squared with the key line wall and wall Y is squared with Z. Any "off square" of the house will be absorbed in the length of wall Y.

Normally, you can square any addition as shown in Figure 20-4.

The measurements from A to D and from B to C must be the same. The corner must be 90 degrees. To check, measure 3 feet from the corner of A on the A line. Then measure 4 feet from the corner of A. The diagonal measurement must be 5 feet.

Pilasters should be provided under all girder framing into the walls in all 6-inch concrete or masonry walls and in all 8-inch hollow block masonry walls.

One way to handle girder support is shown in Figure 20-5.

Pillars or Piers

Pillars should be no more than 8 feet on centers and be a minimum of 10 inches in diameter. The bottom of the pillars should extend beyond the frost line and have a bearing area of about 2 square feet. They should extend a minimum of 8 inches above the finish grade. Figures 20-6 through 20-10 illustrate pillar and foundation construction.

Slabs

Above grade slabs should have a proper vapor barrier and be 4 inches thick, with the top of the slab at least 8 inches above the exterior finish grade. Figure 20-11 illustrates the layout of a ground supported slab. On-slab construction with good slab height is shown in Figure 20-12.

When adding on, be sure to determine the height of the existing floor you intend to join. A good idea is to remove a small section of the exterior wall at the point where the addition will join. This lets you see the existing finish floor. *Caution:* The floor level of the existing house may be off. Place your floor even with the existing floor in the joining access area (door, hallway).

A termite shield goes between the wood framing and the masonry. (See Figure 20-13A and B.)

Framing

Now you are ready to start framing. In our case we've built a concrete block solid foundation and pillars for crawl space construction. You need a sill plate on top of the block foundation wall, and girders for the center. Plate and girder material requirements are based on linear feet. A building 24 feet wide and 48 feet long would have 144 linear feet all the way around. The girder length would be 48 feet. If you will be using one girder down the center made of a double 2 x 12, 96 linear feet of 2 x 12's are required.

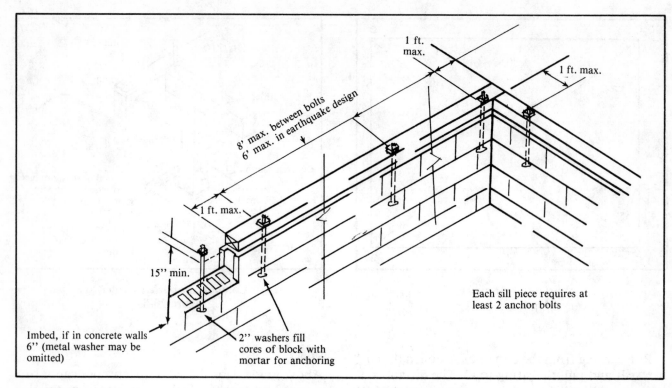

Sill Anchorage to Masonry/Concrete Walls
Figure 20-6

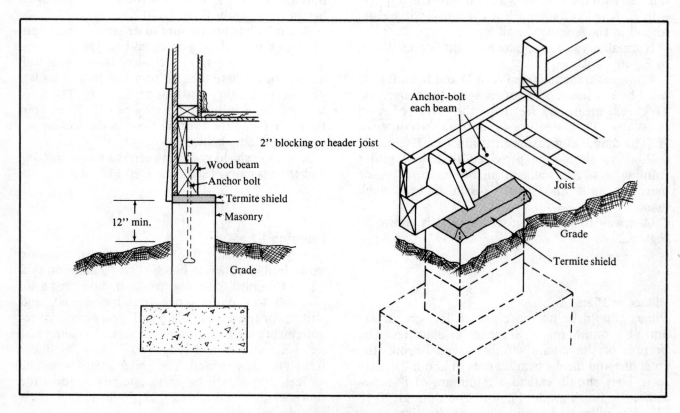

Free Standing Exterior Pier
Figure 20-7

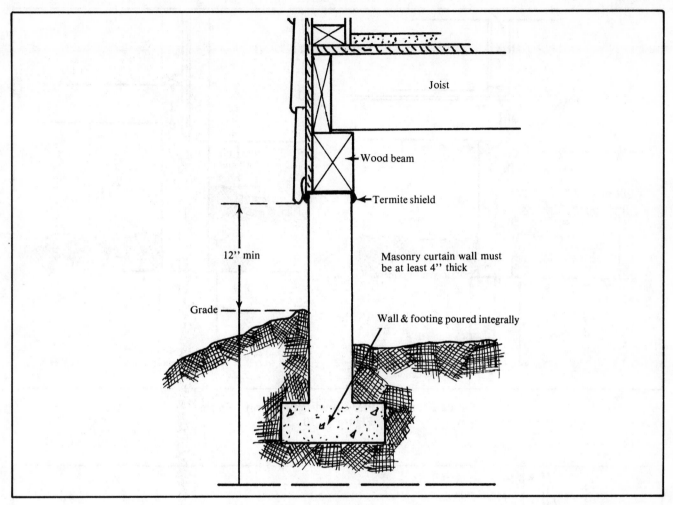

Masonry Curtain Wall
Figure 20-8

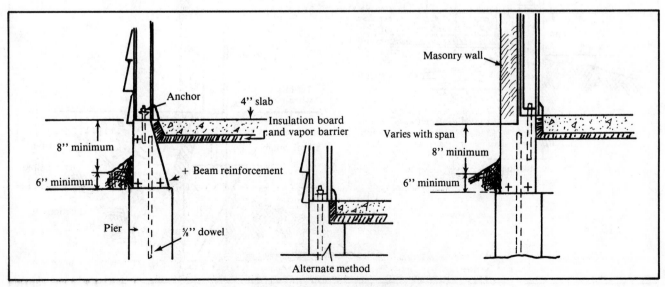

Grade Beam and Piers - Ground Support Slabs
Figure 20-9

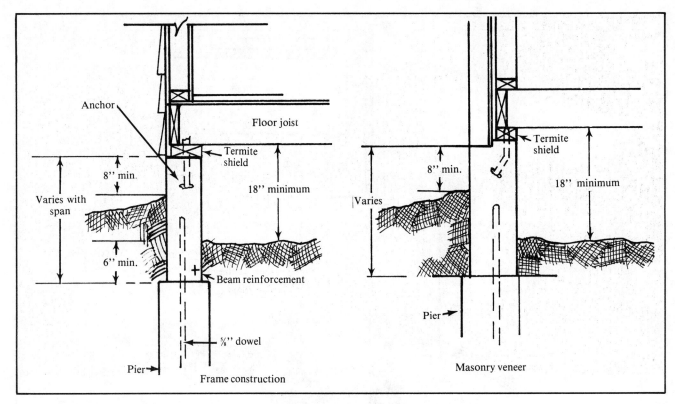

Grade Beam and Pier Construction Crawl Space
Figure 20-10

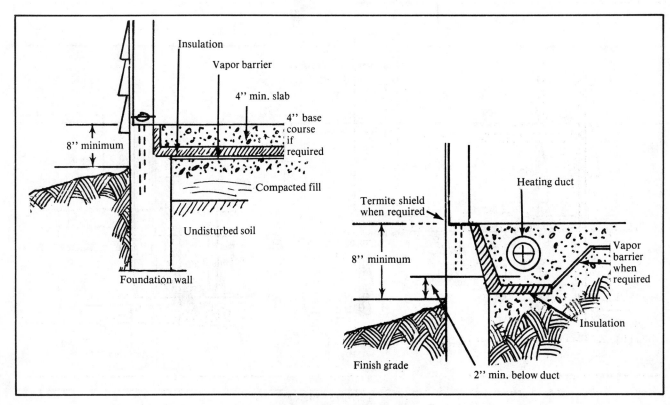

Ground Supported Slabs
Figure 20-11

On Slab Construction
Figure 20-12

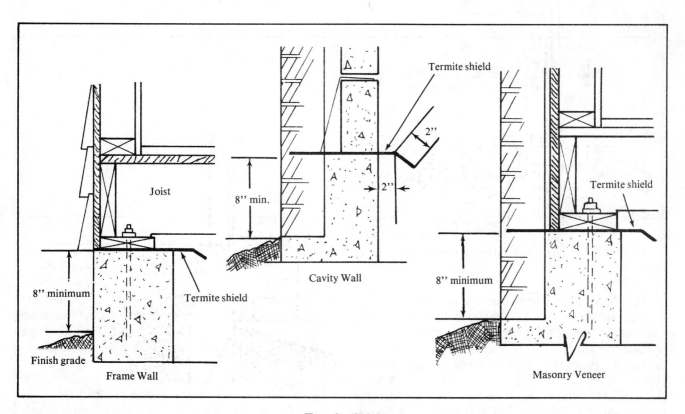

Termite Shields
Figure 20-13A

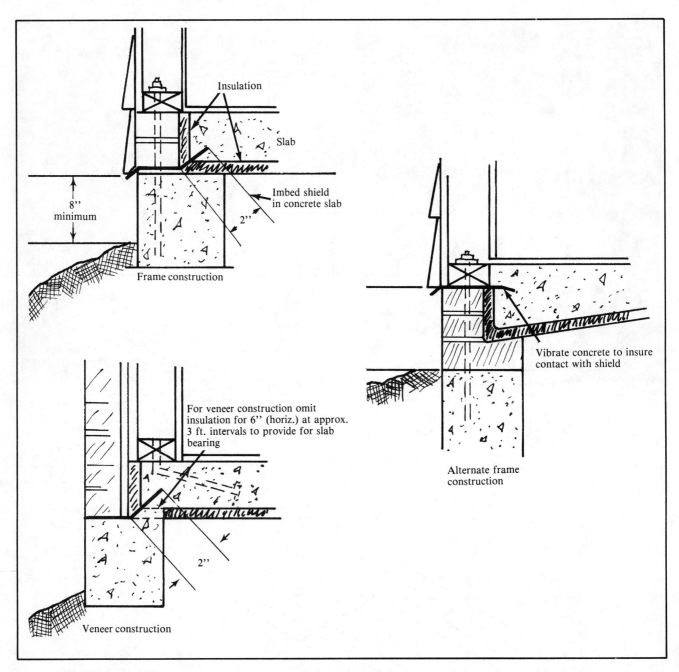

Termite Shields for Slab on Grade Construction
Figure 20-13B

Lumber is sold by the board foot. A board foot is one square foot of wood one-inch thick. The number of board feet in a piece of lumber is determined by multiplying the width times the thickness in inches, times the length in feet and then dividing by 12. A 16-foot 2 x 12, for example, contains 32 board feet. Figure 20-14 gives the board feet for various sizes of lumber. The actual size of a 2 x 4 is 1½ by 3½ inches, but the board footage is based on the name or "nominal" size.

Joist headers and stringers are also determined by linear feet. A building 24 x 48 feet would require 96 linear feet of headers and 48 linear feet of stringers.

Figure 20-15 shows three ways to tie joists to girders. C is the most popular method.

Figure 20-16 shows ways to run joists over the top of girders and bearing partitions.

Size Inches	8'	10'	12'	14'	16'	18'	20'
2 x 4	5⅓	6⅔	8	9⅓	10⅔	12	13⅓
2 x 6	8	10	12	16	18	18	20
2 x 8	10⅔	13⅓	16	18⅔	21⅓	24	26⅔
2 x 10	13⅓	16⅔	20	23⅓	26⅔	30	33⅓
2 x 12	16	20	24	28	32	36	40
4 x 4	10⅔	13⅓	16	18⅔	21⅓	24	26⅔
4 x 6	16	20	24	28	32	36	40
4 x 8	21⅓	26⅔	32	37⅓	42⅔	48	53⅓

Number of Board Feet Length of Lumber
Figure 20-14

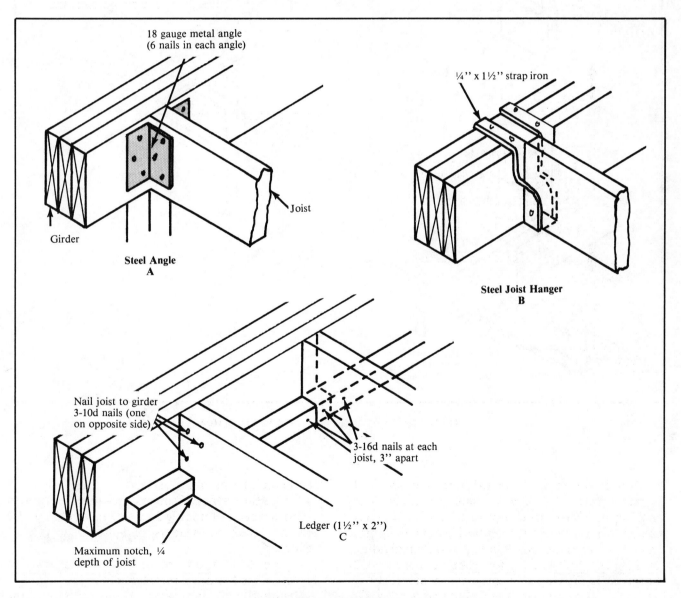

Framing to Wood Girders
Figure 20-15

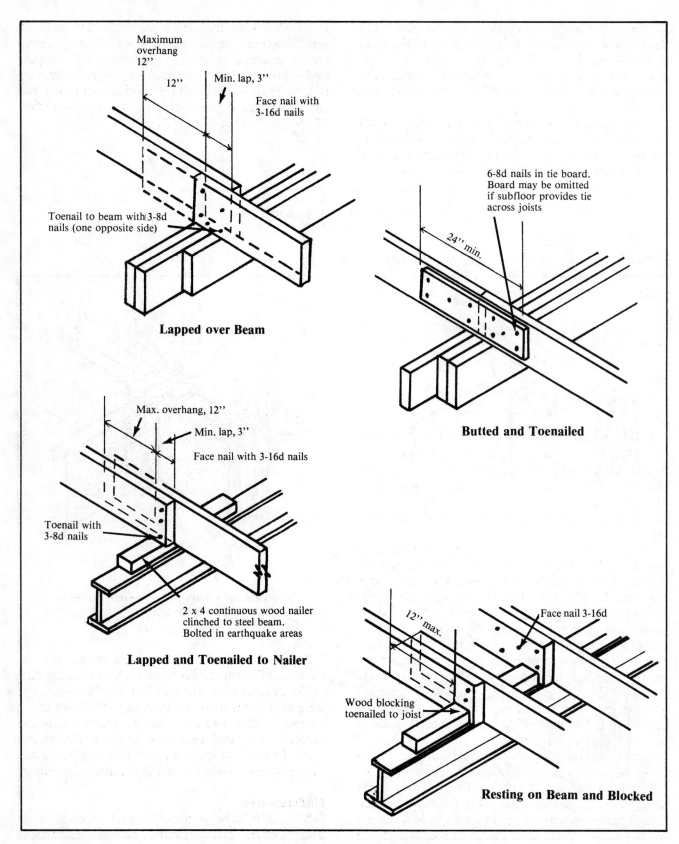

Lapped over Beam

Maximum overhang 12"

12"

Min. lap, 3"

Face nail with 3-16d nails

Toenail to beam with 3-8d nails (one opposite side)

Butted and Toenailed

6-8d nails in tie board. Board may be omitted if subfloor provides tie across joists

24" min.

Lapped and Toenailed to Nailer

Max. overhang, 12"

Min. lap, 3"

Face nail with 3-16d nails

Toenail with 3-8d nails

2 x 4 continuous wood nailer clinched to steel beam. Bolted in earthquake areas

Resting on Beam and Blocked

12" max.

Face nail 3-16d

Wood blocking toenailed to joist

Framing over Girders and Bearing Partitions
Figure 20-16

When estimating material, remember that your dealer probably stocks only even-number lengths: 8, 12, 14, 16, 18, etc. If the joists are to be 11 feet long, you have to buy 12-foot lengths, or 22-foot lengths which can be cut in half for two 11-foot pieces.

To determine how many joists are needed in our 24 x 48 addition, divide the length of the building by the spacing of the joists, center to center in feet, and add one joist for the end. Thus:

48 divided by 4/3   = 36
add 1 for the end   1
_____

Total   37

That's for one side—a girder goes down the center—so we'll have a run of 12-foot 2 x 10 floor joists on each side of the girder for a total of 74 joists.

A 12-foot 2 x 10 has 20 board feet. 74 joists equal:

74
x   20
_____

1480 Board Feet

Don't forget that the joist under partitions should be doubled. Determine where the partitions will be located and add a double joist there. Add these to the 74 and then add 15 percent more. Some of the lumber will not be fit to use for joists unless you go down to the yard and select it yourself. That takes time.

After the floor joists are in place, take the ones you discarded and cut them for solid bridging midway between the girder and foundation wall. There are three 2 x 10's left. If all things are running true to course, they are probably only fit to use as nailers (for ceiling materials) on top of the room partitions, parallel to the ceiling joists.

After the floor is framed, you're ready for the subfloor, also known as sheathing or deck. 1/2-inch sheathing grade plywood with exterior glue is a good choice where the joists are on 16-inch centers and 5/8-inch-thick particle board is going to be used as underlayment.

Georgia-Pacific offers a structural composite plywood panel called "Stable-X". Other mills offer similar products. These can be used as a single layer flooring system. They have a solid core with T & G edges and can be used as a single layer structural floor if applied with glue. The panels measure 4 x 8 feet and are available in 5/8-, 3/4-, 19/32-, and 23/32-inch thicknesses. These panels have an oriented flakeboard core with a face and back of southern pine veneer. The solid, homogeneous core provides high impact resistance, and tongue-and-groove edges produce stronger, sturdier floors. Sealed edges reduce moisture entering the core, and nailing lines assist in proper application of the panel. (See Figure 20-17.)

Structural Composite Plywood Panel
Figure 20-17

Figure 20-18 illustrates the installation of combined subfloor-underlayment such as "Stable-X."

The regular plywood subflooring is laid with the long side vertical to the joist run. Nail the edges every 6 inches and at 8 to 10 inches across the center of the panel with 8d box nails or ring-shank nails. Leave 1/16-inch spacing between butt ends of panels and 1/8-inch spacing at panel edge joints.

Underlayment
While 5/8-inch particle board meets most underlayment requirements, plywood underlayment is recommended for floor coverings that need adhesive holding power. Stagger joints over the

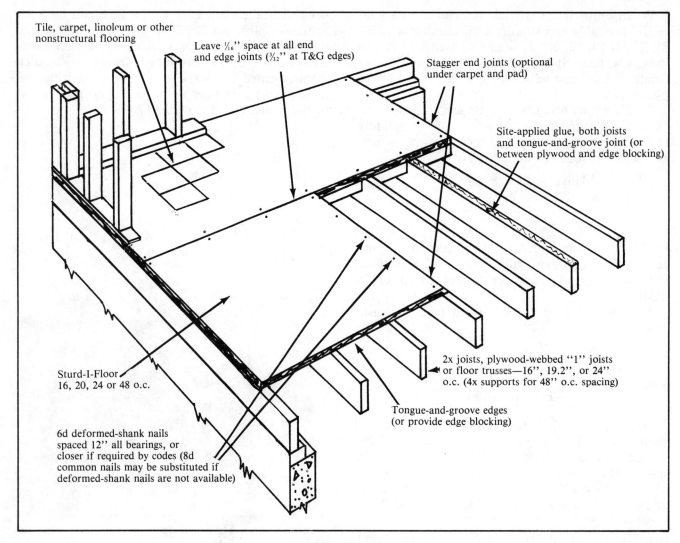

Tile, carpet, linoleum or other nonstructural flooring

Leave ⅟₁₆" space at all end and edge joints (³⁄₃₂" at T&G edges)

Stagger end joints (optional under carpet and pad)

Site-applied glue, both joists and tongue-and-groove joint (or between plywood and edge blocking)

Sturd-I-Floor 16, 20, 24 or 48 o.c.

2x joists, plywood-webbed "I" joists or floor trusses—16", 19.2", or 24" o.c. (4x supports for 48" o.c. spacing)

Tongue-and-groove edges (or provide edge blocking)

6d deformed-shank nails spaced 12" all bearings, or closer if required by codes (8d common nails may be substituted if deformed-shank nails are not available)

APA Glued Floor System
Figure 20-18

subfloor and nail to the joist with ring-shank nails. Space nails at the edges 6 inches apart and 8 inches intermediate.

Underlayment grades of plywood are preferred for some installations. They have a solid, touch-sanded surface for direct application of nonstructural finish flooring such as carpeting, linoleum or tile. Special inner-ply construction resists dents and punctures from concentrated loads. Plywood is also the most dimensionally stable underlayment panel and eliminates most swelling and buckling or humps around nails. Any unevenness is easily bridged by plywood underlayment. Keep in mind that where 1/2-inch plywood subflooring is used, plywood underlayment should be thick enough to eliminate "give" in the floor. In any underlay-

ment, nails must be flush with the floor, particularly where vinyl, linoleum or any resilient tile is to be used. Figure 20-19 shows how to put down plywood underlayment.

Exterior Wall Framing
Wall framing includes studs, corner posts, partition "T"'s, headers, bracing, and plates. The studs are either 2 x 4's or 2 x 6's. It saves time to use pre-cut studs. Place studs on 16-inch or 24-inch centers. Nail the studs to the sole plate with three 10d or four 8d nails, or two 16d nails from underneath. Where structural sheathing overlaps the sole plate, nail the sheathing to the sole plate with 8d nails, at 8 inches o.c.

Where the sheathing is not nailed to the sole

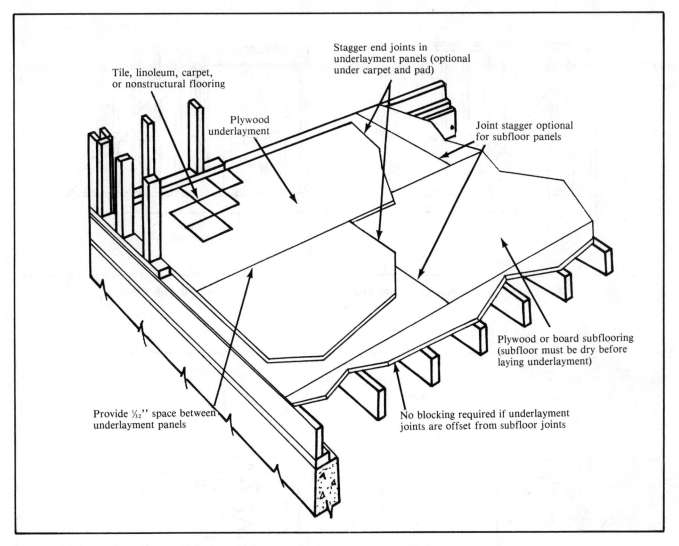

Tile, linoleum, carpet, or nonstructural flooring

Plywood underlayment

Stagger end joints in underlayment panels (optional under carpet and pad)

Joint stagger optional for subfloor panels

Plywood or board subflooring (subfloor must be dry before laying underlayment)

Provide 1/32" space between underlayment panels

No blocking required if underlayment joints are offset from subfloor joints

Plywood Underlayment Installation
Figure 20-19

plate as described above, provide (when called for in high wind areas) 18 gauge galvanized steel straps at every other stud with four 6d nails in each end of the strap, or use metal anchors designed for this purpose.

Corner bracing may be let-in or may be 4 x 8-foot 1/2-inch-thick plywood nailed on 4-inch spacing with 8d nails. Figure 20-20 illustrates let-in bracing with 4-inch material.

One method of estimating the number of studs 16 inches o.c. is to allow 1 stud per linear foot of wall. This should include studs for doubling at windows and doors and for making corners and "T"s. But that rule of thumb doesn't allow for crooked or bad lumber which can be used for firestops, bracing and short studs. If you figure one stud per foot of wall and add 20 percent, you'll be closer to

the mark.

A rule of thumb for a quick estimate is to figure one stud for every three square feet of floor space. A 2100-square-foot house will require 700 studs.

A corner post uses three studs and blocking as shown in Figure 20-21. A partition "T" also requires three studs and blocking as shown in the figure.

Nail the corner post and "T"s together with 16d nails not more than 24 inches o.c. in each wide face, with at least 3 nails into each filler block.

Double studs are required at all window, door and other openings with the jamb (or jack stud) supporting the header. The jamb stud should extend in one piece from the header to the sole plate. Use 16d nails in framing header openings and jamb studs. Use good lumber that has plenty of strength

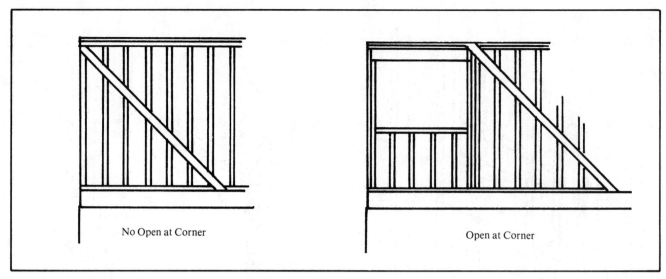

No Open at Corner

Open at Corner

Let-In on Approximately 45° on Outside and Catch Sole Plate and Top Plate
Figure 20-20

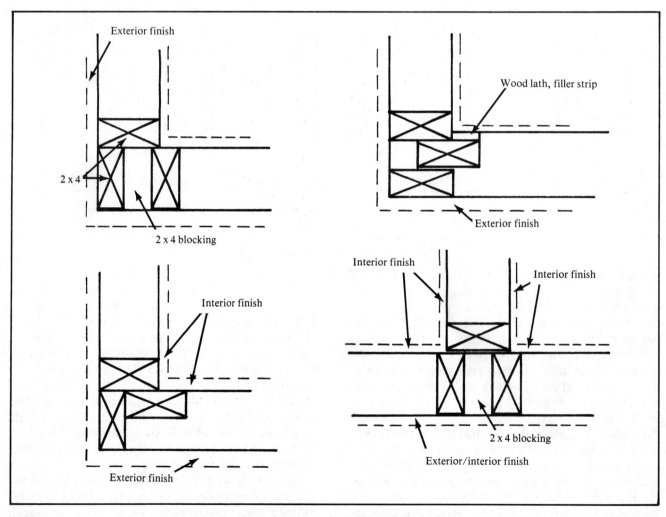

Exterior finish

Wood lath, filler strip

2 x 4

2 x 4 blocking

Exterior finish

Interior finish

Interior finish

Interior finish

2 x 4 blocking

Exterior finish

Exterior/interior finish

Corner Posts and Partition Post
Figure 20-21

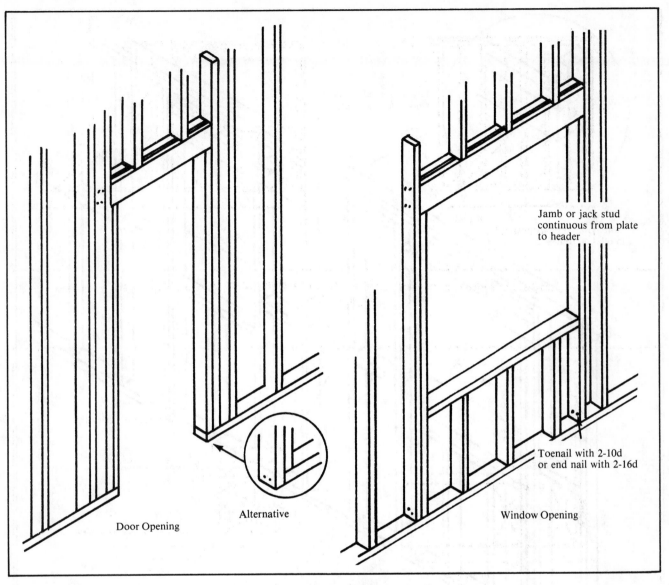

Jamb or jack stud
continuous from plate
to header

Toenail with 2-10d
or end nail with 2-16d

Door Opening

Alternative

Window Opening

Framed Openings in Exterior Wall
Figure 20-22

for headers. My practice is to use doubled 2 x 10's for all weight-bearing exterior and interior walls. A 2 x 4 bottom plate under the 2 x 10's makes the header the correct height for window and door heights. This eliminates any concern about adequate support. Figures 20-22 and 20-23 show correct exterior wall framing.

The best way to frame a wall is on the floor. Nail it all together and raise it as a complete unit. The sole plate is nailed to the studs from the bottom side with two 16d nails. The first 2 x 4 member of the top plate is nailed from the top side with two 16d nails. Lap the top 2 x 4 member of the top plate over the first 2 x 4 member at corners and intersec-

ting walls. Use at least two 16d nails in overlaps. Three are better.

Here are some points to remember when nailing the top 2 x 4 to the first 2 x 4 of the top plate:
• When the structural sheathing is not tied to either top plate member, or in areas subject to high winds, anchorage or tie-in should be provided by 18 gauge, 1-inch galvanized steel straps nailed to every other stud and to the roof and ceiling framing with four 6d nails in each end strap.
• If the plates are cut more than one-half their width for ductwork or piping, they should be reinforced with 18 gauge steel straps.
• When structural sheathing is tied to the upper

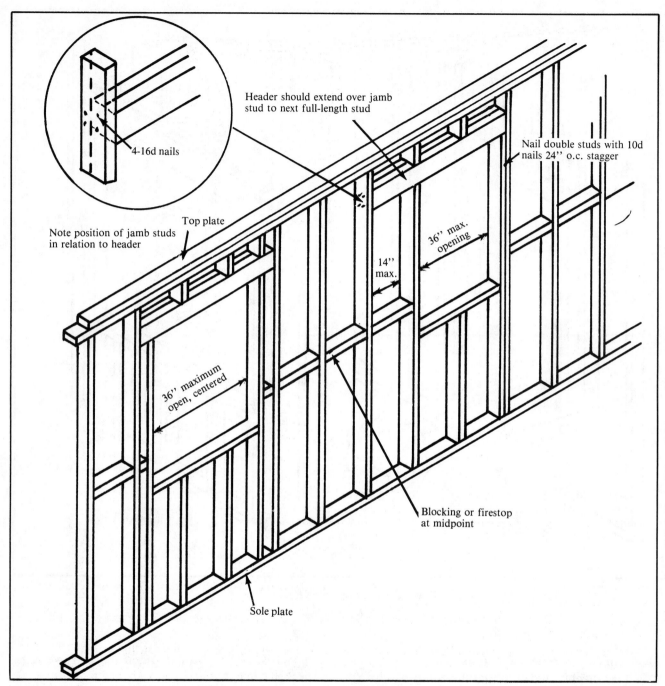

Various Single Framing - Opening Methods
Figure 20-23

member of the plate with 8d nails, 8 inches o.c., nail the upper 2 x 4 to the lower one with 16d nails at 16 inches o.c.

• When the structural sheathing is tied to the lower 2 x 4 member of the plate with 8d nails 8 inches o.c., nail the upper 2 x 4 member to the lower one with 16d nails, 8 inches o.c.

Wall Sheathing
Wall sheathing is required except where exterior finish material does not require a solid backing. Where masonry veneer is used, install fiberboard or other insulating sheathing directly to the studs. It is a good idea to use sheathing on the exterior wall in all cases for its insulating value. Figure 20-24 illustrates the typical installation method.

Figure 20-24 shows rigid fiberboard. The nailing

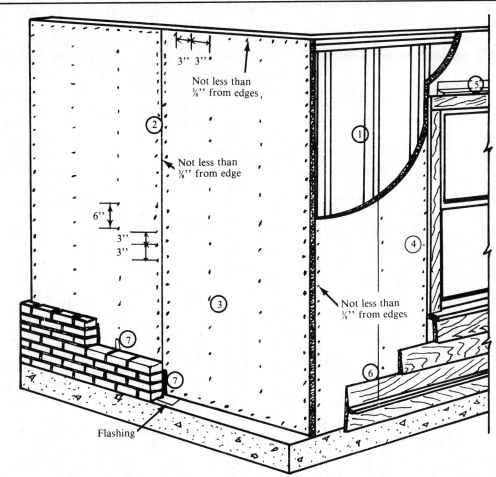

Recommended nails or staples. 1½'' No. 11 gauge galvanized roofing nails with ⁷⁄₁₆'' head, or 6d common nails. In stapling application, where corner bracing is omitted, use 1½'' No. 16 gauge galvanized wire staples with ⁷⁄₁₆'' crown and divergent chisel points.

① Erect framing 16'' or 24'' on center. Sheathing units should extend from sill to top plate (provide headers if and where ends of units will be unsupported).

② Sheathing shall be applied vertically to the framing allowing ⅛'' space between adjoining units. Sheathing is cut scant for this allowance.

③ In application without diagonal corner bracing, nail or staple sheathing first to intermediate framing members at 6'' centers; then fasten along all edges at 3'' centers and not less than ⅜'' from edge. Drive nails flush. Apply staples so that their crowns will slightly depress sheathing surface, and with length of their crowns parallel to direction of edges of sheathing and long dimension of framing members. 4' wide panels show stud-locator nail markings at intermediate 16'' o.c. studs.

④ Bring sheathing units to moderate contact with framing around windows and doors--never force units into place.

⑤ Provide sheet metal flashing over head casing of all windows and doors.

⑥ Apply wood siding in manner to insure that ends fall over centerlines of studs.

⑦ Allow 1'' air space between masonry veneer and sheathing. Under masonry veneer, apply non-corrosive metal ties at rate of one tie to every 160-260 square inches of wall area, driving nails through sheathing into studs to 1'' depth.

Sheathing Installation Methods
Figure 20-24

specs shown would be appropriate where diagonal corner bracing is not used. Nail spacing may be increased to 8 inches at intermediate framing members and 4 inches at edges of boards at stud locations.

Sole plates are nailed to the subfloor and joists with 16d nails at each joist.

Ceiling Joists

Ceiling joists should be of sufficient size to offer rigid ceiling support. (See Figure 5-3B in Chapter 5). Calculate the number of ceiling joists required for an addition the same way you calculated floor joists. Put them on 16-inch centers, and toenail them to the exterior wall plate with three 10d nails. When the rafters are installed next to the joists, secure the joists to the rafters with three 16d nails.

The ends of ceiling joists should be lapped over a bearing partition, and toenailed to the plate with two 10d nails. Three 16d nails through the overlap section of the joists will secure the joists against rafter thrust.

After all the ceiling joists are up, anchor them in

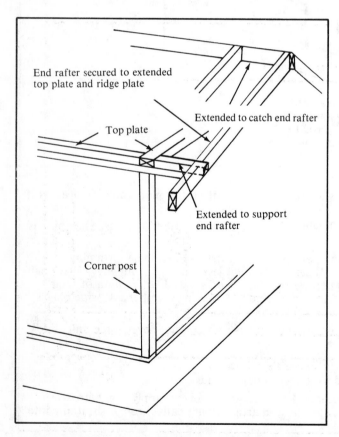

End rafter secured to extended top plate and ridge plate

Extended to catch end rafter

Top plate

Extended to support end rafter

Corner post

End Rafters Secured to Plate and Ridge Extenders
Figure 20-25

position with a "trough" running vertically over the top of the joists at a position about 1/3 the building width from the exterior wall. The "trough" is a 2 x 6 laid flat over the top of the ceiling joists and nailed to each with two 16d nails. A second 2 x 6 is placed on its edge and joined to the flat 2 x 6 on the side toward the center of the house. The edge-up 2 x 6 is anchored to the flat 2 x 6 with 16d nails spaced 12 inches o.c. and toenailed to each joist with one 10d nail. 2 x 4 braces will rest on the "trough" and support the rafter purlin. (See Figure 4-10 in Chapter 4).

The ceiling joists are cut on the top at the exterior plate area to fit flush with the top of the rafter run. This is done before the ceiling joists are put in place.

Rafters

Figure 5-3C in Chapter 5 gives you the species and size of lumber for various rafter spans.

Charts and conversion diagrams are available which show the rafter length requirements of various buildings when you know the building width and pitch or the width and height of the roof at the ridge plate. The length given will be from the ridge plate to the outside edge of the top plate of the wall. The overhang length is added to the length specified in the chart or conversion. The rafter length requirement can also be worked out on your carpenter's square. The rule is:

To find the length of a common rafter, multiply the "length given in the table" by the number of feet of run.

"Length given in the table" refers to the table on your carpenter square. It is the first line, indicated as "Length Common Rafters per Foot Run."

There are 17 of these tables, beginning at the 2-inch mark and continuing to the 18-inch mark.

Position the square to face you, as shown in Figure 20-26.

Now, let's see how long the rafters for our 24 x 48-foot addition will have to be for a 5-inch rise per foot run (5/12 pitch). First, find on the inch line on the top edge of the square the figure that is equal to the rise of the roof, in this case 5. Look at the 5-inch mark on the body of the square. On the first line under the 5-inch mark, find 13, which is the length of the rafter in inches "per foot run" for a 5/12 pitch.

The building is 24 feet wide. Therefore, the run of the rafter will be half that width: 24 divided by 2 equals 12 feet.

Since you know that the length of the rafter per

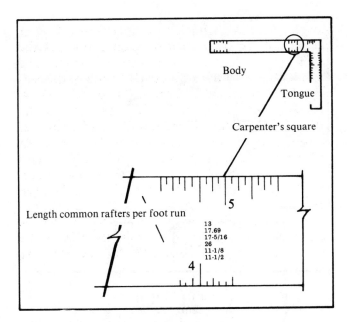

Position the Square to Face You
Figure 20-26

"one foot run" equals 13 inches, the total length of the rafter will be 13 inches multiplied by 12 feet, which equals 156 inches or 13 feet. Thus the rafter from the center of the ridge plate to the outside of the exterior wall plate is 13 feet.

Add to the 13 feet the length of your overhang, and you have the length of material required for the rafter. (The overhang is the tail or eave of the rafter.)

After you've established the total length of the rafter, mark both ends. Remember to allow for half the thickness of the ridge plate.

Determine both the ridge angle and the bird mouth (seat cut) by placing the square so that the 12-inch mark on the body and the 5-inch mark on the tongue (the rise) will be at the edge of the board (rafter).

Let's work this out on our rafter. (See Figure 20-27.)

Points A and B are the ends of the rafter. To obtain the bottom or seat cut, take the 12-inch mark

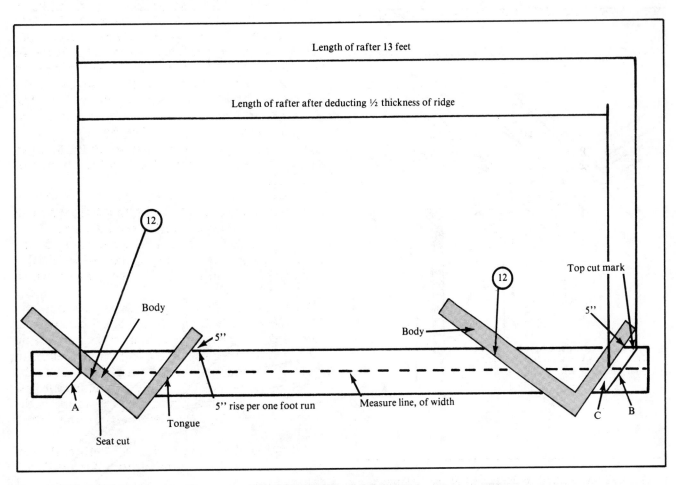

Using the Carpenter Square
Figure 20-27

on the body of the square and the 5-inch mark on the tongue. Lay the square on the rafter so that the body will coincide with point A or the lower end of the rafter. Mark along the body and cut.

Now measure the deduction for half the thickness of the ridge. Half the thickness of our ridge plate is 3/4 inch. Deduct the 3/4 inch at right angles to the top cut mark or plumb line (point C). Then draw a line parallel to the top cut mark and make the cut. Notch the bird mouth to half the width of the rafter.

There are other methods of determining rafter length. You probably have your own best method. If it works, stick with it.

Next, nail the rafters up. They can be braced from the "trough" as previously discussed, or as shown in Figure 20-28. The rafters are toenailed to the plate with four 10d nails. Spike the rafter to the

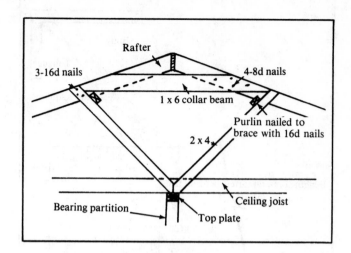

Rafter Bracing
Figure 20-28

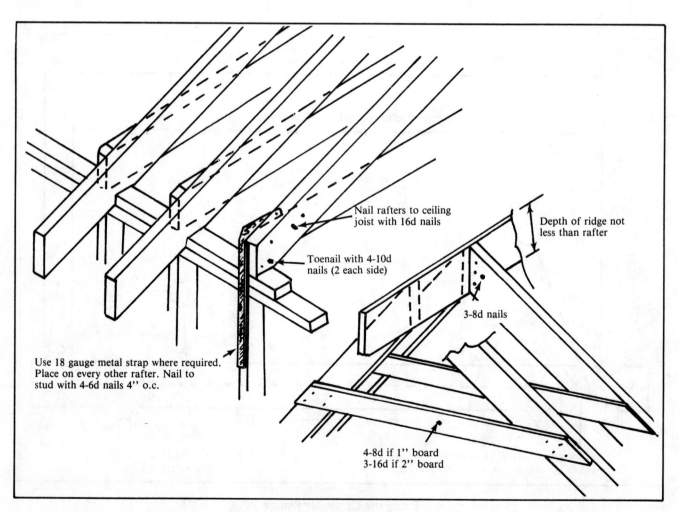

Securing Rafters
Figure 20-29

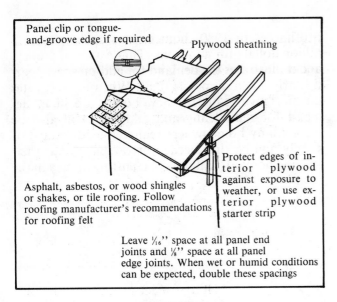

Panel clip or tongue-and-groove edge if required

Plywood sheathing

Asphalt, asbestos, or wood shingles or shakes, or tile roofing. Follow roofing manufacturer's recommendations for roofing felt

Protect edges of interior plywood against exposure to weather, or use exterior plywood starter strip

Leave 1/16" space at all panel end joints and 1/8" space at all panel edge joints. When wet or humid conditions can be expected, double these spacings

Panel Clip
Figure 20-30

ceiling joists with three 16d nails for a solid lock-in. Nail at the ridge plate as shown in Chapter 5, Figure 5-11, or toenail to the ridge with three 8d nails.

Nail up 1 x 6-inch collar boards, maximum spacing, 4 feet o.c. Nail them to each rafter with four 8d nails. If you're using a 2 x 4 collar, nail to each rafter with three 16d nails.

Figure 20-29 illustrates the method of securing rafters to the ridge and plate.

Rafters will bow, turn, and twist at the tail. A straightener of the same material as the rafter nailed to the ends of the rafters will eliminate this problem and provide a solid backing for the fascia board and gutter mounts. Use two 16d nails to face nail the straightener to the rafter ends.

Stud the gable ends. Remember what's been said about proper attic ventilation. Use adequate vents.

1/2-inch plywood makes an excellent deck for 16-inch o.c. spaced rafters. Apply the long side ver-

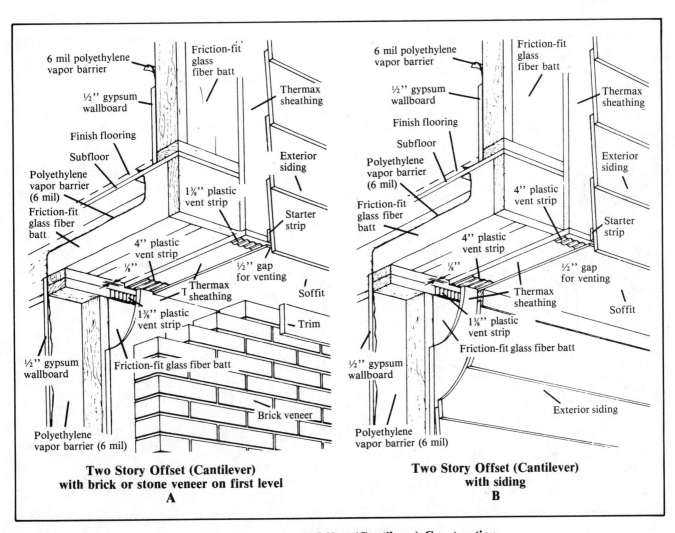

6 mil polyethylene vapor barrier

Friction-fit glass fiber batt

Thermax sheathing

1/2" gypsum wallboard

Finish flooring

Subfloor

Exterior siding

Polyethylene vapor barrier (6 mil)

Friction-fit glass fiber batt

1⅜" plastic vent strip

Starter strip

4" plastic vent strip

1/8"

Thermax sheathing

1/2" gap for venting

Soffit

1⅜" plastic vent strip

Trim

1/2" gypsum wallboard

Friction-fit glass fiber batt

Brick veneer

Polyethylene vapor barrier (6 mil)

Two Story Offset (Cantilever)
with brick or stone veneer on first level
A

6 mil polyethylene vapor barrier

Friction-fit glass fiber batt

Thermax sheathing

1/2" gypsum wallboard

Finish flooring

Subfloor

Polyethylene vapor barrier (6 mil)

Exterior siding

Friction-fit glass fiber batt

4" plastic vent strip

4" plastic vent strip

1/8"

Starter strip

1/2" gap for venting

Thermax sheathing

Soffit

1⅜" plastic vent strip

Friction-fit glass fiber batt

1/2" gypsum wallboard

Exterior siding

Polyethylene vapor barrier (6 mil)

Two Story Offset (Cantilever)
with siding
B

Details for Two-story Offset (Cantilever) Construction
Figure 20-31

tical to the rafters. Stagger the joints. Nail 6 inches o.c. along the intermediate members, using 8d nails. Figure 20-30 shows roof decking with panel clips.

Figure 20-31A and B gives details for two-story offset (cantilever) construction.

Summary

Whether you go down, out or up, additions can be profitable. In 1980, homeowners spent almost 8 billion dollars for additions. That's a lot of money, and it illustrates the demand for add-on space, particularly when money is tight. When interest rates on mortgage loans for new homes are high, demand for remodeling and additions will also be high. Many homeowners realize the advantage of simply adding that needed extra space onto their existing home. It's almost like putting money in the bank.

21

Basic Plumbing and Wiring

Plumbing and wiring are two areas that are best left to professionals. In fact, most areas allow only licensed people to do this work. That's a good policy because both require special methods and know-how. And codes are very strict in these areas.

It would take several books to cover these two subjects completely. We'll just discuss the basics to give you some idea of what to look for.

Basic Plumbing

There is a water source, either private, such as a well, or public, such as the city system. The water enters the house, and one branch goes directly to the various fixtures and appliances. This is the cold water supply. A second branch goes to the hot water heater, and from there to the various fixtures and appliances. This is the hot water supply. After the water is used, it goes out the waste line to a septic tank (privately owned system), or to a sewer (publicly owned system).

That's all there is to it except for the lines, turns, stub-outs, cut-offs, vents, plugs, clean-outs, traps, and air chambers. There are iron pipes, galvanized pipes, copper pipes and plastic pipes. Also gas lines, steam heat lines and such. A useful reference for the professional remodeler is *Basic Plumbing with Illustrations,* available from Craftsman Book Company, 6058 Corte del Cedro, Carlsbad, California, 92008, or through the order form in the back of this manual.

Plastic pipe is economical and easy to install. If the pipes are installed too tightly against wood members in a wall or floor, there will usually be some noise when the pipes contract and expand under temperature change.

Piping should be installed without damaging structural members. This may be hard to get across to some plumbers. But make sure they leave the existing members intact.

Basic Wiring

Wiring is another strict code job that you will sub out unless you're a licensed electrician.

In today's homes a 150-amp service is recommended. Most homes are equipped with an electric range, high speed dryer, electric water heater, central heating/cooling units in addition to general lighting and small appliances. Number 1/0 copper 3-wire service entrance cable (type RHW insulation) or a comparable-capacity aluminum entrance cable is required for 150-amp service.

General purpose circuits are wired with No. 12 copper (12/2 with ground) for a 15 or 20 amp load. For kitchen appliance circuits it's best to go to the No. 10 wire for a 20 amp load circuit. That is, No. 10 with ground. The same is used for the laundry circuit. A No. 10 wire is actually a 30-amp capacity line while a No. 12 is a 20-amp capacity wire. But, when you use a No. 12 wire for a 15-amp load and No. 10 wire for a 20-amp load you've got yourself an extra margin of safety. The wires never heat up before the fuse or circuit breaker goes. Well, that's the way I do it anyway. I have a deep respect for electricity. Once the wires are covered up behind the walls it's too late to change them.

The maximum carrying capacity of a 20-amp circuit with a No. 12 wire is 2400 watts.

To determine the amperage, divide wattage by voltage. Conversely, to find the wattage, multiply the amperage by the voltage. Thus, a 5000-watt central air conditioning unit would be:

5000 watts divided by 220 volts, equals 23 amps.

A wall outlet is required for every 12 feet of wall space. Locate the bottom of the outlet anywhere from 6 to 12 inches from the floor. The 12-inch placement is usually considered standard, though some people prefer it lower or higher. I locate the wall switches with the bottom of the box 48 inches from the floor. The code might specify a different height. But if you put the switch much lower it will be within the reach of small children. And you know how kids love to play with light switches.

In kitchens, one outlet for every 4 feet of counter space is considered essential for the various kitchen appliances now commonly used.

An outside light fixture with interior wall switch control should be installed at each exterior door.

Permanent lighting fixtures, wall switch controlled, should be installed in kitchens, dining rooms, bathrooms and halls.

In other habitable rooms, including living rooms and bedrooms, permanent lighting fixtures must be wall-switch controlled. Wall-switch controlled outlets should also be provided.

Permanent lighting fixtures are required over stairways connecting habitable rooms, or halls at different elevations. Such fixtures should have several switches, one at each elevation.

A permanent lighting fixture should be installed to illuminate basement stairs and should be controlled by a switch at the head of the stairs. Multiple switch controls should be installed at both elevations when the basement has other exit doorways.

Permanent fixtures should be installed in utility rooms, garages, closets, etc.

Don't put a switch or outlet within reach of a tub or shower. In fact, don't put *anything* electrical near the tub or shower! If you do, you're inviting disaster.

Ground-type outlets and fixtures are recommended throughout the house.

At least two 15-amp circuits serving all lighting outlets, and those receptacle outlets not in the kitchen, dining and laundry areas, are required for the first 500 square feet of floor space. At least one circuit is needed for each additional 500 square feet or fraction thereof.

The service entrance panel should be of sufficient size to permit at least one future circuit.

A rule of thumb for estimating the amount of wire for general purpose circuits is one linear foot of wire for each square foot of floor space.

Summary

When bidding on an addition, get firm prices from electrical and plumbing subcontractors. Most working builders or contractors are quite capable of plumbing or wiring a house or addition but lack the technical knowledge or background to pass a licensing examination. Don't insist on doing everything yourself. Concentrate on what you do best, and leave the rest to others. Doing the job right—with all the joints tight and all slips shod—is a demanding enough job.

22

Job Estimating

The material estimate is not actually an estimate. Instead, think of it as a list of every piece of lumber, every pound of nails and every other item that will be used in a job. Then get the price for each item and add it all up. This is your material cost. The important thing is to make certain that everything needed to do the job is included in the list. Figure 22-1 is a list of items that should be included.

A good method is to itemize the materials in sections. For example, list the footing, foundation wall, and piers together. If forming for masonry is required, include that material also.

Then list all the floor framing material by sizes and quantities. Next is the studding material and wall sheathing. Then come the ceiling joists and roof framing material, including the decking, paper and shingles.

All windows and doors are listed for drying in. This "dry in" material is listed separately and in the proper work category as work progresses. The itemized list identifies each step and each length of lumber for its intended use. This ensures that a certain length 2 x 6 ordered for a gable roof valley rafter won't be used for something else. A 12-foot 2 x 6 ordered to be used as two 6-foot valley rafters may well end up being used for another purpose.

The lumber should be stacked by size on the site. If you happen to forget what a particular board is for, the list will remind you. Remind your crew members that a lot of material can be wasted when you grab the nearest board and cut off 2 feet every

time you need some odd board size. Try to use the sizes ordered for their intended purpose.

The time required to itemize the material by size and job is well invested. Having the material stacked by size or in order of use is essential if work and material are to go smoothly.

This is the nuts and bolts of building. A well-prepared list giving the materials by job category is the first essential. You can always look at the work in progress to determine how much is completed. You can also look at the stacks of materials and tell how much more work remains to be done for each particular category.

There are no short cuts to figuring or itemizing material. You can't know what the materials will cost if you don't know what materials are required. So list everything!

Labor

Now that you've listed everything and have an item-by-item price, what about manhours? What formula do you use to find the total manhours required to finish the job?

In remodeling a house about all you can rely on is experience. There are so many variables involved, and each remodeling job is different. Each has its own problems. It's through experience that you learn to determine the number of manhours. There is no single formula that applies to all situations. It

Excavation
Backfilling
Clearing the site
Compacting
Dump fee
Equipment rental
Equipment transport
Establishing new grades
General excavation
Hauling to dump
Pit excavation
Pumping
Relocating utilities
Removing obstructions
Shoring
Stripping topsoil
Trenching

Demolition
Cabinet removal
Ceiling finish removal
Concrete cutting
Debris box
Door removal
Dump fee
Dust partition
Electrical removal
Equipment rental
Fixtures removal
Flooring removal
Framing removal
Hauling to dump
Masonry removal
Plumbing removal
Roofing removal
Salvage value allowance
Siding removal
Slab breaking
Temporary weather protection
Wall finish removal
Window removal

Concrete
Admixtures
Anchors
Apron
Caps
Cement
Columns
Crushed stone
Curbs
Curing
Drainage
Equipment rental
Expansion joints
Fill
Finishing
Floating
Footings
Foundations
Grading
Gutters
Handling
Mixing
Piers
Ready mix
Sand

Screeds
Slabs
Stairs
Standby time
Tamping
Topping
Vapor barrier
Waterproofing

Forms
Braces
Caps
Cleaning for reuse
Columns
Equipment rental
Footings
Foundations
Key joints
Layout
Nails
Piers
Salvage value
Slab
Stair
Stakes
Ties
Walers
Wall

Reinforcing
Bars
Handling
Mesh
Placing
Tying

Masonry
Arches
Backing
Barbecues
Cement
Ceramic tile
Chimney
Chimney cap
Cleaning
Clean-out doors
Dampers
Equipment rental
Fireplace
Fireplace form
Flashing
Flue
Foundation
Glass block
Handling
Hearths
Laying
Lime
Lintels
Mantels
Marble
Mixing
Mortar
Paving
Piers
Repair
Reinforcing

Repointing
Sand
Sandblasting
Sills
Steps
Stonework
Tile
Veneer
Vents
Wall ties
Walls
Waterproofing

Rough Carpentry
Area walls
Backing
Beams
Blocking
Bracing
Bridging
Building paper
Columns
Cornice
Cripples
Door frames
Dormers
Entrance hoods
Fascia
Fences
Flashing
Framing clips
Furring
Girders
Gravel stop
Grounds
Half timber work
Hangers
Headers
Hip jacks
Insulation
Jack rafters
Joists, ceiling
Joists, floor
Ledgers
Nails
Outriggers
Pier pads
Plates
Porches
Posts
Rafters
Ribbons
Ridges
Roof edging
Roof trusses
Rough frames
Rough layout
Scaffolding
Sheathing, roof
Sheathing, wall
Sills
Sleepers
Soffit
Stairs
Straps
Strong backs
Studs

Subfloor
Timber connectors
Trimmers
Valley flashing
Valley jacks
Vents
Window frames

Finish Carpentry
Baseboard
Bath accessories
Belt course
Built-ins
Cabinets
Casings
Caulking
Ceiling tile
Closet doors
Closets
Corner board
Cornice
Counter tops
Cupolas
Door chimes
Door hardware
Door jambs
Door stop
Door trim
Doors
Drywall
Entrances
Fans
Flooring
Frames
Garage doors
Hardware
Jambs
Linen closets
Locksets
Louver vents
Mail slot
Mantels
Medicine cabinets
Mirrors
Molding
Nails
Paneling
Rake
Range hood
Risers
Roofing
Room dividers
Sash
Screen doors
Screens
Shelving
Shutters
Siding
Sills
Sliding doors
Stairs
Stops
Storm doors
Threshold
Treads
Trellis
Trim

Material Estimate List
Figure 22-1

Vents
Wallboard
Watertable
Window trim
Wardrobe closets
Weatherstripping
Windows

Flooring
Adhesive
Asphalt tile
Carpet
Cork tile
Flagstone
Hardwood
Linoleum
Marble
Nails
Pad
Rubber tile
Seamless vinyl
Slate
Tack strip
Terrazzo
Tile
Vinyl tile
Wood flooring

Plumbing
Bathtubs
Bar sink
Couplings
Dishwasher
Drain lines
Dryers
Faucets
Fittings
Furnace hookup
Garbage disposers
Gas service lines
Hanging brackets
Hardware
Laundry trays
Lavatories
Medicine cabinets
Pipe
Pumps
Septic tank
Service sinks
Sewer lines
Sinks
Showers
Stack extension
Supply lines
Tanks
Valves
Vanity cabinets
Vent stacks
Washers
Waste lines
Water closets
Water heaters
Water meter
Water softeners
Water tank
Water tap

Heating
Air conditioning
Air return
Baseboard
Bathroom
Blowers
Collars
Dampers

Ducts
Electric service
Furnaces
Gas lines
Grilles
Hot water
Infrared
Radiant cable
Radiators
Registers
Relocation of system
Thermostat
Vents
Wall units

Roofing
Adhesive
Asbestos
Asphalt shingles
Built-up
Canvas
Caulking
Concrete
Copper
Corrugated
Downspouts
Felt
Fiberglass shingles
Flashing
Gravel
Gutters
Gypsum
Hip units
Insulation
Nails
Ridge units
Roll roofing
Scaffolding
Shakes
Sheet metal
Slate
Tile
Tin
Vents
Wood shingles

Sheet Metal
Access doors
Caulking
Downspouts
Ducts
Flashing
Gutters
Laundry chutes
Roof flashing
Valley flashing
Vents

Electrical Work
Air conditioning
Appliance hook-up
Bell wiring
Cable
Ceiling fixtures
Circuit breakers
Circuit load adequate
Clock outlet
Conduit
Cover plates
Dimmers
Dishwashers
Dryers
Fans
Fixtures
Furnaces

Garbage disposers
High voltage line
Hood hook-up
Hook-up
Lighting
Meter boxes
Ovens
Panel boards
Plug outlets
Ranges
Receptacles
Relocation of existing lines
Service entrance
Switches
Switching
Telephone outlets
Television wiring
Thermostat wiring
Transformers
Vent fans
Wall fixtures
Washers
Water heaters
Wire

Plastering
Bases
Beads
Cement
Coloring
Cornerite
Coves
Gypsum
Keene's cement
Lath
Lime
Partitions
Sand
Soffits

Painting and Decorating
Aluminum paint
Cabinets
Caulking
Ceramic tile
Concrete
Doors
Draperies
Filler
Finishing
Floors
Masonry
Paperhanging
Paste
Roof
Sandblasting
Shingle stain
Stucco
Wallpaper removal
Windows
Wood

Glass and Glazing
Breakage allowance
Crystal
Hackout
Insulating glass
Mirrors
Obscure
Ornamental
Plate
Putty
Reglaze
Window glass

Indirect Costs
Barricades
Bid bond
Builder's risk insurance
Building permit fee
Business license
Cleaning floor
Cleaning glass
Clean-up
Completion bond
Debris removal
Design fee
Equipment floater insurance
Equipment rental
Estimating fee
Expendable tools
Field supplies
Job shanty
Job phone
Job signs
Liability insurance
Local business license
Maintenance bond
Patching after subcontractors
Payment bond
Plan checking fee
Plan cost
Protecting adjoining property
Protection during construction
Removing utilities
Repairing damage
Sales commission
Sales taxes
Sewer connection fee
State contractor's license
Street closing fee
Street repair bond
Supervision
Survey
Temporary electrical
Temporary fencing
Temporary heating
Temporary lighting
Temporary toilets
Temporary water
Transportation equipment
Travel expense
Watchman
Water meter fee
Waxing floors

Administrative Overhead
Accounting
Advertising
Automobiles
Depreciation
Donations
Dues and subscriptions
Entertaining
Interest
Legal
Licenses and fees
Office insurance
Office phone
Office rent
Office salaries
Office utilities
Pensions
Postage
Profit sharing
Repairs
Small tools
Taxes
Uncollectable accounts

Material Estimate List
Figure 22-1 (continued)

will take two men "X" number of days to remove a partition, refinish the walls and redo the floor. But it's a big "X".

For additions and house construction, however, there are a number of formulas. Some are worthless. Some are less helpful. But, just because a formula works for another builder, don't expect it to work for you. No two builders work alike or at the same rate. Your own formula may not work the same way on two different jobs.

One formula determines manhour requirements per 1000 board feet of lumber. It provides 22 manhours for nailing 1000 board feet of 2 x 8 floor joists.

One 2 x 8, 10 feet long contains 13⅓ board feet. There would be around 1000 board feet for 77 2 x 8 floor joists. That's 17 minutes for picking up, measuring, sawing and nailing in place one floor joist. You might do it slower or faster than that. To most builders, 17 minutes for one joist is pretty slow.

Every builder develops his own system of determining manhours. He bases it on experience, the peculiarities of the job, the weather, the number of Mondays, the number of Fridays and many other variables.

When figuring manhours for a basementless structure, a rule of thumb is:

One manhour per square foot of floor space. (2/3 skilled labor, 1/3 semi-skilled labor)

This is a *base-line* rule. Each builder must adjust it to his own job and methods.

Let's look at it more closely. A 1000 square-foot addition (no basement) will require 1000 manhours, 666 of which are skilled, 333 of which are semi-skilled. This means that you, a skilled carpenter (or builder), and a semi-skilled helper can do the job in 1000 manhours. If you sub out part of the work, you reduce your own manhours by that much. If you base labor on a $20-per-square-foot basis, the labor cost is $20,000.

Total Cost

There are various formulas you can use to estimate the total cost of a house or addition. Some are vague; some are fairly accurate.

The following rule of thumb can often be applied:

The total square-foot cost of a house (no land, basement or exotic treatment) is three times the carpenter hourly union rate.

Thus, if the union scale for a carpenter is $30 an hour, a 1500 square foot house would cost $135,000, or $90 per square foot.

This won't work in areas where labor or materials are out of proportion. It applies to standard construction. For intricate, complicated work the price goes up accordingly.

But it all comes back to the individual builder. You have to use your own method whether it's based on the position of the moon or carefully kept job records. If you prepare a bid, you must decide on a price. Your records and experience are your best guide. You know how much work you can do in the winter in your area. You also know what effect the hot summer sun has on your crew. But, you have to be competitive or someone else will get the job. Carefully itemize and properly categorize every material item. That will help you be accurate *and* competitive.

Site Preparation

Before you prepare a bid or cost sheet, you have to determine what is necessary to ready the site for construction: how much dirt has to be removed, how many trees, etc. Underground lines may have to be relocated. You may have to strip, store and re-spread the top soil. If a basement is involved, how much will it cost to dig and backfill? Be sure to include trenching and pit excavating for draining. Sometimes these below ground items cost as much as the house.

Site preparation for most additions presents few problems. It is usually only a task of removing some shrubbery. Where it is extensive, most small builders sub the job out since few of us have the necessary equipment for such work.

You're probably like most small builders working out of a "pick-up truck office." When you need heavy equipment, call on your old backhoe, front-end loader or bulldozer buddy. You show him the job site, tell him what has to be done, and let him give you his price for the job. You may have to watch him to make sure he doesn't push everything into the next block or county. But you know he'll get the job done.

Profits? To most small builders that means no more than making a living. Let's say that you bid on a 600-square-foot addition. Using our rule of thumb, the base line figure would be 600 manhours to do the job. The union hourly rate for a carpenter, say, is $30 an hour. The total cost for the 600-square-foot addition is:

$$600 \times \$30 \times 3 = \$54,000$$

That is, the square-foot cost for the complete addition is three times the carpenter union hourly rate. This is a base line rule of thumb.

The base line is then adjusted for your particular

area and any special treatment for the job. Anything other than standard construction raises the base figure. You may also have to adjust the base to include taxes, insurance, fringe benefits, and contributions.

Overhead

Most states levy S.U.I. (unemployment insurance tax) on employers. This is based on the total payroll for each calendar quarter. The actual tax percentage is usually based on the employer's history of unemployment claims and may vary from less than 1 percent of the payroll to 4 percent or more. The Federal Government levies an unemployment insurance tax based on the payroll (F.U.T.A.). The tax runs about 0.7 percent of payroll. The Federal Government also collects Social Security (F.I.C.A.) and Medicare taxes. Together they come to about 8 percent of the payroll, depending on the earnings of each employee, and are collected from the employer each calendar quarter.

States generally require employers to maintain Worker's Compensation Insurance. Your cost is based on some percentage of payroll for each trade. Most light construction trades have a rate between 5 and 8 percent. Some trades, such as roofers, have higher rates. The actual cost of Worker's Compensation Insurance varies from one area to the next and from one year to the next, depending on the history of injuries for the previous period.

You should also carry liability insurance to protect yourself in the event of an accident. Liability insurance is based on the total payroll and is about 1½ percent of the payroll. The higher the liability limits, the higher the cost.

What does all this add up to? Approximate percentages look like this:

State Unemployment Insurance............4.0%
F.I.C.A. and Medicare...................7.0%
F.U.T.A. 0.7%
Worker's Compensation Insurance....5.0-15.0%
Liability Insurance.....................2.0%

Total   15.7 to 25.7%

Your base line cost should be adjusted to include these *operating costs*.

Even though your pickup is your office, you still have other overhead expenses. Your truck, tools, tool replacement, equipment maintenance, fire insurance, surety bonds, telephone and advertising are overhead. Anything you need to get in business, stay in business, and do the business, is overhead.

The time spent on government paperwork is overhead. This includes directives from the local, state and federal level telling you what you can and can't do, what you must and must not do, and what you'd better do.

Operating costs and other overhead are *not* profit. That's what you pay off the top to stay in business. If you do five jobs a year, the expense can be "assigned" proportionately to each job.

Profit is the percentage assigned to each job as a *contractor's fee* or as *profit*. It's determined after you compute all labor and material costs, all operating costs and all other overhead.

Your base line figure can absorb these costs. In many areas such costs are already included in the base line. In any event, it should be adjustable to fit your particular need.

You may want to let someone else handle your bookkeeping, but most small builders are capable of doing their own paperwork or bookkeeping. Your spouse may be happy to assume this chore. The bookkeeper knows best where the money goes and how much is available at any moment. Larger builders usually have office personnel to handle this work.

Summary

This isn't a book about accounting or bookkeeping. If you can count to ten and keep your records and figures straight, you shouldn't have any trouble.

What are your chances of expanding your business to where you can work from a desk instead of a pickup? Who knows? Some contractors prefer to stay small. They avoid all the problems of the "pencil" builder and the office-type contractor. Keeping a big crew busy isn't easy. Just look around you. How many small contractors have been in business for ten years and are still going strong? How many larger contractors make a big splash and then fall by the wayside?

Some remodelers will grow and become major contracting firms. Some will stay small because that's how they feel comfortable. It's an individual thing. Some have a knack for expanding and holding their own. Some don't. You can make that decision when the time comes.

If you advertise, use taste. A little "quiet" advertising is fine. "Loud" advertising hurts more than it helps. If you let your last job be your best recommendation, advertising will never be a problem. Your customers will give you all the advertising you'll need. If your last job won't recommend you, no amount of advertising will help.

Remember this rule:

If the customer gets what he wants, he'll be back. If he gets what he doesn't want, he'll never return. Neither will his friends.

We've covered a lot of ground in this book. Mr. Brown now has a dandy house—a showplace of sorts. The word will get around. If you don't have another job lined up on his street, you will eventually. That's how it works. Sometimes you get stuck on a street so long that it seems like home to you.

Contracting can be, and for many is, a creative, satisfying and rewarding career. Your reward will be proportionate to the skill, diligence and professionalism you bring to the work you handle.

Index

A

Access patterns, unacceptable, 29
Accessories, solid vinyl, 205-206
Adding outward, 365
Adding partitions, 134-136
Adding second story, 361
Additions, squaring, 13
Add-on, 359, 365
Adhesive, nail-on, 139
Adhesive, vinyl sheet, 123
Adhesives, 146
Advertising, 393
A-frame roof, 217
Agreement, memorandum, 114
Air conditioning, 19
Air infiltration, categories, 227
Air movement, 227-234
Air panel, removing, 299
Air system duct, 298
Air system hook-up, 298
Airway, insulation, 219
Algae discoloration, roof, 271
Alkyd binders, paint, 329
Aluminum paint, 331
Aluminum primer, 227
Aluminum sheet metal, 50-51
Aluminum siding, 208
Anchorage, sill, 368
Angle firestop, 296
APA glued floor, 376
Application, cap flashing, 260
Application, roofing, 249-251
Application, starter strips, 249
Application, step flashing, 255-256
Application, strip shingles, 245
Application, underlayment, 246-247
Application, valley shingles, 252-254
Applying individual shingles, 266-267
Applying roll roofing, 267-273
Applying roofing over roofing, 264-266
Applying sheathing, 381
Applying shingles, 248
Applying shingles, low slopes, 260
Applying shingles, steep slopes, 261-262
Applying wood shingles, 274
Area/rake conversion table, 243
Areas, moisture content, 49
Areaways, window, 355
Ash dump, 303
Asphalt roofing, 235
Asphalt shingles, 237
Asphalt tile, 128
Attaching counter top, 98
Attic ceiling height, 27
Attic ceilings, 359
Attic conversion, 359
Attic skylights, 361
Attic ventilation, 229-232
Auxiliary sill, 37
Aviation snips, 196

B

Backer board, 140-141
Barrier, vapor, 219-224
Base and wall cabinets, 27
Base cabinets, 95, 97-98
"Base Line" formula, 392-393
Base molding, 166
Basement construction, 353
Basement conversion, 353
Basement installation, fireplace, 289
Basement insulation, 224
Basement paneling, 144-150
Basement space, 27
Basement wall dampproofing, 356
Basements, 10
Basic plumbing, 387
Basic wiring, 387
Bath, half, 77, 80
Bathroom addition, 81
Bathroom checklist, 82-86
Bathroom fittings/hardware, 76
Bathroom, minimum space, 25
Bathroom, plans/specifications, 78
Bathroom vanities/cabinets, 76
Bathrooms, 55-56
Baths, whirlpool, 68-72
Beam and post, 31
Beam details, 373-374
Beam, flush, 134
Beam support, 134
Bearing partition, 11
Bedroom, minimum space, 25
Bidding on additions, 388
Bidets, 68
Blistering and peeling paint, 339
Block wall, 11
Blocked joist, 374
Blocking between joists, 135
Board feet chart, 373
Board footage table, 143
Board insulation, 214-215
Border, ceiling tile, 152-153
Border tile layout, 162
Boxing around beams, 164
Boxing around stairways, 165
Boxing around windows, 164
Bracing rafters, 384
Bracing wall, 378, 381
Brick, fire, 280
Brick veneer application, 381
Brick wall, 11
Bright paint, 335
Building, squaring, 366-367
Built-in heat circulator, 283
Built-up girder, 11-12
Butted joist, 374
Butting, vinyl panels, 201
By-pass door, 173

C

Cabinet, base, 95

C (continued)

Cabinet, bath, 79
Cabinet, cut-out, 97
Cabinet, plan/elevation, 115
Cabinet, proportions, 97
Cabinet, specifications, range, 96
Cabinet, wall, 95, 98
Cabinets, 17
Cabinets, kitchen, 95-107
Calculating paint coverage, 333
Cantilever construction, 385
Cap flashing, chimney, 259
Carpenter ants, 20
Carpenter's square, 382-383
Carpet, 130
Carpet yardage calculator, 130
Carports, minimum space, 33
Cast iron tubs, 56-57
Caulking, 190, 225, 340
Caulks, 225
Ceiling, 133
Ceiling, attic, 359
Ceiling heights, 27
Ceiling insulation, 218-219
Ceiling joists, 382
Ceiling joist lumber, 47
Ceiling joist support, 134
Ceiling materials chart, 160-161
Ceiling molding, 165-166
Ceiling tile, 150-157
Cement blocks, 366
Cement methods, 151
Cement paints, 329
Ceramic tile, 77
Chalk line, roof, 246
Chalking, paint, 342
Chamber, smoke, 280
Chart, board feet, 373
Chart, exterior paint, 332
Chase, chimney, 298
Checking and cracking, 341
Checklist, bathroom, 82-84
Chimney base, 298
Chimney, conventional, 302-305
Chimney cricket, 257-259
Chimney diagram, 303
Chimney elbow, 295
Chimney height, masonry, 303
Chimney height, minimum, 296-297
Chimney, masonry, 302-305
Chimney offset, masonry, 304
Chimney roof measurement, 297
Chimney, smoke pipe, 305
Chimney support, 295
Chimney through roof, 297
Chimney top construction, 303, 305
Chimney top housing, 297-298
Chimney top, masonry, 303, 305
Chimneys, 10
Chimneys fireplaces, 277
Chimney walls, masonry, 304-305

C (continued 2)

Choosing primer/finish coat, 344
Circulator, built-in, 283
Circulators, fireplace, 306-308
Clip, plywood panel, 385
Cleaning/preparation center, 92
Clean-out door, 303, 305
Clogged drains, 10
Closed cut valley, 254
Closet dimensions, 26
Closet, minimum space, 25
Clusters, skylights, 323-324
Coarse aggregate, 45
Coat closet, 27
Collars, fireplace duct, 299
Color and texture, 327
Color effects, 328
Color for moods, 328
Color guide, shingles, 239
Commode, 67, 79
Common roof styles, 238
Common wall, 13
Common wire nails, 54
Comparative R-values, 214
Comparison, paint binders, 330
Complex roof, 217
Computing stair run, 312
Concealed nail method, 270-271
Concrete estimating, 45
Concrete mixing, 45
Concrete pedestal, 42
Concrete proportions, 45
Concrete stairs, 311
Concrete subfloors, 127
Condensation problems, 226
Construction grade lumber, 46-48
Construction notes, panels, 207-208
Construction, stairs, 312
Contract, 88-89
Contractor, general, 5
Conventional chimneys, 302-305
Conversion, basement, 353
Corbelling, chimney, 305
Cork tile, 129
Corners, 180-181
Corner guard molding, 165-166
Corner posts, 378
Corner treatment, 162
Cooking/serving center, 92
Cooktops, electric, 107
Cooktops, gas, 106
Correcting floor squeaks, 39-40
Correcting foundations, 42
Correcting masonry pillars, 42
Correcting paint problems, 338, 340-343
Correcting rafter sag, 40-41
Correcting sagging girders, 38-39
Correcting sagging roofs, 40-41
Corridor kitchen, 93
Cost rule, 392
Cost total, 392

Counter/cabinet details, 98
Counter height, 97
Counter top area, 27
Counter tops, 98
Counters, kitchen, 94
Crawl space access, 33
Crawl space plenum, 222
Cricket, 257-259, 305
Cross tees, 158-159
Coverage and exposure, roofing, 236
Cracked walls, 10
Crown, eliminating, 13
Crumbling mortar, 42
Curb mount skylights, 316-319
Curtain wall, 9
Curtain wall, masonry, 369
Customer notice, 90
Customers, 7
Cutting chimney hole, roof, 296
Cutting rafters, 382-384
Cutting siding, 179
Cutting vinyl sheet, 121

D

Damaged panel replacement, 203-204
Damaged sill, 36-37
Damper, 10
Damper, fireplace, 278-280
Dampproofing, 354
Deck preparation, 245
Defects, structural, 35
Design, exterior, 34
Destructive insects, 20
Details, 2-story offset, 385
Determining fireplace location, 284
Determining pitch/slope, 241
Determining stair run, 312
Diagonal brace, 378
Diffuser, skylights, 316, 319
Dimensions, bathroom fixtures, 26
Dimensions, closet, 26
Dimensions, fireplace, 278-279
Dimensions, fireplace opening, 300
Dimensions, rooms, 25
Door casing, 165-166, 175
Door clearances, 171
Door frame construction, 170
Door framing, 379
Door locations, 28-29
Door, prehung, 169
Door side jambs, 170
Door sill arrangements, 172
Door sizes, 32
Door, sliding, 173
Door stop clearance, 174
Doors, 14, 167, 171
Doors and windows, 167
Dormer, 359-360
Dormer roof valley, 252-254
Double coverage roof, 272-273
Double dome skylight, 315-316
Double sliding door, 173
Drain tile, 354
Drawer area, 27
Dressing rooms, 56, 80
Drip caps, door, 170
Drip edges, 245
Drop ceiling, 97
Drop-in range, 98, 101
Drywall painting, specifications, 337
Drywall stud spacing, 137
Drywall thickness, 137
Ductless range hood, 109

E

Eave flashing, 246-247
Edge strips, roofing, 269
Effects of color, 328
Eggshell paint, 334-337
Elbow, chimney, 295
Electric cooktops, 107
Electric heating systems, 19
Electric outlets, bath, 76
Electric stove, free-standing, 99
Electric wall oven, 102
Electrical, 18
Electrical connection, fireplace, 299
Eliminating floor squeaks, 39-40
Ells, chimney, 295
Enamel, 331
Entrance stairs, 310
Epoxy binders, paint, 329
Estimate list, material, 390-391
Estimating ceiling tile, 151
Estimating concrete, 45
Estimating job, 389
Estimating paint coverage, 333
Estimating paint needs, 331
Estimating roof areas, 239-240
Estimating roofing, 239
Estimating sheet, 197

Estimating siding requirements, 182, 191
Evaluating, 7-8
Evaluation checklist, 21-23
Examination, building, 8
Excessive chalking, 342
Exhaust fan, 57
Experience, 389
Exposed beam support, 134
Exposed nail method, 268
Exposed nails, 179
Extenders, plate/ridge, 382
Exterior doors, 167, 171
Exterior free-standing pier, 368
Exterior paint chart, 332
Exterior painting, 345-348
Exterior siding, 179
Exterior stairs, 310
Exterior surface preparation, 331
Exterior wall framing, 376-380

F

Factory-built fireplaces, 282
Factory-built inserts, 306-308
Factory-built stairs, 312
Fascia treatment, vinyl, 203-204
Fasteners, roof, 244, 250
Fastening wood to concrete, 51
Faucet controls, 64
Fiberglass shower cove, 58-63
Fiberglass tubs, 56-57
F.I.C.A. tax, 393
Figure "Base Line", 392-393
Figuring paint coverage, 345
Finish coat, 344
Finish interior walls, 136
Finish treatment, beams, 164
Finishing fireplaces, 300-302
Finishing hardboard siding, 181
Firebrick, 280
Fireplace and chimneys, 277
Fireplace, basement installation, 289
Fireplace, built-in, 282-302
Fireplace circulator, 302-306
Fireplace dimensions, 278-279
Fireplace dimensions (factory), 290
Fireplace electrical connection, 299
Fireplace, factory built-in, 282
Fireplace finish wall, 301
Fireplace footing, masonry, 280
Fireplace framing, 300-302
Fireplace grille, 300
Fireplace hearth extension, 301-302
Fireplace location, 284
Fireplace, minimum clearance, 290
Fireplace, mobile home, 287
Fireplace offset dimensions, 291-292
Fireplace opening, 300
Fireplace, proper construction, 278
Fireplace repair, 277
Fireplace research data, 281-282
Fireplace sealing details, 302
Fireplace, stacking two, 288
Fireplace surround, 301
Fireplaces, 277
Firestop angle, 296
Firestop spacers, 293-294
Fitting vinyl panels, 199-203
Fitting vinyl sheets, 119-126
Fittings, bathroom, 76-77
Fixture spacing, bathroom, 79
Fixtures, bathroom, 26
Flashing, 381
Flashing around chimneys, 257
Flashing, dormer walls, 360
Flashing roof, 255-256
"Floating angle" installation, 138-139
Floor cover removal, 126
Floor covering, bath, 77
Floor framing, 12
Floor insulation, 221, 224
Floor joist spans, 46
Floor joist support, 373-374
Floor joists, 46
Floor plan, kitchen, 115
Floor reference points, 119-120
Floor squeaks, 39-40
Flooring, 117
Flooring, nonresilient, 129-130
Flooring, room preparation, 119
Flooring sheet vinyl, 119-126
Flooring wood, 117-119
Flue lining, chimney, 303-305
Flue size, chimney, 280
Flue supports, 296
Flush beam, 134
Foam plastic sheathing, 179
Foams, insulation, 214
Foil-back gypsum board, 179
Footing, 9, 11
Footings, 365-372

Form, kitchen preference, 110
Forms, fireplace circulator, 306-308
Foundation drains, 354
Foundation squaring, 367
Foundation vents, 229
Foundation walls, 355-356
Foundations, 8
Frame nailing, 53
Framing, 367
Framing chimney roof hole, 296
Framing, cluster installation, 323-324
Framing dimensions, skylights, 321
Framing doors/windows, 379-380
Framing fireplace, 300-302
Framing floor, 12
Framing flush ceiling, 134
Framing gable dormer, 360
Framing lumber, 45-48
Framing opening, skylights, 317, 321-324
Framing over girders, 374
Framing over partitions, 374
Framing, roof, 13-14
Framing shed dormer, 360
Framing skylight shaft, 325
Framing, soil pipe, 136
Framing square, 383
Framing tub/shower, 58-61
Framing, vent stack, 136
Framing wall, 12, 376-380
Framing wood girder, 373
Free standing piers, 368
Free standing range, 103
Furring, 151, 180, 186
Furring ceilings, 140
Furring for panels, 144
Furring for strip method, 151
Furring strips, 137
F.U.T.A. tax, 393

G

Gable dormers, 360
Gable outlet ventilators, 233
Galvanized sheet metal, 50
Garages, minimum, 33
Gas cook tops, 106
Gas free-standing ranges, 105
Gas ranges, 103
Gas slide-in ranges, 104
General exterior design, 34
General space requirements, 24-25
Girder, 373-374
Girder, built-up, 11
Girder/joist details, 373-374
Girder sizes, 11
Girder span, 11
Girders, repositioning, 35
Girders, sagging, 34
Girders, steel, 12
Girders, wood, 12
Glass, 50-51
Glued floor, 376
Grab bars, 57, 79
Grade beam and piers, 369-370
Greenery, bathroom, 77
Ground cover, barrier, 222
Ground support slabs, 370-372
Guide, window materials, 176
Guide, window selection, 175
Gutter peeling, 340
Gypsum board, 138
Gypsum board application, 138-139

H

Habitable rooms, 354
Handgrips, 57
Handrail, stairs, 309-310
Hanger wire, 159
Half bath, 78
Halls, minimum space, 25
Hard flooring, 129
Hardboard lap siding, 192-196
Hardboard panels, 144
Hardboard shakes, 181, 195-196, 198
Hardboard siding, 179
Hardware, bathrooms, 76-77
Headers, stairs, 310, 312
Headers, windows/doors, 176, 379-380
Headroom, stairs, 309
Heating and cooling, 19
Heating duct, 370
Heating, forced air, 19
Heating, gravity, 19
Heating, radiant, 19
High gloss paint, 334-337
Hinges, door, 171
Hip and ridges, 260-261
Hip and ridges, roofing, 269-271, 273
Hip/valley conversion table, 244
Home evaluation checklist, 21-23
Horizontal wall furring, 137
Horizontal wallboard application, 138-139

Hot water system, 19
House design, 19
House jacks, 36-37
House paint, 331

I

Ice dams, 234
Individual "hex" shingles, 267
Inlet ventilators, 232
Inside corners, 180-181, 198
Inspecting roof, 14
Installation, door hinges, 171
Installation, door trim, 174
Installation, drop-in range, 98, 101
Installation, fiberglass shower, 59-63
Installation, fiberglass tub, 58
Installation, fireplace, 285-288
Installation, fireplace circulator, 306-308
Installation, hardboard panels, 181-190
Installation, lap siding, 192-196
Installation, range hood, 108
Installation, spas, 73-76
Installation, strip flooring, 117-118
Installation, whirlpool baths, 68-72
Installation, window frame, 177
Installing air system, 298
Installing backer board, 140-141
Installing basement paneling, 144-150
Installing built-in fireplace, 282-302
Installing ceiling tile, 151-157
Installing chimney sections, 297
Installing chimney supports, 295
Installing fireplace duct, 299
Installing firestop spacers, 293
Installing floor tile, 127-129
Installing insulation, 219-224
Installing panels, 207
Installing rafters, 382-383
Installing sheathing, 381
Installing suspended ceiling, 158-165
Installing the chimney, 284-288
Installing the termination, 299
Insulating ceiling/walls, 218-224
Insulating floors, 221, 224
Insulation, 17, 140, 211
Insulation application, siding, 200
Insulation materials, 211-214
Insulation recommendations, 226
Insurance percentages, 393
Itemized materials list, 390-391
Itemizing materials, 389
Interior doors, 167, 171
Interior stairs, 309-310
Interior surface preparation, 333
Interior trim, 165-166
Interior wall finish, 136
Interlocking shingles, 235

J

J channel, 200
Jack, house, 36-37
Jack post, 38-39
Job estimating, 389
Joint molding, 180
Joints, 180
Joist, ceiling, 382
Joist/girder details, 373-374
Joist support, 134

K

Keyline layout, 366
Kitchen cabinet proportions, 97
Kitchen cabinets, 95-107
Kitchen cabinets, stock, 97
Kitchen cabinets, wall, 97
Kitchen change, 112
Kitchen, corridor, 93
Kitchen counter, 94
Kitchen, L-shaped, 93
Kitchen, old, 112
Kitchen plan/elevation, 115
Kitchen planning, 109
Kitchen planning steps, 113
Kitchen preference form, 110-111
Kitchen, sidewall, 94
Kitchen storage, 27, 94
Kitchen, U-shaped, 92-93
Kitchen ventilation, 108
Kitchen "Work Triangle", 91-93
Kitchens, 91
Knee-wall insulation, 224
Knee-walls, attic, 359
Knock-downs, 97

L

Labor, 389
Lacquer, 331
Landings, stairs, 309-311
Lap over beams, 374
Lap siding, hardboard, 192-196
Latex paints, 329

Laundry area, minimum, 25
Lavatories, 58, 64-67
Lavatory spacing, 79
Leaks, 16
Ledgers, girder, 373
Let-in bracing, 378
Liability insurance, 393
Light direction, skylights, 325
Lighting, bathrooms, 76
List, material estimate, 390-391
Load-bearing partitions, 30
Locating chimney centerpoint, 296
Low lustre paint, 334-336
Low slope roofing, 260
L-shaped kitchen, 93
Lumber, ceiling joists, 47
Lumber, floor joists, 46
Lumber, framing, 45-48
Lumber grades, 45
Lumber, rafters, 48
Lumber species, 45-48

M

Main runner layout, 158-159
Main runners, 158
Main stairs, 311
Malco snaplock punch, 200
Manhour rule-of-thumb, 392
Manhours per SF, 392
Manufactured fireplaces, 280-284
Masonry chimneys, 302
Masonry construction, 180
Masonry curtain walls, 369
Masonry fireplace footings, 280
Masonry, paint and repainting, 337
Masonry painting specifications, 335
Masonry systems, 216-217
Masonry walls, 11
Material estimate, 389
Material estimate list, 390-391
Materials, window, 176
M.D.O. plywood, 207
Memorandum of Agreement, 114
Metal corners, 180
Metal trim, 180
Metal-furred underlayment, 217
Microwave ovens, 95-97
Minimum clearance, fireplace, 290
Minimum material thickness, 137
Minimum pitch/slope, 238
Minimum room sizes, 25
Minimum size bathroom, 81
Mirrors, 64
Mirror height, bath, 79
Mobile home fireplace, 287
Modern bathrooms, 56
Modern luxury bathrooms, 80
Moisture content areas, 49
Moisture content, lumber, 49
Moisture problems, paint, 339-340
Molding, 166
Mud, sheetrock, 141
Multiple switch controls, 388

N

Nail sizes, 54
Nail spacing, paneling, 190, 208
Nailer, steel girder, 374
Nailing at corners, 140
Nailing schedule, 53
Nailing sheathing, 381
Nails, roofing, 244, 250
Nails, wallboard, 138
New partition framing, 135
Non-load bearing partition, 30
Non-load bearing partition framing, 135
Non-moisture blistering, 340
Non-resilient flooring, 129
Notebook, 55
Notice to Customer, 90

O

Odd shapes, painting, 333
Offset, chimney, masonry, 304
Offset construction, 385
Oil binders, paint, 329
Oil stains, 331
Old kitchens, 91
One story footings, 9
On-slab construction, 367-372
Open riser, stairs, 311
Open roof valleys, 252
Open stringers, stairs, 310
Operating costs, 393
Optional shafts, skylights, 323, 325
Outside air system, 298
Outside corners, 180-181, 198
Outside finishes, 15
Overhead, 393

P

Paint and enamel, 331
Paint and painting, 327
Paint binders, 330
Paint blistering, 339
Paint checking, 341
Paint cracking, 341
Paint estimating, 345
Paint failure, 15, 204
Paint materials, 329
Paint peeling, 337
Paint problems, correcting, 338, 340-343
Paint scaling, 339
Paint selection chart, 332
Paint types, 330
Paint wrinkling, 342-343
Painting doors, 350-351
Painting masonry, 332
Painting metal, 332
Painting new wood, 331
Painting outside house, 345-348
Painting siding, 196
Painting specifications, 334-337
Painting trim, 350-351
Painting walls/ceilings, 348-350
Painting windows, 349-350
Paints, 334-338
Panel clips, 385
Panel fitting, vinyl, 199-203
Panel installation, hardboard, 181-190
Panel siding nails, 190
Paneling a basement, 144-150
Paneling, wood and fiberboard, 150
Paper holder, bath, 79
Parging basement walls, 354
Particle board tile, 119
Partition, adding, 134-136
Partition, basement, 148
Partition, bearing, 11
Partition, non-bearing, 11
Partition post, 378
Partition, removal, 133-134
Payroll percentages, 393
Pedestal, 12
Pedestal, concrete, 42
Peeling gutters, 340
Peeling paint, 337-339
Penetrating paint, 334
Permanent light fixtures, 388
Piers, exterior, 368
Pillars, 9, 11
Pillars, installing new, 36
Pillars or piers, 367
Pitch, roof, 382-383
Pitch/slope calculation, 241
Plan, floor/elevation, 115
Planning, 24
Planning kitchens, 109
Planning steps, kitchen, 113
Plans, bathroom, 77-81
Plaster, 16
Plaster/drywall paints, 336-337
Plastic vent strip, 385
Plate extenders, 382
Plywood, 50
Plywood and hardboard, 144
Plywood cross sections, 208
Plywood nailing schedule, 208
Plywood panel clips, 385
Plywood underlayment, 126, 377
Plumbing, 18
Plumbing, basic, 387
Polyethylene film, 179
Porches, 15
Post, 11
Post jack, 38-39
Post, steel, 12
Post support, 42
Post, wood, 9, 42
Posts, corner, 378
Posts, partition, 378
Powder post beetles, 20
Powder rooms, 56
Prefabricated fireplace, 306-308
Prefinished metal corners, 180
Prehung doors, 168-169
Price escalation clause, 78
Primer, 344
Primer, paint/enamel, 331
Procedures, installing vinyl, 121-126
Profit, 392-393
Proportions, kitchen cabinets, 97
Proposal and contract, 88-89
Purlin, 384
PVC molding, 165

Q

Quarry tile, 130
Quiet conditioning, 142

R

Rafter bracing, 384
Rafter cutting, 383-384
Rafter/joist bracing, 40-41
Rafter lumber, 48
Rafter measuring, 383-384
Rafter nailing, 52-53
Rafter run, 383-384
Rafter sag, 40-41
Rafter spacing, skylights, 315, 321
Rafter support, rake, 382
Rafter tie-in, 385
Rafters, 382-386
Railings, stairs, 309-310
Rake overhang support, 382
Range, drop-in, 98, 101
Range, free-standing, 99
Range, gas free-standing, 103, 105
Range, gas slide-in, 104
Range hood installation, 108-109
Range, slide-in, 100
Reading the square, 383
Ready-mix cement, 45
Ready-mix compound, 41
Recommended R-values, 225
Reflective insulation, 213
Refrigerator/storage center, 91
Reinforced beam, 134
Relocating partitions, 29-30
Remodeled bath, 80
Removing air panel, 299
Removing floor tile, 126
Removing old roofing, 263
Removing partitions, 29-30, 133-134
Removing vinyl flooring, 126
Repainting masonry, 332
Repainting plaster/drywall, 337
Repainting wood, 331
Repairing structural defects, 35
Replacing damaged panels, 204
Replacing sills, 35-36, 39
Reroofing, 263
Resanding floors, 17
Restaining, 332
Ridge extender, 382
Ridge plate brace, 40
Ridges and hips, 260-261, 271, 273
Right materials, 44
Rigid board insulation, 213
Rise, stairs, 309-312
Rod, closet, 26
Roll roofing, 236, 267, 273
Roof double coverage, 272-273
Roof framing, 13
Roof horizontal area, 240
Roof inspecting, 14
Roof overhang, 11
Roof pitch, 238, 382-383
Roof pitch calculation, 241
Roof pitch/slopes, 240
Roof step flashing, 255
Roof valley, 252-254
Roofing, 15
Roofing, built-up, 16
Roofing, metal, 16
Roofing over roofing, 264-266
Room add-on, 365
Room dimensions, minimum, 25
Room divider, 95
Room layout, 30-31
Room layout chart, 157
Rough opening, window, 177
Rubber base paint, 329
Rubber tile, 129
Rule-of-thumb, manhours, 392
Rule-of-thumb, SF cost, 392
Run, stairs, 309-312
R-value, materials, 212

S

Sag, rafter, 40-41
Sagging girders, 36, 38-39
Sagging sheathing, 41
Satellite concept, addition, 365
Satin paint, 336-337
Saturated felt, 236
Scaling paint, 337, 339
Scissor truss system, 217
Seam sealer, vinyl, 123
Second story addition, 361-362
Securing rafters, 384
Self-flashing skylight, 319-320
Semi-gloss paint, 336
Shaft control lighting, 325
Shaft, skylight, 323, 325
Shakes, hardboard, 181, 198
Shallow beam, 31
Shallow post, 31
Sheathing boards, 35, 50

Sheathing installation methods, 381
Sheathing, "Thermax", 214-217
Shed dormer, 359
Sheet vinyl flooring, 119-126
Sheetrock, 138
Sheetrock compound, 141
Sheetrock footage table, 143
Sheetrock taping, 141
Shims, 137
Shingle application methods, 248-250
Shingle application, sloped roofs, 261-262
Shingle color guide, 239
Shingle first course, 248-249
Shingles, 235
Shingles, wood, 16, 274
Shower coves/cabinets, 57
Shower heads, 57, 79
Shutters, 346
Sidewall, kitchen, 94
Siding, 14
Siding, exterior, 179
Siding panel installation, 181-190
Siding requirement, 191
Siding, vinyl, 196, 199-204
Siding, wood, 204
Sill, 11
Sill anchorage, 368
Sill, auxiliary, 37
Sill, damaged, 37
Sill, door, 170
Sills, repositioning, 35
Sills, replacing, 36
Simple roofs, 240
Single dome skylight, 315-316
Single framing method, 380
Site preparation, 392
Skylights, 315
Skylights, attic, 361
Skylight "bubble", 322
Skylight clusters, 322-324
Skylight curb mount, 316-319
Skylight insulation values, 316
Skylight light distribution, 315, 325
Skylight, self-flashing, 319-320
Skylight shaft treatment, 323, 325
Slabs, 367-372
Sliding door, 173
Small bathrooms, 8, 80
Smoke chamber, 280
Smoke pipe protection, 306
Soap dish, 79
Social Security tax, 393
Soffit channel, 203
Soffit directional change, 203
Soffit ventilators, 232, 234
Soft gloss paint, 334-335
Solid bridging, 40
Solid vinyl accessories, 205-206
Soot pocket, chimney, 305
Space requirements, 24-25
Span, maximum, 11
Spas, 68, 73-76
Special framing, 136
Square, carpenter, 383
Square foot cost, 392
Squaring a ceiling, 154-155
Squeaks, floor, 39-40
"Stable-X" plywood, 375
Stacking chimneys, built-in, 288
Stain, 331, 334-336
Staining new wood, 331
Staining siding, 196
Staining specifications, wood, 334, 336
Staples, roofing, 244-245
Stapling shingles, 251
Stapling vinyl flooring, 125
Starter strip, 180-181, 248
Starter strip, vinyl, 200
Stairs, concrete, 311
Stairs, construction, 312
Stairs, entrance, 310
Stairs, exterior, 310
Stairs, interior, 309, 311
Stairs, nosing, 311
Stairs, specifications, 309-311
Stairs, width, 309-311
Stairs, wood, 310, 312
Stairways, 309
State unemployment insurance, 393
Steel circulators, fireplace, 306-308
Steel joist hanger, 373
Steel reinforced beam, 134
Steel tubs, 56-57
Stepped footing, 366
Stock cabinets, kitchen, 97
Stone wall, 11
Storage, 27, 76
Storm doors, 168
Storm windows, 168

Stringers, stairs, 310, 312
Strip shingles, 235
Structural defects, 35
Structural plywood, 50
Structural sheathing, 381
Structural supports, 11
"Stud driver", 51, 54, 145
"Sturd-I-Floor", 376
Subcontracting, 55
Subcontractors, 388
Subfloor, concrete, 127
Subfloor underlayment, 126
Supports, flue, 296
Supporting partition, 11
Surface discoloration, paint, 343-344
Surface mounted fixture, 164
Surface preparation, painting, 333
Surface problems, paint, 337
Suspended ceilings, 155
Suspended ceiling installation, 158-165

T

Table, fireplace dimensions, 279
Taping, sheetrock, 141
Ten foot rule, 297
Terminations, installing, 299
Termite areas, 54
Termite protection, 51
Termite shields, 367-368, 371-372
Termites, 20
Terrazzo floor, 130
Thermax sheathing, 214-216
Three-tab shingles, 251
Threshold, door, 170
Thresholds, 15
Tile, asphalt, 128
Tile, ceiling, 150
Tile, cork, 129
Tile removal, 126
Tile, rubber, 129
Tile, vinyl-asbestos, 127
Toilet, 67, 79
Tongue-and-groove plywood, 376
Top housing, chimney, 297
Top plate extension, 382
Total cost, 392
Total rise/run, 312
Towel bars, 79
Traffic flow, 28
Tread-to-riser relationship, 312

Trim, 17
Trim and shutters, 346
Trim interior, 165-166
Trough bracing, 40-41
Tub, bath, 79
Tub, glass doors, 57
Tub metal hanger, 57
Tubs, 56-58
Two story footings, 9
Two story offset, 385
Two-tab shingles, 251
Types of paint, 330-331
Typical duct installation, 299

U

Unacceptable access patterns, 29
Under-eave peeling, 337
Underlayment, 375-376
Underlayment, existing floor, 127
Underlayment, low slope, 261
Underlayment, roof deck, 245
Underlayment, subfloor, 126
Unemployment insurance, 393
Unlocking tool, 204, 206
Unprimed hardboard shakes, 198
Urethane finishes, 330
Use of lumber, 49-50
Use of plywood, 50
U-shaped kitchens, 92-93
Using a brush, 346-347
Using carpenter's square, 382-383
Utility runs, walls, 136

V

Valley flashing, 247
Valley tin, 50-51
Valley underlayment, 247-248
Valleys, open, 247
Vapor barrier, 140, 144, 180, 219-224, 228-229, 370
Vapor resistant coating, 227
Vanities, bathroom, 76
Varnish, 331
VDC base flashing, 200
VDDC dual cap, 200
Vent strips, 385
Vented range hood, 109
Ventilation, 17, 227-234
Ventilation areas, roofs, 230-232
Ventilation, crawl space, 229

Ventilation, kitchen, 108
Ventilation requirements, 230-231
Ventilator, 219, 230
Vertical sidewall flashing, 256
Vertical siding, vinyl, 200, 202
Vertical wallboard application, 138-139
VFT finishing trim, 200
Vinyl-asbestos tile, 127
Vinyl cutting procedure, 199
Vinyl fascia detail, 204
Vinyl nailing procedure, 199
Vinyl sheet, 119-126
Vinyl sheet cutting, 121-122
Vinyl siding, 196, 199-204
Vinyl soffit application, 202-203
Vinyl trim, 202
VJ channels, 202
VSD starter, 202

W

Wallboard fasteners, 138
Wallboard footage table, 143
Wall bracing, 378, 381
Wall cabinet, 27, 97
Wall construction, 179
Wall framing, 12, 376-378, 380
Wall framing lumber, 45
Wall insulation, 220, 223-224
Wall molding, 158
Wall outlets, 19
Wall oven, electric, 102
Wall switch, 19
Walls and ceilings, 16, 133
Walls, bath, 77
Walls, foundation, 356
Walls, interior finish, 136
Walls, sound resistant, 142
Walls, utility runs, 136
Waterproofed basement, 10
Waterproofed walls, 354
Water resistant wallboard, 140-141
Weatherstripping and caulking, 225
Weep holes, weep joints, 11
"Wet look" paints, 327
Where to install molding, 166
Whirlpool baths, 68-72
Wide overhangs, 203
Wind zone areas, 52
Winders, stairs, 309, 311

Window and door headers, 176
Window and door trim, vinyl, 202
Window arrangement, 32
Window boxing, 164
Window casing, 165-166
Window flashing, 381
Window framing, 379-380
Window headers, 379-380
Window installation, 177
Window opening, 379-380
Window rough opening, 177
Window sills, 14
Window valance, 164
Windows, 168
Windows and doors, 167
Windows, inspecting, 14
Windows, materials, 176
Windows, selection, 175
Wire fasteners, 54
Wiring, basic, 387
Wiring, damage, 19
Wood and fiberboard paneling, 150
Wood block flooring, 119
Wood brackets, 134
Wood corners, shakes, 198
Wood flooring, 117-119
Wood furring, ceiling, 140
Wood painting specifications, exterior, 334
Wood painting specifications, interior, 336
Wood post, 42
Wood shakes, 274-275
Wood shingles, 16, 274
Wood siding, 204
Wood staining specifications, interior, 334
Wood strip flooring, 117
Wood surface preparation, 331-333
Wooden posts, 9
Work island, 95
"Work triangle", 91-93
Worker's compensation insurance, 393
Working the square, 367
Woven and closed valleys, 248
Woven roof valley, 255
Wrinkling, paint, 342-343
Wythe, chimney, 303, 305

XYZ

Your services, kitchen, 113
Zone wind velocities, 52

Other Practical References

Running Your Remodeling Business
Everything you need to know about operating a remodeling business, from making your first sale to insuring your profits: how to advertise, write up a contract, estimate, schedule your jobs, arrange financing (for both you and your customers), and when and how to expand your business. Explains what you need to know about insurance, bonds, and liens, and how to collect the money you've earned. Includes sample business forms for your use. **272 pages, 8½ x 11, $21.00**

Contractor's Survival Manual
How to survive hard times in construction and take full advantage of the profitable cycles. Shows what to do when the bills can't be paid, finding money and buying time, transferring debt, and all the alternatives to bankruptcy. Explains how to build profits, avoid problems in zoning and permits, taxes, time-keeping, and payroll. Unconventional advice includes how to invest in inflation, get high appraisals, trade and postpone income, and how to stay hip-deep in profitable work. **160 pages, 8½ x 11, $16.75**

Wood-Frame House Construction
From the layout of the outer walls, excavation and formwork, to finish carpentry, and painting, every step of construction is covered in detail with clear illustrations and explanations. Everything the builder needs to know about framing, roofing, siding, insulation and vapor barrier, interior finishing, floor coverings, and stairs. . . complete step by step "how to" information on what goes into building a frame house. **240 pages, 8½ x 11, $14.25. Revised edition**

Estimating Tables for Home Building
Produce accurate estimates in minutes for nearly any home or multi-family dwelling. This handy manual has the tables you need to find the quantity of materials and labor for most residential construction. Includes overhead and profit, how to develop unit costs for labor and materials and how to be sure you've considered every cost in the job. **336 pages, 8½ x 11, $21.50**

Building Layout
Shows how to use a transit to locate the building on the lot correctly, plan proper grades with minimum excavation, find utility lines and easements, establish correct elevations, lay out accurate foundations and set correct floor heights. Explains planning sewer connections, leveling a foundation out of level, using a story pole and batterboards, working on steep sites, and minimizing excavation costs. **240 pages, 5½ x 8½, $11.75**

Handbook of Construction Contracting Vol. 1
Volume 1: Everything you need to know to start and run your construction business; the pros and cons of each type of contracting, the records you'll need to keep, and how to read and understand house plans and specs to find any problems before the actual work begins. All aspects of construction are covered in detail, including all-weather wood foundations, practical math for the jobsite, and elementary surveying. **416 pages, 8½ x 11, $21.75**

Remodeler's Handbook
The complete manual of home improvement contracting: Planning the job, estimating costs, doing the work, running your company and making profits. Pages of sample forms, contracts, documents, clear illustrations and examples. Chapters on evaluating the work, rehabilitation, kitchens, bathrooms, adding living area, re-flooring, re-siding, re-roofing, replacing windows and doors, installing new wall and ceiling cover, re-painting, upgrading insulation, combating moisture damage, estimating, selling your services, and bookkeeping for remodelers. **416 pages, 8½ x 11, $18.50**

Spec Builder's Guide
Explains how to plan and build a home, control your construction costs, and then sell the house at a price that earns a decent return on the time and money you've invested. Includes professional tips to ensure success as a spec builder: how government statistics help you judge the housing market, cutting costs at every opportunity without sacrificing quality, and taking advantage of construction cycles. Every chapter includes checklists, diagrams, charts, figures, and estimating tables. **448 pages, 8½ x 11, $24.00**

Rough Carpentry
All rough carpentry is covered in detail: sills, girders, columns, joists, sheathing, ceiling, roof and wall framing, roof trusses, dormers, bay windows, furring and grounds, stairs and insulation. Many of the 24 chapters explain practical code approved methods for saving lumber and time without sacrificing quality. Chapters on columns, headers, rafters, joists and girders show how to use simple engineering principles to select the right lumber dimension for whatever species and grade you are using. **288 pages, 8½ x 11, $16.00**

Contractor's Guide to the Building Code
Explains in plain English exactly what the Uniform Building Code requires and shows how to design and construct residential and light commercial buildings that will pass inspection the first time. Suggests how to work with the inspector to minimize construction costs, what common building short cuts are likely to be cited, and where exceptions are granted. **312 pages, 5½ x 8½, $16.25**

Residential Electrical Design
Explains what every builder needs to know about designing electrical systems for residential construction. Shows how to draw up an electrical plan from the blueprints, including the service entrance, grounding, lighting requirements for kitchen, bedroom and bath and how to lay them out. Explains how to plan electrical heating systems and what equipment you'll need, how to plan outdoor lighting, and much more. If you are a builder who ever has to plan an electrical system, you should have this book. **194 pages, 8½ x 11, $11.50**

Stair Builders Handbook
If you know the floor to floor rise, this handbook will give you everything else: the number and dimension of treads and risers, the total run, the correct well hole opening, the angle of incline, the quantity of materials and settings for your framing square for over 3,500 code approved rise and run combinations—several for every 1/8 inch interval from a 3 foot to a 12 foot floor to floor rise. **416 pages, 8½ x 5½, $13.75**

Plumbers Handbook Revised

This new edition shows what will and what will not pass inspection in drainage, vent, and waste piping, septic tanks, water supply, fire protection, and gas piping systems. All tables, standards, and specifications are completely up-to-date with recent changes in the plumbing code. Covers common layouts for residential work, how to size piping, selecting and hanging fixtures, practical recommendations and trade tips. This book is the approved reference for the plumbing contractors exam in many states. **240 pages, 8½ x 11, $18.00**

How to Sell Remodeling

Proven, effective sales methods for repair and remodeling contractors: finding qualified leads, making the sales call, identifying what your prospects really need, pricing the job, arranging financing, and closing the sale. Explains how to organize and staff a sales team, how to bring in the work to keep your crews busy and your business growing, and much more. Includes blank forms, tables, and charts. **240 pages, 8½ x 11, $17.50**

National Construction Estimator

Current building costs in dollars and cents for residential, commercial and industrial construction. Prices for every commonly used building material, and the proper labor cost associated with installation of the material. Everything figured out to give you the "in place" cost in seconds. Many time-saving rules of thumb, waste and coverage factors and estimating tables are included. **544 pages, 8½ x 11, $19.50. Revised annually.**

Estimating Home Building Costs

Estimate every phase of residential construction from site costs to the profit margin you should include in your bid. Shows how to keep track of manhours and make accurate labor cost estimates for footings, foundations, framing and sheathing finishes, electrical, plumbing and more. Explains the work being estimated and provides sample cost estimate worksheets with complete instructions for each job phase. **320 pages, 5½ x 8½, $17.00**

Roof Framing

Frame any type of roof in common use today, even if you've never framed a roof before. Shows how to use a pocket calculator to figure any common, hip, valley, and jack rafter length in seconds. Over 400 illustrations take you through every measurement and every cut on each type of roof: gable, hip, Dutch, Tudor, gambrel, shed, gazebo and more. **480 pages, 5½ x 8½, $22.00**

Builder's Guide to Accounting Revised

Step-by-step, easy to follow guidelines for setting up and maintaining an efficient record keeping system for your building business. Not a book of theory, this practical, newly-revised guide to all accounting methods shows how to meet state and federal accounting requirements, including new depreciation rules, and explains what the tax reform act of 1986 can mean to your business. Full of charts, diagrams, blank forms, simple directions and examples. **304 pages, 8½ x 11, $17.25**

Basic Plumbing with Illustrations

The journeyman's and apprentice's guide to installing plumbing, piping and fixtures in residential and light commercial buildings: how to select the right materials, lay out the job and do professional quality plumbing work. Explains the use of essential tools and materials, how to make repairs, maintain plumbing systems, install fixtures and add to existing systems. **320 pages, 8½ x 11, $17.50**

Home Wiring: Improvement, Extension, Repairs

How to repair electrical wiring in older homes, extend or expand an existing electrical system in homes being remodeled, and bring the electrical system up to modern standards in any residence. Shows how to use the anticipated loads and demand factors to figure the amperage and number of new circuits needed, and how to size and install wiring, conduit, switches, and auxiliary panels and fixtures. Explains how to test and troubleshoot fixtures, circuit wiring, and switches, as well as how to service or replace low voltage systems. **224 pages, 5½ x 8½, $15.00**

BUSINESS REPLY MAIL

FIRST CLASS   PERMIT NO. 271   CARLSBAD, CA

POSTAGE WILL BE PAID BY ADDRESSEE

Craftsman Book Company
6058 Corte Del Cedro
P. O. Box 6500
Carlsbad, CA 92008—0992

BUSINESS REPLY MAIL

FIRST CLASS   PERMIT NO. 271   CARLSBAD, CA

POSTAGE WILL BE PAID BY ADDRESSEE

Craftsman Book Company
6058 Corte Del Cedro
P. O. Box 6500
Carlsbad, CA 92008—0992

BUSINESS REPLY MAIL

FIRST CLASS   PERMIT NO. 271   CARLSBAD, CA

POSTAGE WILL BE PAID BY ADDRESSEE

Craftsman Book Company
6058 Corte Del Cedro
P. O. Box 6500
Carlsbad, CA 92008—0992